Random Matrices and the Six-Vertex Model

Volume 32

CRM MONOGRAPH SERIES

Centre de Recherches Mathématiques
Montréal

Random Matrices and the Six-Vertex Model

Pavel Bleher
Karl Liechty

American Mathematical Society
Providence, Rhode Island USA

The production of this volume was supported in part by the Fonds de recherche du Québec–Nature et technologies (FRQNT) and the Natural Sciences and Engineering Research Council of Canada (NSERC).

2010 *Mathematics Subject Classification.* Primary 60B20; Secondary 82B23.

For additional information and updates on this book, visit
www.ams.org/bookpages/crmm-32

Library of Congress Cataloging-in-Publication Data

Bleher, Pavel, 1947– author.
Random matrices and the six-vertex model / Pavel Bleher, Karl Liechty.
pages cm. — (CRM monograph series / Centre de Recherches Mathematiques ; 32)
Includes bibliographical references.
ISBN 978-1-4704-0961-6 (alk. paper)
1. Random matrices. I. Liechty, Karl, 1981– author. II. Title.

QA188.B54 2013
519.2—dc23

2013032106

∞ The paper used in this book is acid-free and falls within the guidelines
established to ensure permanence and durability.
Visit the AMS home page at http://www.ams.org/

10 9 8 7 6 5 4 3 2 1   19 18 17 16 15 14

Contents

Introduction

The theory of random matrices has proven to have a wide reach into many areas of mathematics, physics, and statistics, and there are many excellent books on the topic. The book of Mehta [61] has become a classic for anyone interested in the subject, and several excellent books on random matrices have appeared in more recent years: [4] by Bai and Silverstein; [3] by Anderson, Guionnet, and Zeitouni; [42] by Forrester; [66] by Pastur and Shcherbina; the CRM volume of lectures [43] edited by Harnad; [73] by Tao; and the Oxford handbook on random matrix theory [1] edited by Akemann, Baik, and Di Francesco. See also the reviews [34] by Di Francesco, Ginsparg, and Zinn-Justin; the ones in the MSRI volume [12], edited by Bleher and Its; [33] by Di Francesco; and the forthcoming book [38] of Eynard. These books and reviews vary in scope and perspective, and they present different approaches to random matrices and their applications to combinatorics, statistics, and physics. In this book we outline a connection from random matrices to the six-vertex model of statistical physics. In particular, this model is related to the unitary matrix ensembles, which are among the most widely studied of the matrix ensembles. For unitary ensembles there is a direct connection to orthogonal polynomials on the real line, and the asymptotics of partition functions as well as local spectral statistics can be studied using the Riemann – Hilbert approach. The focus of this book is a description of the Riemann – Hilbert method for both continuous and discrete orthogonal polynomials, and applications of this approach to matrix models as well as to the six-vertex model.

The Riemann – Hilbert approach to ensembles of random matrices was initiated in the late 1990s in the papers [10] by Bleher and Its, and [29, 30] by Deift, Kriecherbauer, McLaughlin, Venakides, and Zhou, and it became a powerful tool in the theory of universality and critical phenomena in random matrices. In particular, the Riemann – Hilbert method allows for an asymptotic analysis of a wide class of orthogonal polynomials, which was a vital ingredient in the proof of universality of scaling limits for correlations of eigenvalues. The main ideas of the Riemann – Hilbert approach to orthogonal polynomials and random matrices are nicely described in the the lectures [27] by Deift. Chapter 2 of this book is adapted from the paper [29].

The six-vertex model dates back to Slater [69] in the early 1940s, and is one of the integrable models of 2-d statistical physics, see [7, 67]. The domain wall boundary conditions considered in this book were introduced by Korepin [50] in 1982. In that paper certain recursions for the partition function were derived. Subsequently these recursions were used by Izergin [46] to give an explicit determinantal formula for the partition function. This formula is the basis for the asymptotic analysis described in this book, and is known as the Izergin – Korepin formula. The relation of the Izergin – Korepin formula to ensembles of random matrices and orthogonal polynomials was discovered and used by Zinn-Justin [78, 79]. For certain values of the parameters, the relevant orthogonal polynomials are classical. In these cases, the Izergin – Korepin formula was used by Colomo and Pronko [23–26] to give expressions for the 1-, 2-, and 3-enumeration of alternating sign matrices. Outside of these special cases the orthogonal polynomials are not classical, and the Riemann – Hilbert approach was employed in a series of papers by Bleher and coauthors [8, 9, 13–15].

The general outline for the book is as follows:

• In Chapter 1 we introduce the unitary matrix ensembles and describe their connections to orthogonal polynomials and integrable systems.

• In Chapter 2 we discuss the Riemann – Hilbert (RH) approach to random matrix ensembles, adapted from the original approach of the paper [29] and the book [27]. We give general formulas for asymptotics of recurrence coefficients for orthogonal polynomials, and give a proof of the universality of the sine and Airy kernels in the bulk and at the edge, respectively, of the spectrum.

• In Chapter 3 we consider an extension of the RH approach to discrete orthogonal polynomials on an infinite lattice, which was originally developed in the book [5] of Baik, Kriecherbauer, McLaughlin, and Miller for discrete orthogonal polynomials on a finite lattice, and then extended to an infinite lattice in the paper [16] by Bleher and Liechty. Again we give general formulas for asymptotics of recurrence coefficients. Universality of the local correlations in the discrete orthogonal polynomial ensemble is discussed, and we give a proof of the scaling limit of the correlation kernel at the point which separates a band from a saturated region.

• In Chapter 4 we introduce the six vertex model with with domain wall boundary conditions.

• In Chapter 5 we derive the Izergin – Korepin formula for the partition function of the six vertex model with with domain wall boundary conditions. The proof is based on the Yang – Baxter equations, and we follow the elegant approach of the papers [51, 55].

• In Chapters 6 – 8 we obtain the large n asymptotic formulas for the partition function in different phase regions on the phase diagram. These chapters follow the works [9, 13, 15]. The methods of Chapters 2 and 3 are applied, and all details of the analysis are presented.

• In Chapter 9 we discuss the asymptotics of the partition function on the critical lines between the phases, as well as the phase transitions. The results of the papers [8, 14] for the partition function on the critical lines are discussed, but we do not present detailed proofs in this book.

Acknowledgements. In part, this book is based on lectures which the first author gave at different universities: Indiana University – Purdue University Indianapolis; Centre de recherches mathématiques, Montréal; Katholieke Universiteit of Leuven, Belgium; and the National University of Singapore, Singapore. The initial work on this book was done during the semester long program *Random Matrix Theory, Interacting Particle Systems and Integrable Systems* at the Mathematical Sciences Research Institute (MSRI) in Berkeley, California, in the fall of 2010. We would like to thank MSRI and the organizers of that program, Jinho Baik, Alexei Borodin, Percy Deift, Alice Guionnet, Craig Tracy, and Pierre van Moerbeke, for allowing us the opportunity to be there. We would also like to thank various people for useful comments and discussions over the years, including Thomas Bothner, Percy Deift,

Bertrand Eynard, John Harnad, Alexander Its, Vladimir Korepin, Arno Kuijlaars, Ken McLaughlin, and Paul Zinn-Justin.

CHAPTER 1

Unitary Matrix Ensembles

1.1. Unitary ensemble with real analytic interaction

Recall that a matrix M is Hermitian if $M = M^*$ (where $M^* = \overline{M^T}$), so that $M_{kj} = \overline{M_{jk}}$. Of course, any Hermitian matrix must have real entries along the diagonal, whereas the entries below the diagonal are completely determined by the entries above the diagonal. It follows that, in order to count the real dimension of the space of $N\times N$ Hermitian matrices, we should count the number of entries along the diagonal, N, and twice the number of entries above the diagonal to account for real and imaginary parts. Thus, if $\mathcal{H}_N$ is the space of $N\times N$ Hermitian matrices, then its real dimension is equal to

$$\dim \mathcal{H}_N = N + 2[1+2+\cdots+N-1] = N + N(N-1) = N^2. \tag{1.1.1}$$

The space $\mathcal{H}_N$ is a real Hilbert space with respect to the scalar product

$$\begin{aligned}(L,M) &= \operatorname{Re}\operatorname{Tr}(LM^*) = \sum_{j,k=1}^{N} \operatorname{Re}\big(L_{jk}\overline{M_{jk}}\big) \\ &= \sum_{j=1}^{N} L_{jj}M_{jj} + 2\sum_{j>k}^{N}[(\operatorname{Re} L_{jk})(\operatorname{Re} M_{jk}) + (\operatorname{Im} L_{jk})(\operatorname{Im} M_{jk})].\end{aligned} \tag{1.1.2}$$

Being a subspace of the space of $N\times N$ matrices with complex entries, $\mathcal{H}_N$ embeds naturally into $\mathbb{C}^{N^2}$. The Euclidean distance inherited from this embedding is given as

$$\begin{aligned}\operatorname{dist}(L,M) = \|L-M\| &= \Big(\sum_{j,k=1}^{N} |L_{jk}-M_{jk}|^2\Big)^{1/2} \\ &= \Big(\sum_{j=1}^{N}|L_{jj}-M_{jj}|^2 + 2\sum_{j>k}^{N}|L_{jk}-M_{jk}|^2\Big)^{1/2}.\end{aligned} \tag{1.1.3}$$

The scalar product (L,M) and the distance $\operatorname{dist}(L,M)$ are invariant with respect to the conjugation by any unitary matrix $U\in \mathrm{U}(N)$,

$$M \to U^{-1}MU, \quad U \in \mathrm{U}(N). \tag{1.1.4}$$

Let $\mathrm{d}M$ be the N^2-dimensional Lebesgue measure,

$$\mathrm{d}M = \prod_{j=1}^{N} \mathrm{d}M_{jj} \prod_{j\neq k}^{N} \mathrm{d}\operatorname{Re} M_{jk}\,\mathrm{d}\operatorname{Im} M_{jk}. \tag{1.1.5}$$

We will consider the probability distribution on $\mathcal{H}_N$ given by

$$\mathrm{d}\mu_N(M) = \frac{1}{Z_N}\,\mathrm{e}^{-N\operatorname{Tr} V(M)}\,\mathrm{d}M, \tag{1.1.6}$$

where $V(x)$ is a real analytic function satisfying the growth condition that

$$\frac{V(x)}{\log(|x|^2+1)} \to +\infty \quad \text{as } |x| \to \infty. \tag{1.1.7}$$

This growth condition is a technical condition for subsequent analysis. Indeed, the reader may simply think of V as a polynomial of even degree and with positive leading coefficient, in which case it is clear what is meant by the matrix $V(M)$ in (1.1.6). If V is not a polynomial, then the matrix $V(M)$ can be understood by applying V to the spectrum of M. That is, M can be diagonalized as $M = U\Lambda U^*$, where U is a unitary matrix and $\Lambda = \operatorname{diag}(\lambda_1, \lambda_2, \ldots, \lambda_N)$ is the matrix of eigenvalues, and we can then define

$$V(M) = U \operatorname{diag}\big(V(\lambda_1), V(\lambda_2), \ldots, V(\lambda_N)\big)U^*. \tag{1.1.8}$$

Notice then that $\operatorname{Tr} V(M)$ is invariant with respect to unitary conjugations given in (1.1.4). Since the distance $\operatorname{dist}(L, M)$ induces the measure $2^{N(N-1)/2}\,\mathrm{d}M$ on $\mathcal{H}_N$, the Lebesgue measure $\mathrm{d}M$ is invariant with respect to unitary conjugations (1.1.4) as well. It follows that the distribution $\mathrm{d}\mu_N(M)$ is invariant with respect to any unitary conjugation (1.1.4), hence the name of the ensemble. The normalizing constant Z_N, called the partition function, is defined such that μ_N is a probability measure. That is, it is the matrix integral

$$Z_N = \int_{\mathcal{H}_N} \mathrm{e}^{-N \operatorname{Tr} V(M)}\,\mathrm{d}M. \tag{1.1.9}$$

Example (Gaussian unitary ensemble). For $V(M) = M^2$, the measure μ_N is the probability distribution of the Gaussian unitary ensemble (GUE). This is the oldest and most well known of the invariant matrix ensembles. In this case,

$$\operatorname{Tr} V(M) = \operatorname{Tr} M^2 = \sum_{j,k=1}^{N} M_{kj}M_{jk} = \sum_{j=1}^{N} M_{jj}^2 + 2\sum_{j>k}|M_{jk}|^2, \tag{1.1.10}$$

hence

$$\mathrm{d}\mu_N^{\mathrm{GUE}}(M) = \frac{1}{Z_N^{\mathrm{GUE}}} \prod_{j=1}^{N}\big(\mathrm{e}^{-NM_{jj}^2}\big) \prod_{j>k}\big(\mathrm{e}^{-2N|M_{jk}|^2}\big)\,\mathrm{d}M, \tag{1.1.11}$$

so that the matrix elements in GUE are independent Gaussian random variables. The partition function of GUE is evaluated as

$$\begin{aligned} Z_N^{\mathrm{GUE}} &= \int_{\mathcal{H}_N} \prod_{j=1}^{N}\big(\mathrm{e}^{-NM_{jj}^2}\big) \prod_{j>k}\big(\mathrm{e}^{-2N|M_{jk}|^2}\big)\,\mathrm{d}M \\ &= \left(\frac{\pi}{N}\right)^{N/2}\left(\frac{\pi}{2N}\right)^{N(N-1)/2} = \left(\frac{\pi}{N}\right)^{N^2/2}\left(\frac{1}{2}\right)^{N(N-1)/2}. \end{aligned} \tag{1.1.12}$$

The GUE is somewhat special in that it lies at the intersection of the invariant ensembles, which are invariant with respect to some sort of matrix conjugation (in this case unitary conjugation), and the Wigner ensembles, for which the matrix entries are independent. If the function $V(x)$ is not quadratic, then the matrix entries become dependent.

1.2. Ensemble of eigenvalues

The central topic in random matrix theory is the distribution of the eigenvalues of a random matrix. We can write a formula for the distribution of eigenvalues of an Hermitian matrix M from distribution (1.1.6) by writing M in terms of its eigenvalues and eigenvectors, so that $M = U\Lambda U^*$, where U is a unitary matrix (of eigenvectors) and $\Lambda = \mathrm{diag}(\lambda_1, \lambda_2, \ldots, \lambda_N)$ is the matrix of eigenvalues. In order to make this map one-to-one, let us consider $\lambda = (\lambda_1, \lambda_2, \ldots, \lambda_N)$ to be ordered, so that $\lambda_1 \le \lambda_2 \le \cdots \le \lambda_N$. In fact, dropping a set of measure zero, we can assume $\lambda_j \ne \lambda_k$ for $j \ne k$, and thus consider λ in the Weyl chamber

$$\lambda_1 < \lambda_2 < \cdots < \lambda_N. \tag{1.2.1}$$

Also, instead of $U \in \mathrm{U}(N)$ we may consider matrix UD, where

$$D = \mathrm{diag}(\mathrm{e}^{\mathrm{i}\vartheta_1}, \ldots, \mathrm{e}^{\mathrm{i}\vartheta_N}) \in \mathrm{D}(N)$$

is any diagonal unitary matrix. Consider therefore the equivalence class

$$\widetilde{U} = \{UD, D \in \mathrm{D}(N)\}, \tag{1.2.2}$$

and the homogeneous space $\widetilde{\mathrm{U}}(N)$ of the equivalence classes $\widetilde{U}$. Then the map

$$(\widetilde{U}, \Lambda) \mapsto M = U\Lambda U^*, \quad U \in \widetilde{U}, \tag{1.2.3}$$

is one-to-one, and we may consider its Jacobian

$$J = \frac{\mathrm{d}M}{\mathrm{d}\widetilde{U}\,\mathrm{d}\lambda}, \tag{1.2.4}$$

where $\mathrm{d}\widetilde{U}$ is the projection of the Haar measure on $\mathrm{U}(N)$ onto $\widetilde{\mathrm{U}}(N)$ and $\mathrm{d}\lambda = \mathrm{d}\lambda_1 \cdots \mathrm{d}\lambda_N$. Since $\mathrm{d}M$ is invariant with respect to the unitary conjugations, and $\mathrm{d}\widetilde{U}$ is invariant with respect to the unitary left shifts, the Jacobian J does not depend on $\widetilde{U}$. Its dependence on $\lambda = (\lambda_1, \ldots, \lambda_N)$ is described as follows.

Proposition 1.2.1 (Weyl's formula). *For some constant $C_N > 0$,*

$$J = C_N \prod_{j<k} |\lambda_k - \lambda_j|^2. \tag{1.2.5}$$

PROOF. Since J does not depend on $\widetilde{U}$, it suffices to evaluate J at $U = I$, i.e., at $\widetilde{U} = \tilde{I} = \mathrm{D}(N)$. In a small neighborhood of $\tilde{I}$, the elements $\widetilde{U}$ are uniquely represented by unitary matrices $U = \mathrm{e}^A$, where $A^* = -A$. As $A \to 0$,

$$U = \mathrm{e}^A = I + A + \mathcal{O}(A^2), \quad U^* = \mathrm{e}^{-A} = I - A + \mathcal{O}(A^2), \tag{1.2.6}$$

and

$$M = U\Lambda U^{-1} = \Lambda + [A, \Lambda] + \mathcal{O}(A^2), \tag{1.2.7}$$

so that

$$M_{ii} = \lambda_i + \mathcal{O}(A^2), \qquad M_{ij} = (\lambda_j - \lambda_i)A_{ij} + \mathcal{O}(A^2), \quad i < j. \tag{1.2.8}$$

This implies that at $A = 0$,

$$\begin{aligned} &\frac{\partial M_{ii}}{\partial \lambda_k} = \delta_{ik}, \qquad &&\frac{\partial M_{ii}}{\partial A_{kl}} = 0, \\ &\frac{\partial M_{ij}}{\partial \lambda_k} = 0, \qquad &&\frac{\partial M_{ij}}{\partial A_{kl}} = (\lambda_j - \lambda_i)\delta_{ik}\delta_{jl}, \quad i < j, \end{aligned} \tag{1.2.9}$$

hence the Jacobian matrix $\nabla_{\lambda,A}M$ is diagonal and its determinant is equal to

$$J = \prod_{j<k} |\lambda_k - \lambda_j|^2. \tag{1.2.10}$$

Since at $A = 0$,

$$\mathrm{d}\widetilde{U} = C_N \,\mathrm{d}A, \quad C_N > 0, \tag{1.2.11}$$

formula (1.2.5) follows. □

It follows that, if we are interested only in eigenvalues, the eigenvectors may be integrated out, and we find that the distribution of eigenvalues of M with respect to the ensemble μ_N is given as

$$\mathrm{d}\mu_N(\lambda) = \frac{1}{\widetilde{Z}_N} \prod_{j>k} (\lambda_j - \lambda_k)^2 \prod_{j=1}^{N} \mathrm{e}^{-NV(\lambda_j)} \,\mathrm{d}\lambda, \tag{1.2.12}$$

where

$$\widetilde{Z}_N = \int \prod_{j>k} (\lambda_j - \lambda_k)^2 \prod_{j=1}^{N} \mathrm{e}^{-NV(\lambda_j)} \,\mathrm{d}\lambda, \quad \mathrm{d}\lambda = \mathrm{d}\lambda_1 \cdots \mathrm{d}\lambda_N. \tag{1.2.13}$$

Since we have considered λ in the Weyl chamber (1.2.1), this gives a measure on ordered eigenvalues, and the integral (1.2.13) is over the Weyl chamber. However, clearly (1.2.12) is symmetric in λ, and therefore the measure can be lifted to an ensemble of unordered eigenvalues, where the integral in (1.2.13) is then understood to be over $\mathbb{R}^N$.

Notice that

$$Z_N = \widetilde{Z}_N \frac{C_N \operatorname{Vol}(\widetilde{\mathrm{U}}(N))}{N!}, \tag{1.2.14}$$

and thus the ratio $\widetilde{Z}_N / Z_N$ does not depend on the potential V. We can calculate this ratio in the case of the GUE, and the result will hold for any unitary ensemble given by (1.1.6). Indeed, for GUE,

$$\mathrm{d}\mu_N^{\mathrm{GUE}}(\lambda) = \frac{1}{\widetilde{Z}_N^{\mathrm{GUE}}} \prod_{j>k} (\lambda_j - \lambda_k)^2 \prod_{j=1}^{N} \mathrm{e}^{-N\lambda_j^2} \,\mathrm{d}\lambda, \tag{1.2.15}$$

where

$$\widetilde{Z}_N^{\mathrm{GUE}} = \int \prod_{j>k} (\lambda_j - \lambda_k)^2 \prod_{j=1}^{N} \mathrm{e}^{-N\lambda_j^2} \,\mathrm{d}\lambda. \tag{1.2.16}$$

The constant $\widetilde{Z}_N^{\mathrm{GUE}}$ is a Selberg integral, and its exact value is

$$\widetilde{Z}_N^{\mathrm{GUE}} = \frac{(2\pi)^{N/2}}{(2N)^{N^2/2}} \prod_{k=1}^{N} k!, \tag{1.2.17}$$

see, e.g., [61]. A proof of formula (1.2.17) from the discrete string equations for orthogonal polynomials is given subsequently in Section 1.3. We therefore have that the partition functions Z_N and $\widetilde{Z}_N$ are related as

$$\frac{\widetilde{Z}_N}{Z_N} = \frac{\widetilde{Z}_N^{\mathrm{GUE}}}{Z_N^{\mathrm{GUE}}} = \frac{1}{\pi^{N(N-1)/2}} \prod_{k=1}^{N} k!. \tag{1.2.18}$$

One of the main problems in random matrix theory is to evaluate the large N asymptotics of the partition function $\widetilde{Z}_N$ and of the correlations between eigenvalues.

From (1.2.12), the joint probability density function for the eigenvalues is given by

$$p_N(x_1,\dots,x_N)=\widetilde{Z}_N^{-1}\prod_{j>k}(x_j-x_k)^2\prod_{j=1}^{N}e^{-NV(x_j)}. \tag{1.2.19}$$

Integrating out $(N-m)$ variables, we obtain the marginal probability density function for m eigenvalues,

$$p_{mN}(x_1,\dots,x_m)=\int_{\mathbb{R}^{N-m}}p_N(x_1,\dots,x_N)\,\mathrm{d}x_{m+1}\cdots\mathrm{d}x_N. \tag{1.2.20}$$

The m-point correlation function is then defined as

$$R_{mN}(x_1,\dots,x_m):=\frac{N!}{(N-m)!}p_{mN}(x_1,\dots,x_m), \tag{1.2.21}$$

see, e.g., [2,3,36,61]. Remarkably, these correlation functions can all be expressed in terms of a system of orthogonal polynomials. Let $\{P_k(x)\}_{k=0}^{\infty}$ be the system of monic orthogonal polynomials defined from the orthogonality condition

$$\int_{-\infty}^{\infty}P_j(x)P_k(x)\mathrm{e}^{-NV(x)}\,\mathrm{d}x=h_k\delta_{jk}, \tag{1.2.22}$$

for some system of normalizing constants $\{h_k\}_{k=0}^{\infty}$. Existence and uniqueness of these polynomials is guaranteed by condition (1.1.7). Define also the functions

$$\psi_k(x)=\frac{1}{h_k^{1/2}}P_k(x)\mathrm{e}^{-NV(x)/2}, \tag{1.2.23}$$

which form an orthonormal basis in $L^2(\mathbb{R}^1)$. We have the following proposition.

Proposition 1.2.2. *The correlation function* (1.2.21) *has the determinantal form*

$$R_{mN}(x_1,\dots,x_m)=\det\big(K_N(x_k,x_l)\big)_{k,l=1}^{m}, \tag{1.2.24}$$

where

$$K_N(x,y)=\sum_{n=0}^{N-1}\psi_n(x)\psi_n(y). \tag{1.2.25}$$

Furthermore, the partition function $\widetilde{Z}_N$ can be written in terms of the orthogonal polynomials (1.2.22) *as*

$$\widetilde{Z}_N=N!\prod_{j=0}^{N-1}h_j. \tag{1.2.26}$$

An ensemble whose correlations can be expressed by such a determinantal formula is called a *determinantal point process*, see [18,20,44,45]. In particular notice that the one point correlation function enables us to write the density of eigenvalues on the real line, which we notate ρ_N in the simple form

$$\rho_N(x):=\frac{R_{1N}(x)}{N}=\frac{K_N(x,x)}{N}. \tag{1.2.27}$$

Before proving Proposition 1.2.2, let us point out some unique properties of the function K_N. Observe that K_N is the kernel of the projection operator onto the N-dimensional space generated by the first N functions ψ_n, $n = 0, \ldots, N-1$. The function $K_N(x, y)$ is called the *reproducing kernel* and it has the following properties:

$$\begin{aligned} &\int_{\mathbb{R}} K_N(x,x)\,\mathrm{d}x = N, \\ &\int_{\mathbb{R}} K_N(x,y)K_N(y,z)\,\mathrm{d}y = K(x,z). \end{aligned} \tag{1.2.28}$$

Indeed, by (1.2.25),

$$\int_{\mathbb{R}} K_N(x,x)\,\mathrm{d}x = \sum_{j=0}^{N-1} \int_{\mathbb{R}} \psi_j(x)^2\,\mathrm{d}x = \sum_{j=0}^{N-1} 1 = N, \tag{1.2.29}$$

and

$$\begin{aligned} \int_{\mathbb{R}} K_N(x,y)K_N(y,z)\,\mathrm{d}y &= \sum_{j,k=0}^{N-1} \int_{\mathbb{R}} \psi_j(x)\psi_j(y)\psi_k(y)\psi_k(z)\,\mathrm{d}y \\ &= \sum_{j=0}^{N-1} \psi_j(x)\psi_j(z) = K(x,z). \end{aligned} \tag{1.2.30}$$

Let us now prove formula (1.2.26) for the partition function. Recall the formula for the Vandermonde determinant,

$$\det\left[x_j^{k-1}\right]_{j,k=1}^N = \prod_{\substack{j,k=1 \\ j<k}}^{N} (x_k - x_j). \tag{1.2.31}$$

The main point in the proof of (1.2.26) is that the function $p_N(x_1, \ldots, x_N)$ in the integrand of (1.2.16) is the product of the square of the Vandermonde determinant and factors which are independent and identical on each of the coordinates x_j. The form of the Vandermonde matrix and multilinearity of the determinant function allow us to replace the jth row of the Vandermonde matrix with any monic polynomial of degree $(j-1)$. In particular, we may use the orthogonal polynomials described in (1.2.22), so that

$$\begin{aligned} \widetilde{Z}_N &= \int_{\mathbb{R}^N} p_N(x_1, \ldots, x_N)\,\mathrm{d}x_1 \cdots \mathrm{d}x_N \\ &= \int_{\mathbb{R}^N} \det \begin{pmatrix} 1 & 1 & \ldots & 1 \\ x_1 & x_2 & \ldots & x_N \\ x_1^2 & x_2^2 & \ldots & x_N^2 \\ \vdots & \vdots & & \vdots \\ x_1^{N-1} & x_2^{N-1} & \ldots & x_N^{N-1} \end{pmatrix}^2 \prod_{j=1}^N \mathrm{e}^{-NV(x_j)}\,\mathrm{d}x_j \end{aligned} \tag{1.2.32}$$

$$= \int_{\mathbb{R}^N} \det \begin{pmatrix} 1 & 1 & \dots & 1 \\ P_1(x_1) & P_1(x_2) & \dots & P_1(x_N) \\ P_2(x_1) & P_2(x_2) & \dots & P_2(x_N) \\ \vdots & \vdots & & \vdots \\ P_{N-1}(x_1) & P_{N-1}(x_2) & \dots & P_{N-1}(x_N) \end{pmatrix}^2 \prod_{j=1}^N \mathrm{e}^{-NV(x_j)}\, \mathrm{d}x_j$$

$$= \int_{\mathbb{R}^N} \Big(\sum_{\sigma\in S_N} (-1)^\sigma \prod_{j=1}^N P_{\sigma(j)-1}(x_j) \Big)^2 \prod_{j=1}^N \mathrm{e}^{-NV(x_j)}\, \mathrm{d}x_j$$

$$= \int_{\mathbb{R}^N} \sum_{\sigma\in S_N} \sum_{\sigma'\in S_N} (-1)^{\sigma+\sigma'} \prod_{j=1}^N P_{\sigma(j)-1}(x_j) P_{\sigma'(j)-1}(x_j) \prod_{j=1}^N \mathrm{e}^{-NV(x_j)}\, \mathrm{d}x_j$$

$$= \sum_{\sigma\in S_N} \sum_{\sigma'\in S_N} (-1)^{\sigma+\sigma'} \prod_{j=1}^N \int_{\mathbb{R}} P_{\sigma(j)-1}(x_j) P_{\sigma'(j)-1}(x_j) \mathrm{e}^{-NV(x_j)}\, \mathrm{d}x_j.$$

The orthogonality condition ensures that only diagonal terms, $\sigma = \sigma'$, are nonzero. Thus we have

$$\widetilde{Z}_N = \sum_{\sigma\in S_N} \prod_{j=1}^N \int_{\mathbb{R}} P^2_{\sigma(j)-1}(x_j) \mathrm{e}^{-NV(x_j)}\, \mathrm{d}x_j = N! \prod_{j=0}^{N-1} h_j. \tag{1.2.33}$$

Let us now prove the formula (1.2.24) for the correlation kernel.

Consider first the case $m = N$, when

(1.2.34)

$$R_{NN}(x_1,\dots,x_N) = N! p_N(x_1,\dots,x_N) = \frac{N!}{\widetilde{Z}_N} \prod_{j>k} (\lambda_j - \lambda_k)^2 \prod_{j=1}^N \mathrm{e}^{-NV(\lambda_j)}$$

$$= \frac{1}{\prod_{j=0}^{N-1} h_j} \det \begin{pmatrix} 1 & 1 & \dots & 1 \\ x_1 & x_2 & \dots & x_N \\ x_1^2 & x_2^2 & \dots & x_N^2 \\ \vdots & \vdots & & \vdots \\ x_1^{N-1} & x_2^{N-1} & \dots & x_N^{N-1} \end{pmatrix}^2 \prod_{j=1}^N \mathrm{e}^{-NV(x_j)}.$$

As in (1.2.32), we obtain that

(1.2.35) $R_{NN}(x_1,\dots,x_N)$

$$= \det \begin{pmatrix} 1 & 1 & \dots & 1 \\ P_1(x_1) & P_1(x_2) & \dots & P_1(x_N) \\ P_2(x_1) & P_2(x_2) & \dots & P_2(x_N) \\ \vdots & \vdots & & \vdots \\ P_{N-1}(x_1) & P_{N-1}(x_2) & \dots & P_{N-1}(x_N) \end{pmatrix}^2 \frac{\prod_{j=1}^N \mathrm{e}^{-NV(x_j)}}{\prod_{j=0}^{N-1} h_j}$$

$$= \det \begin{pmatrix} \psi_0(x_1) & \psi_0(x_2) & \dots & \psi_0(x_N) \\ \psi_1(x_1) & \psi_1(x_2) & \dots & \psi_1(x_N) \\ \psi_2(x_1) & \psi_2(x_2) & \dots & \psi_2(x_N) \\ \vdots & \vdots & & \vdots \\ \psi_{N-1}(x_1) & \psi_{N-1}(x_2) & \dots & \psi_{N-1}(x_N) \end{pmatrix}^2,$$

hence

$$
\begin{aligned}
(1.2.36)\quad & R_{NN}(x_1,\dots,x_N) \\
&= \det \begin{pmatrix} \psi_0(x_1) & \psi_1(x_1) & \dots & \psi_{N-1}(x_1) \\ \psi_0(x_2) & \psi_1(x_2) & \dots & \psi_{N-1}(x_2) \\ \vdots & \vdots & & \vdots \\ \psi_0(x_N) & \psi_1(x_N) & \dots & \psi_{N-1}(x_N) \end{pmatrix} \begin{pmatrix} \psi_0(x_1) & \psi_0(x_2) & \dots & \psi_0(x_N) \\ \psi_1(x_1) & \psi_1(x_2) & \dots & \psi_1(x_N) \\ \vdots & \vdots & & \vdots \\ \psi_{N-1}(x_1) & \psi_{N-1}(x_2) & \dots & \psi_{N-1}(x_N) \end{pmatrix} \\
&= \det\big(K_N(x_k,x_l)\big)_{k,l=1}^N .
\end{aligned}
$$

We will prove (1.2.24) by induction in m, starting from $m = N$ down, but we first need to prove the following inductive lemma.

Lemma 1.2.3. *For any* $1 < m \le N$,

$$
(1.2.37)\quad \int_{\mathbb{R}} \det\big(K_N(x_j,x_k)\big)_{j,k=1}^m \, dx_m = (N-m+1)\det\big(K_N(x_j,x_k)\big)_{j,k=1}^{m-1}.
$$

PROOF. Let us expand $\det\big(K_N(x_j,x_k)\big)_{j,k=1}^m$ with respect to the last column:

$$
\begin{aligned}
(1.2.38)\quad & \det\big(K_N(x_j,x_k)\big)_{j,k=1}^m \\
&= \sum_{i=1}^m (-1)^{m+i} K_N(x_i,x_m) \det\big(K_N(x_j,x_k)\big)_{\substack{1\le j\le m, j\ne i \\ 1\le k\le m-1}}.
\end{aligned}
$$

When we integrate the last term, $i = m$, in this formula, we obtain, by (1.2.28),

$$
(1.2.39)\quad \int_{\mathbb{R}} K_N(x_m,x_m)\det\big(K_N(x_j,x_k)\big)_{j,k=1}^{m-1} dx_m = N\det\big(K_N(x_j,x_k)\big)_{j,k=1}^{m-1}.
$$

When we integrate the ith term in (1.2.38), we expand $\det\big(K_N(x_j,x_k)\big)_{\substack{1\le j\le m, j\ne i \\ 1\le k\le m-1}}$ with respect to the last row:

$$
\begin{aligned}
(1.2.40)\quad & \det\big(K_N(x_j,x_k)\big)_{\substack{1\le j\le m, j\ne i \\ 1\le k\le m-1}} \\
&= \sum_{l=1}^{m-1} (-1)^{m-1+l} K_N(x_m,x_l) \det\big(K_N(x_j,x_k)\big)_{\substack{1\le j\le m, j\ne i \\ 1\le k\le m-1, k\ne l}}
\end{aligned}
$$

By (1.2.28), the integration of the ith term in (1.2.38) gives

$$
(1.2.41)\quad \sum_{l=1}^{m-1} (-1)^{i+l-1} K_N(x_i,x_l) \det\big(K_N(x_j,x_k)\big)_{\substack{1\le j\le m, j\ne i \\ 1\le k\le m-1, k\ne l}},
$$

which coincides with the expansion of $-\det\big(K_N(x_j,x_k)\big)_{j,k=1}^{m-1}$ with respect to the ith row. Thus,

$$
\begin{aligned}
(1.2.42)\quad & \int_{\mathbb{R}} \det\big(K_N(x_j,x_k)\big)_{j,k=1}^m \, dx_m \\
&= N\det\big(K_N(x_j,x_k)\big)_{j,k=1}^{m-1} - \sum_{i=1}^{m-1} \det\big(K_N(x_j,x_k)\big)_{j,k=1}^{m-1} \\
&= (N-m+1)\det\big(K_N(x_j,x_k)\big)_{j,k=1}^{m-1}.
\end{aligned}
$$

Lemma 1.2.3 is proved. □

Assume now that formula (1.2.24) is already proven for some $m \leq N$. Let us prove it for $m-1$. From (1.2.21) we have that

$$R_{m-1,N}(x_1,\dots,x_{m-1}) = \frac{1}{N-m+1}\int_{\mathbb{R}} R_{mN}(x_1,\dots,x_{m-1},x_m)\,\mathrm{d}x_m, \tag{1.2.43}$$

hence by the inductive assumption and Lemma 1.2.3,

$$\begin{aligned} R_{m-1,N}(x_1,\dots,x_{m-1}) &= \frac{1}{N-m+1}\int_{\mathbb{R}} \det\big(K_N(x_j,x_k)\big)_{j,k=1}^{m}\,\mathrm{d}x_m \\ &= \det\big(K_N(x_j,x_k)\big)_{j,k=1}^{m-1}. \end{aligned} \tag{1.2.44}$$

This proves formula (1.2.24) by induction.

The kernel K_N can be expressed in terms of ψ_{N-1} and ψ_N only, due to the Christoffel–Darboux formula, which is presented in the next section. We also consider recurrence and differential equations for the functions ψ_n.

1.3. Recurrence equations and discrete string equations for orthogonal polynomials

An important feature of orthogonal polynomials on the real line is the *three-term recurrence relation*, see, e.g., [72],

$$\begin{aligned} xP_n(x) &= P_{n+1}(x) + \beta_n P_n(x) + \gamma_n^2 P_{n-1}(x), \\ \gamma_n &= \left(\frac{h_n}{h_{n-1}}\right)^{1/2} > 0, \quad n \geq 1; \qquad \gamma_0 = 0. \end{aligned} \tag{1.3.1}$$

For the functions ψ_n it reads as

$$x\psi_n(x) = \gamma_{n+1}\psi_{n+1}(x) + \beta_n\psi_n(x) + \gamma_n\psi_{n-1}(x). \tag{1.3.2}$$

This allows the calculation

$$\begin{aligned} (x-y)&\sum_{n=0}^{N-1}\psi_n(x)\psi_n(y) \\ &= \sum_{n=0}^{N-1}\big[\big(\gamma_{n+1}\psi_{n+1}(x) + \beta_n\psi_n(x) + \gamma_n\psi_{n-1}(x)\big)\psi_n(y) \\ &\qquad\qquad - \psi_n(x)\big(\gamma_{n+1}\psi_{n+1}(y) + \beta_n\psi_n(y) + \gamma_n\psi_{n-1}(y)\big)\big] \\ &= \gamma_N[\psi_N(x)\psi_{N-1}(y) - \psi_{N-1}(x)\psi_N(y)], \end{aligned} \tag{1.3.3}$$

as the first line of (1.3.3) gives a telescoping sum. Hence,

$$K_N(x,y) = \sum_{n=0}^{N-1}\psi_n(x)\psi_n(y) = \gamma_N\frac{\psi_N(x)\psi_{N-1}(y) - \psi_{N-1}(x)\psi_N(y)}{x-y}, \tag{1.3.4}$$

which is the *Christoffel–Darboux formula*. For the eigenvalue density function $K_N(x,x)/N$ we take the limit as $y \to x$, obtaining that

$$\rho_N(x) = \frac{K_N(x,x)}{N} = \frac{\gamma_N}{N}[\psi_N'(x)\psi_{N-1}(x) - \psi_{N-1}'(x)\psi_N(x)]. \tag{1.3.5}$$

We now use the recurrence (1.3.2) derive the discrete string equations for orthogonal polynomials. To do so, we will consider two operators on $L^2(\mathbb{R})$. The first operator is that of multiplication by x. If we think of $L^2(\mathbb{R})$ in the basis

$\{\psi_n(x)\}$, then, according to the three-term recurrence relation, this operator has the symmetric tridiagonal Jacobi matrix,

$$Q = \begin{pmatrix} \beta_0 & \gamma_1 & 0 & 0 & \dots \\ \gamma_1 & \beta_1 & \gamma_2 & 0 & \dots \\ 0 & \gamma_2 & \beta_2 & \gamma_3 & \dots \\ 0 & 0 & \gamma_3 & \beta_3 & \dots \\ \vdots & \vdots & \vdots & \vdots & \ddots \end{pmatrix}. \tag{1.3.6}$$

The other operator we will consider is that of differentiation. Let $P = (P_{nm})_{n,m=0}^{\infty}$ be a matrix of the operator $f(z) \mapsto f'(z)$ in the basis $\{\psi_n(z)\}_{n=0}^{\infty}$, so that

$$\psi_n'(x) = \sum_{m=0}^{\infty} P_{nm}\psi_m(x), \tag{1.3.7}$$

or equivalently

$$P_{nm} = \int_{-\infty}^{\infty} \psi_n'(x)\psi_m(x)\,\mathrm{d}x. \tag{1.3.8}$$

Integrating by parts gives that $P_{mn} = -P_{nm}$. Since $P_n'(z)$ is a polynomial of degree strictly less than n, we have

$$\begin{aligned} \psi_n'(z) &= -\frac{NV'(z)}{2}\psi_n(z) + \frac{P_n'(z)}{\sqrt{h_n}}\mathrm{e}^{-NV(z)/2} \\ &= -\frac{NV'(z)}{2}\psi_n(z) + \frac{n}{\gamma_n}\psi_{n-1}(z) + \cdots, \end{aligned} \tag{1.3.9}$$

hence

$$\left[P + \frac{NV'(Q)}{2}\right]_{nm} = 0 \quad \text{for } m \geq n. \tag{1.3.10}$$

Due to the antisymmetry of the matrix P, equation (1.3.10) gives a full characterization of the matrix P in terms of the matrix $V'(Q)$. More specifically, we have

$$P_{nm} = \begin{cases} \left[-\dfrac{NV'(Q)}{2}\right]_{nm} & \text{for } m \geq n \\ \left[\dfrac{NV'(Q)}{2}\right]_{nm} & \text{for } m \leq n. \end{cases} \tag{1.3.11}$$

Since $P_{nn} = 0$, we obtain that

$$[V'(Q)]_{nn} = 0. \tag{1.3.12}$$

In addition,

$$\left[-P + \frac{NV'(Q)}{2}\right]_{n,n-1} = 0, \quad \left[P + \frac{NV'(Q)}{2}\right]_{n,n-1} = \frac{n}{\gamma_n}, \tag{1.3.13}$$

hence

$$\gamma_n[V'(Q)]_{n,n-1} = \frac{n}{N}. \tag{1.3.14}$$

Thus we have the *discrete string equations* for the recurrence coefficients,

$$\begin{cases} \gamma_n [V'(Q)]_{n,n-1} = \dfrac{n}{N}, \\ [V'(Q)]_{nn} = 0. \end{cases} \tag{1.3.15}$$

The string equations can be brought to a variational form.

Proposition 1.3.1. *Define the infinite Hamiltonian,*

$$H(\gamma,\beta) = N \operatorname{Tr} V(Q) - \sum_{n=1}^{\infty} n \log \gamma_n^2, \quad \gamma = (\gamma_0, \gamma_1, \dots),\ \beta = (\beta_0, \beta_1, \dots). \tag{1.3.16}$$

Then equations (1.3.15) *can be written as*

$$\frac{\partial H}{\partial \gamma_n} = 0, \quad \frac{\partial H}{\partial \beta_n} = 0; \qquad n \ge 1, \tag{1.3.17}$$

which are the Euler – Lagrange equations for the Hamiltonian H.

PROOF. We have that

$$\frac{\partial H}{\partial \gamma_n} = N \operatorname{Tr}\left(V'(Q) \frac{\partial Q}{\partial \gamma_n} \right) - \frac{2n}{\gamma_n} = 2N[V'(Q)]_{n,n-1} - \frac{2n}{\gamma_n}, \tag{1.3.18}$$

and

$$\frac{\partial H}{\partial \beta_n} = N \operatorname{Tr}\left(V'(Q) \frac{\partial Q}{\partial \beta_n} \right) = N[V'(Q)]_{nn}, \tag{1.3.19}$$

hence equations (1.3.15) are equivalent to (1.3.17). □

Example (The even quartic model). Let us take

$$V(M) = \frac{t}{2} M^2 + \frac{g}{4} M^4, \tag{1.3.20}$$

for $g > 0$, $t \in \mathbb{R}$. In this case, since V is even, $\beta_n = 0$, and we have one string equation,

$$\gamma_n^2 (t + g\gamma_{n-1}^2 + g\gamma_n^2 + g\gamma_{n+1}^2) = \frac{n}{N}, \tag{1.3.21}$$

with the initial conditions

$$\gamma_0 = 0, \quad \gamma_1 = \frac{\int_{-\infty}^{\infty} z^2 \mathrm{e}^{-NV(z)}\, \mathrm{d}z}{\int_{-\infty}^{\infty} \mathrm{e}^{-NV(z)}\, \mathrm{d}z}. \tag{1.3.22}$$

The Hamiltonian is

$$H(\gamma) = \sum_{n=1}^{\infty} \left[\frac{N}{2} \gamma_n^2 (2t + g\gamma_{n-1}^2 + g\gamma_n^2 + g\gamma_{n+1}^2) - n \log \gamma_n^2 \right]. \tag{1.3.23}$$

The minimization of the functional H gives a useful procedure for a numerical solution of the string equations, see [11,22]. The initial value problem (1.3.21) and (1.3.22) for the string equations is difficult to solve numerically, as the recursion is very unstable. However, the minimization of H with $\gamma_0 = 0$ and some boundary conditions at $n = N$, say $\gamma_N = 0$, works very well. In fact, the boundary condition at $n = N$ creates a narrow boundary layer near $n = N$, and it does not affect significantly the main part of the graph of γ_n^2. Figure 1.1 presents a computer solution, $y = \gamma_n^2$, of the string equation for the quartic model: $g = 1$, $t = -1$,

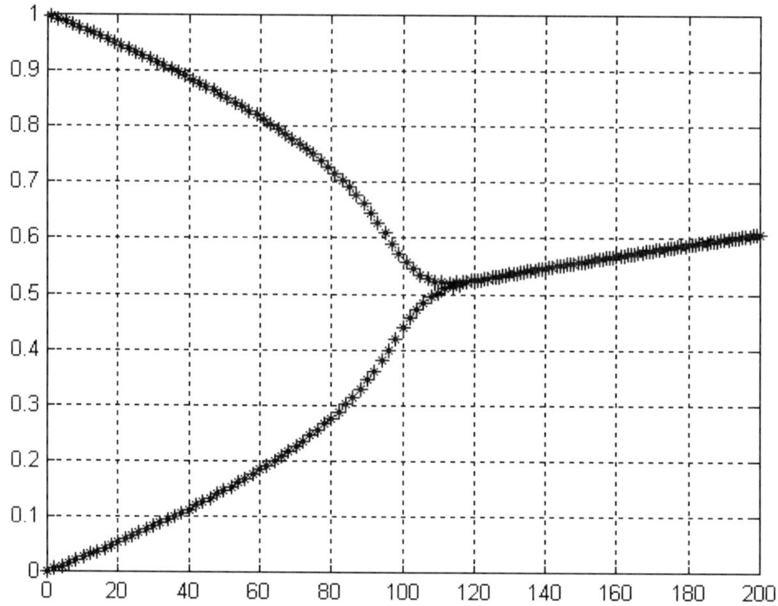

FIGURE 1.1. A computer solution, $y = \gamma_n^2$, of the string equation for the quartic model: $g = 1$, $t = -1$, $N = 400$.

$N = 400$. For this solution, it is shown in [10, 11] that there is a critical value, $\lambda_c = \frac{1}{4}$, so that for any $\varepsilon > 0$, as $N \to \infty$,

$$\gamma_n^2 = R\left(\frac{n}{N}\right) + \mathcal{O}(N^{-1}), \quad \text{if } \frac{n}{N} \geq \lambda_c + \varepsilon, \tag{1.3.24}$$

and

$$\gamma_n^2 = \begin{cases} R\left(\frac{n}{N}\right) + \mathcal{O}(N^{-1}), & n = 2k+1, \\ L\left(\frac{n}{N}\right) + \mathcal{O}(N^{-1}), & n = 2k, \end{cases} \qquad \text{if } \frac{n}{N} \leq \lambda_c - \varepsilon. \tag{1.3.25}$$

The functions R for $\lambda \geq \lambda_c$ and R, L for $\lambda \leq \lambda_c$ can be found from the string equation (1.3.21):

$$R(\lambda) = \frac{1 + \sqrt{1 + 12\lambda}}{6}, \quad \lambda > \lambda_c, \tag{1.3.26}$$

and

$$R(\lambda), L(\lambda) = \frac{1 \pm \sqrt{1 - 4\lambda}}{2}, \quad \lambda < \lambda_c. \tag{1.3.27}$$

PROOF OF FORMULA (1.2.17). We can use the the discrete string equations to obtain the recurrence coefficients for the Hermite polynomials, and therefore the formula (1.2.17). Setting $t = 2$ and $g = 0$ in (1.3.20) we obtain $V(M) = M^2$, the Gaussian unitary ensemble. Formula (1.3.21) reduces to

$$\gamma_n^2 = \frac{n}{2N}, \tag{1.3.28}$$

which are the recurrence coefficients of the Hermite polynomials. We obtain from (1.3.1) that

$$\frac{h_n}{h_{n-1}} = \frac{n}{2N}. \tag{1.3.29}$$

Since

$$h_0 = \int_{-\infty}^{\infty} \mathrm{e}^{-Nx^2}\,\mathrm{d}x = \sqrt{\frac{\pi}{N}}, \tag{1.3.30}$$

(1.3.29) gives that

$$h_n = h_0 \prod_{k=1}^{n} \frac{k}{2N} = \frac{\sqrt{\pi}\,n!}{2^n N^{(2n+1)/2}}. \tag{1.3.31}$$

Applying now formula (1.2.26), we obtain that

$$\widetilde{Z}_N^{\mathrm{GUE}} = N! \prod_{n=0}^{N-1} h_n = N! \prod_{n=0}^{N-1} \frac{\sqrt{\pi}\,n!}{2^n N^{(2n+1)/2}} = \frac{(2\pi)^{N/2}}{(2N)^{N^2/2}} \prod_{n=1}^{N} n!. \tag{1.3.32}$$

which gives a proof to Selberg's formula (1.2.17). □

We will discuss in the next chapter how to justify asymptotics (1.3.24), (1.3.25), and their extension for a general V.

1.4. Deformation equations for the recurrence coefficients

In this section we prove some differential identities for the recurrence coefficients of a system of orthogonal polynomials. These identities are in fact quite general, and apply to systems whose measure of orthogonality may not be absolutely continuous with respect to Lebesgue measure. We will consider the setting in which $\{P_n(x)\}_{n=0}^{\infty}$ is a system of monic polynomials satisfying the orthogonality condition

$$\int_{\mathbb{R}} P_n(x)P_m(x)\mathrm{e}^{-Nu_kx^k}\,\mathrm{d}\mu(x) = h_n\delta_{mn}, \tag{1.4.1}$$

where $\mathrm{d}\mu(x)$ is any measure on $\mathbb{R}$ such that the system of orthogonal polynomials exists. We consider deformations of this system with respect to the parameter u_k. Let us rewrite the recurrence equation (1.3.1), explicitly noting the dependence of each recurrence coefficient on the parameter u_k:

$$xP_n(x) = P_{n+1}(x)+\beta_n(u_k)P_n(x)+\gamma_n^2(u_k)P_{n-1}(x), \quad \gamma_n^2(u_k) = \frac{h_n(u_k)}{h_{n-1}(u_k)}. \tag{1.4.2}$$

We have the following theorem (see [37,41]).

Theorem 1.4.1. *Let Q be the semi-infinite matrix defined in* (1.3.1). *As functions of the parameter u_k, the recurrence coefficients in equation* (1.4.2) *satisfy the following differential identities*:

$$h_n'(u_k) = -N[Q^k]_{nn}h_n(u_k) \tag{1.4.3}$$

$$\gamma_n'(u_k) = \frac{N\gamma_n(u_k)}{2}([Q^k]_{n-1,n-1} - [Q^k]_{nn}) \tag{1.4.4}$$

$$\beta_n'(u_k) = N(\gamma_n(u_k)[Q^k]_{n-1,n} - \gamma_{n+1}(u_k)[Q^k]_{n,n+1}). \tag{1.4.5}$$

PROOF. Recall that

$$P_n(x) = x^n + p_{n,n-1}x^{n-1} + p_{n,n-2}x^{n-2} + \cdots, \tag{1.4.6}$$

where each of the coefficients $p_{n,j}$ depend on the parameter u_k as well. Differentiating (1.4.1) for $n = m$ with respect to u_k gives

$$\int_{\mathbb{R}} \left[-Nx^k P_n(x)P_n(x) + 2P_n(x)\left(\frac{\mathrm{d}}{\mathrm{d}u_k}P_n(x)\right)\right] \mathrm{e}^{-Nu_kx^k}\,\mathrm{d}\mu(x) = h_n'(u_k). \tag{1.4.7}$$

According to (1.4.6)), $(\mathrm{d}/\,\mathrm{d}u_k)P_n(x)$ is a polynomial of degree at most $n-1$, thus

$$\int_{\mathbb{R}} P_n(x)\left(\frac{\mathrm{d}}{\mathrm{d}u_k}P_n(x)\right)\mathrm{e}^{-Nu_kx^k}\,\mathrm{d}\mu(x) = 0,$$

and we have

$$h_n'(u_k) = \int_{\mathbb{R}} -Nx^k P_n(x)P_n(x)\mathrm{e}^{-Nu_kx^k}\,\mathrm{d}\mu(x), \tag{1.4.8}$$

which is (1.4.3).

If we differentiate the identity

$$\gamma_n^2(u_k) = \frac{h_n(u_k)}{h_{n-1}(u_k)}$$

with respect to u_k, then apply the differential identity (1.4.3), we get

$$\begin{aligned} 2\gamma_n(u_k)\gamma_n'(u_k) &= \frac{h_n'(u_k)h_{n-1}(u_k) - h_n(u_k)h_{n-1}'(u_k)}{h_{n-1}(u_k)^2} \\ &= \frac{Nh_n(t)}{h_{n-1}(t)}([Q^k]_{n-1,n-1} - [Q^k]_{nn}), \end{aligned} \tag{1.4.9}$$

which simplifies to (1.4.4).

If we differentiate (1.4.1) for $m = n-1$ with respect to u_k, we find that

$$\begin{aligned} 0 &= \int_{\mathbb{R}} \left[-Nx^k P_{n-1}(x)P_n(x) + P_{n-1}(x)\left(\frac{\mathrm{d}}{\mathrm{d}u_k}P_n(x)\right)\right] \mathrm{e}^{-Nu_kx^k}\,\mathrm{d}\mu(x) \\ &= -\frac{N}{\gamma_n}[Q^k]_{n-1,n}h_n(u_k) + h_{n-1}(u_k)p_{n,n-1}'(u_k), \end{aligned} \tag{1.4.10}$$

which is

$$p_{n,n-1}'(u_k) = N\gamma_n[Q^k]_{n-1,n}. \tag{1.4.11}$$

By matching the coefficient of the x^n-term in the three term recurrence equation (1.4.2), we find that

$$\beta_n(u_k) = p_{n,n-1}(u_k) - p_{n+1,n}(u_k). \tag{1.4.12}$$

Combining (1.4.11) with (1.4.12) gives (1.4.5). □

In particular if we consider deformations with respect to the linear term in the potential, we can apply Theorem 1.4.1 with $k = 1$, obtaining

$$\frac{\partial}{\partial u_1}h_n = -N\beta_n h_n, \tag{1.4.13}$$

$$\frac{\partial}{\partial u_1}\gamma_n = \frac{N\gamma_n}{2}(\beta_{n-1} - \beta_n), \tag{1.4.14}$$

$$\frac{\partial}{\partial u_1}\beta_n = N(\gamma_n^2 - \gamma_{n+1}^2). \tag{1.4.15}$$

Also, if we consider deformations with respect to the quadratic term in the potential, we can apply Theorem 1.4.1 with $k = 2$, obtaining

$$\frac{\partial}{\partial u_2} h_n = -N h_n(\gamma_n^2 + \beta_n^2 + \gamma_{n+1}^2), \tag{1.4.16}$$

$$\frac{\partial}{\partial u_2} \gamma_n = \frac{N\gamma_n}{2}(\gamma_{n-1}^2 - \gamma_{n+1}^2 + \beta_{n-1}^2 - \beta_n^2), \tag{1.4.17}$$

$$\frac{\partial}{\partial u_2} \beta_n = N(\gamma_n^2 \beta_{n-1} + \gamma_n^2 \beta_n - \gamma_{n+1}^2 \beta_n - \gamma_{n+1}^2 \beta_{n+1}). \tag{1.4.18}$$

We can now use these deformation equations to derive some deformation equations for the partition function of the ensemble of eigenvalues in a random matrix ensemble. Recall the formula for the partition function given in (1.2.33):

$$\widetilde{Z}_N = N! \prod_{j=0}^{N-1} h_j. \tag{1.4.19}$$

We have the following theorem.

Theorem 1.4.2. *The partition function $\widetilde{Z}_N$ satisfies the identities*

$$\frac{\partial^2}{\partial u_1^2}(\log \widetilde{Z}_N) = N^2 \gamma_N^2 \tag{1.4.20}$$

$$\frac{\partial^2}{\partial u_2^2}(\log \widetilde{Z}_N) = N^2 \gamma_N^2(\gamma_{N+1}^2 + \gamma_{N-1}^2 + \beta_{N-1}^2 + \beta_N^2 + 2\beta_{N-1}\beta_N). \tag{1.4.21}$$

PROOF. Differentiating (1.4.19) with respect to u_1 and using the differential identity (1.4.13) gives

$$\frac{\partial}{\partial u_1} \widetilde{Z}_N = -N\widetilde{Z}_N \sum_{j=0}^{N-1} \beta_j. \tag{1.4.22}$$

Differentiating again with respect to u_1 and applying (1.4.13) and (1.4.15) gives

$$\frac{\partial^2}{\partial u_1^2} \widetilde{Z}_N = N^2 \left(\frac{\partial}{\partial u_1} \widetilde{Z}_N\right) \left(\sum_{j=0}^{n-1} \beta_j\right)^2 - N\left(\frac{\partial}{\partial u_1} \widetilde{Z}_N\right) \sum_{j=0}^{n-1} N(\gamma_j^2 - \gamma_{j+1}^2). \tag{1.4.23}$$

Using (1.4.22) and noting the telescoping sum in (1.4.23), we can write

$$\begin{aligned} \frac{\partial^2}{\partial u_1^2} \widetilde{Z}_N &= \frac{(\partial \widetilde{Z}_N / \partial u_1)^2}{\widetilde{Z}_N} + N^2 \widetilde{Z}_N \gamma_n^2 \\ &= \frac{(\partial \widetilde{Z}_N / \partial u_1)^2}{\widetilde{Z}_N} + N^2 \widetilde{Z}_N \frac{h_n}{h_{n-1}}. \end{aligned} \tag{1.4.24}$$

Equivalently, we can write

$$\frac{\partial^2 \widetilde{Z}_N / \partial u_1^2}{\widetilde{Z}_N} - \left(\frac{\partial \widetilde{Z}_N / \partial u_1}{\widetilde{Z}_N}\right)^2 = N^2 \frac{\widetilde{Z}_{N+1} \widetilde{Z}_{N-1}}{\widetilde{Z}_N^2} \tag{1.4.25}$$

or

$$\frac{\partial^2}{\partial u_1^2}(\log \widetilde{Z}_N) = N^2 \frac{\widetilde{Z}_{N+1} \widetilde{Z}_{N-1}}{\widetilde{Z}_N^2}.$$

This equation is clearly equivalent to (1.4.20), and is called the *Toda differential equation.*

To prove (1.4.21), we can write (1.4.16) as

$$\frac{\partial}{\partial u_2}(\log h_j) = -N(\gamma_j^2 + \beta_j^2 + \gamma_{j+1}^2). \tag{1.4.26}$$

Differentiating this expression once again with respect to u_2 and applying (1.4.17) and (1.4.18), it is straightforward to see that

$$\frac{\partial^2}{\partial u_2^2}(\log h_j) = -N^2(I_j - I_{j+1}), \tag{1.4.27}$$

where

$$I_j = \gamma_j^2(\gamma_{j+1}^2 + \gamma_{j-1}^2 + \beta_{j-1}^2 + \beta_j^2 + 2\beta_{j-1}\beta_j). \tag{1.4.28}$$

It follows that

$$\begin{aligned} \frac{\partial^2}{\partial u_2^2}\left(\log \widetilde{Z}_N\right) &= \sum_{j=0}^{N-1} \frac{\partial^2}{\partial u_2^2}\left(\log h_j\right) \\ &= -N^2 \sum_{j=0}^{N-1} I_j - I_{j+1} \\ &= -N^2(I_0 - I_N) \\ &= N^2 I_N, \end{aligned} \tag{1.4.29}$$

since $I_0 = \gamma_0 = 0$. This proves (1.4.21). □

1.5. Differential equations and Lax pair for the ψ-functions

Define

$$\vec{\Psi}_n(z) = \begin{pmatrix} \psi_n(z) \\ \psi_{n-1}(z) \end{pmatrix}. \tag{1.5.1}$$

We have the following theorem (see [37]).

Theorem 1.5.1. *The vector $\vec{\Psi}_n(z)$ satisfies the ODE*

$$\vec{\Psi}_n'(z) = NA_n(z)\vec{\Psi}_n(z), \tag{1.5.2}$$

where

$$A_n(z) = \begin{pmatrix} -V'(z)/2 - \gamma_n u_n(z) & \gamma_n v_n(z) \\ -\gamma_n v_{n-1}(z) & V'(z)/2 + \gamma_n u_n(z) \end{pmatrix} \tag{1.5.3}$$

and

$$u_n(z) = [W(Q,z)]_{n,n-1}, \quad v_n(z) = [W(Q,z)]_{nn}, \tag{1.5.4}$$

where

$$W(Q,z) = \frac{V'(Q) - V'(z)}{Q - z}. \tag{1.5.5}$$

Observe that $\operatorname{Tr} A_n(z) = 0$.

The three-term recurrence relation (1.3.1) can be written as

$$\vec{\Psi}_{n+1}(z) = U_n(z)\vec{\Psi}_n(z), \tag{1.5.6}$$

where

$$U_n(z) = \begin{pmatrix} \gamma_{n+1}^{-1}(z-\beta_n) & -\gamma_{n+1}^{-1}\gamma_n \\ 1 & 0 \end{pmatrix} \tag{1.5.7}$$

Differential equation 1.5.2 and recurrence equation (1.5.6) form the Lax pair for discrete string equations (1.3.15). This means that the compatibility condition of (1.5.2) and (1.5.6),

$$U_n' = N(A_{n+1}U_n - U_nA_n), \tag{1.5.8}$$

when written for the matrix elements, implies (1.3.15).

Example (Even quartic model). For $V(M) = (t/2)M^2 + (g/4)M^4$, the matrix $A_n(z)$ is

$$A_n(z) = \begin{pmatrix} -\frac{1}{2}(tz+gz^3) - g\gamma_n^2 z & \gamma_n(gz^2+\theta_n) \\ -\gamma_n(gz^2+\theta_{n-1}) & \frac{1}{2}(tz+gz^3) + g\gamma_n^2 z \end{pmatrix} \tag{1.5.9}$$

where

$$\theta_n = t + g\gamma_n^2 + g\gamma_{n+1}^2. \tag{1.5.10}$$

CHAPTER 2

The Riemann – Hilbert Problem for Orthogonal Polynomials

In this chapter, we present a powerful method for computing asymptotic properties of orthogonal polynomials. As in the previous chapter, we will denote by $\{P_n(z)\}_{n=0}^{\infty}$ the orthogonal polynomials defined by the orthogonality

$$\int_{\mathbb{R}} P_n(x)P_m(x)w(x)\,\mathrm{d}x = h_n\delta_{mn} \quad w(x) = \mathrm{e}^{-NV(x)}, \tag{2.0.1}$$

for V real analytic satisfying (1.1.7). The idea of this asymptotic method is to write a certain matrix-valued Riemann – Hilbert problem whose solution is constructed from the orthogonal polynomials. Independent of the solution in terms of orthogonal polynomials, the Riemann – Hilbert problem can be studied in the large N limit, which in turn gives us large N asymptotic properties of orthogonal polynomials.

2.1. The Cauchy transform and its properties

The *Cauchy transform* of the product $P_n(x)w(x)$ is defined as

$$C(P_nw)(z) = \frac{1}{2\pi\mathrm{i}}\int_{-\infty}^{\infty} \frac{P_n(u)\,w(u)\,\mathrm{d}u}{u-z}, \qquad z \in \mathbb{C}\setminus\mathbb{R}. \tag{2.1.1}$$

For $z \in \mathbb{R}$, the above integral is clearly singular, but $C(P_nw)(z)$ takes limiting values as z approaches the real line from the upper and lower half planes. Define

$$C(P_nw)_{\pm}(x) = \lim_{\substack{z\to x\\ \pm\,\mathrm{Im}\,z>0}} C(P_nw)(z), \qquad -\infty < x < \infty. \tag{2.1.2}$$

Then the well-known Plemelj – Sokhotski formula (see, e.g., [64]) for the jump of the Cauchy type integral gives that

$$C(P_nw)_+(x) - C(P_nw)_-(x) = w(x)P_n(x). \tag{2.1.3}$$

Expanding $(u-z)^{-1}$ as a geometric series, we see that the asymptotics of $C(P_nw)(z)$ as $z \to \infty$, $z \in \mathbb{C}$, are given as

$$\begin{aligned} C(P_nw)(z) &= \frac{1}{2\pi\mathrm{i}}\int_{-\infty}^{\infty} \frac{w(u)P_n(u)\,\mathrm{d}u}{u-z} \\ &\sim -\frac{1}{2\pi\mathrm{i}}\int_{-\infty}^{\infty} w(u)P_n(u)\sum_{j=0}^{\infty}\frac{u^j}{z^{j+1}}\,\mathrm{d}u. \end{aligned} \tag{2.1.4}$$

Due to the orthogonality, the first n terms in this expansion are zero, and we have that

$$C(P_nw)(z) \sim -\frac{h_n}{2\pi\mathrm{i}z^{n+1}} + \sum_{j=n+2}^{\infty}\frac{\alpha_j}{z^j}, \tag{2.1.5}$$

for some constants α_j. The sign $\sim$ in (2.1.4) and (2.1.5) means an asymptotic expansion, so that for any $k \ge n+2$, there exists a constant $C_k > 0$ such that for all $z \in \mathbb{C}$,

$$C(P_n w)(z) - \left(-\frac{h_n}{2\pi \mathrm{i} z^{n+1}} + \sum_{j=n+2}^{k} \frac{\alpha_j}{z^j}\right) \le \frac{C_k}{(1+|z|)^{k+1}}. \tag{2.1.6}$$

There can be some doubts about the uniformity of this asymptotics near the real axis, but we assume that the weight $w(z)$ is analytic in a strip $\{z : |\operatorname{Im} z| \le a\}$, $a > 0$, hence the contour of integration in (2.1.4) can be shifted, and (2.1.6) holds uniformly in the complex plane.

2.2. The Riemann – Hilbert problem

Introduce now the matrix-valued function,

$$\mathbf{Y}_n(z) = \begin{pmatrix} P_n(z) & C(P_n w)(z) \\ c\,P_{n-1}(z) & c\,C(P_{n-1}w)(z) \end{pmatrix}, \tag{2.2.1}$$

where the constant,

$$c = -\frac{2\pi \mathrm{i}}{h_{n-1}}, \tag{2.2.2}$$

is chosen in such a way that

$$c\,C\big(P_{n-1}w)(z)\big) \sim \frac{1}{z^n} + \dots, \tag{2.2.3}$$

see (2.1.5). The function $\mathbf{Y}_n$ solves the following Riemann–Hilbert problem (RHP):

(1) $\mathbf{Y}_n(z)$ is analytic on $\mathbb{C}^+ \equiv \{\operatorname{Im} z \ge 0\}$ and $\mathbb{C}^- \equiv \{\operatorname{Im} z \le 0\}$ and is two-valued on $\mathbb{R} = \mathbb{C}^+ \cap \mathbb{C}^-$.

(2) For any real x,

$$\mathbf{Y}_{n+}(x) = \mathbf{Y}_{n-}(x) j_Y(x), \quad j_Y(x) = \begin{pmatrix} 1 & w(x) \\ 0 & 1 \end{pmatrix}. \tag{2.2.4}$$

(3) As $z \to \infty$,

$$\mathbf{Y}_n(z) \sim \left(I + \sum_{k=1}^{\infty} \frac{\mathbf{Y}_k}{z^k}\right) \begin{pmatrix} z^n & 0 \\ 0 & z^{-n} \end{pmatrix} \tag{2.2.5}$$

where $\mathbf{Y}_k$, $k = 1, 2, \dots$, are some constant 2×2 matrices.

Observe that (2.2.4) follows from (2.1.3), while (2.2.5) from (2.2.3). The RHP (1)–(3) has some nice properties.

First of all, (2.2.1) is the only solution of the RHP. Let us sketch a proof of the uniqueness. It follows from (2.2.4) that

$$\det \mathbf{Y}_{n+}(x) = \det \mathbf{Y}_{n-}(x), \tag{2.2.6}$$

hence $\det \mathbf{Y}_n(z)$ has no jump on the real axis, and thus $\det \mathbf{Y}_n(z)$ is an entire function. At infinity, by (2.2.5),

$$\det \mathbf{Y}_n(z) \sim 1 + \cdots, \tag{2.2.7}$$

hence by the Liouville theorem

$$\det \mathbf{Y}_n(z) \equiv 1. \tag{2.2.8}$$

In particular, $\mathbf{Y}_n(z)$ is invertible for any z. Suppose that $\widetilde{\mathbf{Y}}_n$ is another solution of the RHP. Then $\mathbf{X}_n = \widetilde{\mathbf{Y}}_n \mathbf{Y}_n^{-1}$ satisfies

$$\mathbf{X}_{n+}(x) = \widetilde{\mathbf{Y}}_{n+}(x)\mathbf{Y}_{n+}(x)^{-1} = \widetilde{\mathbf{Y}}_{n-}(x)j_Y(x)j_Y(x)^{-1}\mathbf{Y}_{n-}^{-1}(x) = \mathbf{X}_{n-}(x), \tag{2.2.9}$$

hence $\mathbf{X}_n$ is an entire matrix-valued function. At infinity, by (2.2.5),

$$\mathbf{X}_n(z) \sim I + \cdots \tag{2.2.10}$$

hence $\mathbf{X}_n(z) \equiv I$, by the Liouville theorem. This implies that $\widetilde{\mathbf{Y}}_n = \mathbf{Y}_n$, the uniqueness.

The recurrence coefficients for the orthogonal polynomials are explicitly given in the expansion of $\mathbf{Y}(z)$ at $z = \infty$ as

$$\gamma_n^2 = [\mathbf{Y}_1]_{21}[\mathbf{Y}_1]_{12}, \tag{2.2.11}$$

and

$$\beta_{n-1} = \frac{[\mathbf{Y}_2]_{21}}{[\mathbf{Y}_1]_{21}} - [\mathbf{Y}_1]_{11}. \tag{2.2.12}$$

Indeed, from (2.2.1), (2.2.5),

$$\mathbf{Y}_n(z)\begin{pmatrix} z^{-n} & 0 \\ 0 & z^n \end{pmatrix} = \begin{pmatrix} z^{-n}P_n(z) & z^n C(P_n w)(z) \\ z^{-n}cP_{n-1}(z) & z^n cC(P_{n-1}w)(z) \end{pmatrix} \sim I + \sum_{k=1}^{\infty} \frac{\mathbf{Y}_k}{z^k}, \tag{2.2.13}$$

hence by (2.1.4), (2.2.2), and (1.3.1)

$$[\mathbf{Y}_1]_{21}[\mathbf{Y}_1]_{12} = \left(-\frac{2\pi \mathrm{i}}{h_{n-1}}\right)\left(-\frac{h_n}{2\pi \mathrm{i}}\right) = \frac{h_n}{h_{n-1}} = \gamma_n^2, \tag{2.2.14}$$

which proves (2.2.11). Also,

$$\frac{[\mathbf{Y}_2]_{21}}{[\mathbf{Y}_1]_{21}} - [\mathbf{Y}_1]_{11} = p_{n-1,n-2} - p_{n,n-1}, \tag{2.2.15}$$

where

$$P_n(z) = \sum_{j=0}^{n} p_{nj} z^j. \tag{2.2.16}$$

By matching coefficients in the z^n term in the recursion (1.3.1) we obtain that

$$p_{n-1,n-2} - p_{n,n-1} = \beta_{n-1}, \tag{2.2.17}$$

hence (2.2.12) follows. The normalizing constant h_n can be found as

$$h_n = -2\pi \mathrm{i}[\mathbf{Y}_1]_{12}, \quad h_{n-1} = -\frac{2\pi \mathrm{i}}{[\mathbf{Y}_1]_{21}}. \tag{2.2.18}$$

The reproducing kernel $K_N(x,y)$ of the eigenvalue correlation functions (see (1.2.24)) is expressed in terms of $\mathbf{Y}_{N+}(x)$ as follows:

$$\begin{aligned} K_N(x,y) \\ = -\mathrm{e}^{-NV(x)/2}\mathrm{e}^{-NV(y)/2}\frac{1}{2\pi \mathrm{i}(x-y)}\begin{pmatrix} 0 & 1 \end{pmatrix} \mathbf{Y}_{N+}^{-1}(y)\mathbf{Y}_{N+}(x)\begin{pmatrix} 1 \\ 0 \end{pmatrix}. \end{aligned} \tag{2.2.19}$$

Indeed, by (1.3.4),

$$\begin{aligned}
K_N(x,y) &= \gamma_N \frac{\psi_N(x)\psi_{N-1}(y) - \psi_{N-1}(x)\psi_N(y)}{x-y} \\
&= \mathrm{e}^{-NV(x)/2}\mathrm{e}^{-NV(y)/2}\frac{\gamma_N}{\sqrt{h_N h_{N-1}}} \\
&\quad \times \frac{P_N(x)P_{N-1}(y) - P_{N-1}(x)P_N(y)}{x-y}.
\end{aligned} \tag{2.2.20}$$

From (2.2.1), (2.2.2) and (2.2.8), we obtain that

$$\begin{aligned}
&-\begin{pmatrix}0 & 1\end{pmatrix} \mathbf{Y}_{N+}^{-1}(y)\mathbf{Y}_{N+}(x)\begin{pmatrix}1\\0\end{pmatrix} \\
&\qquad = \frac{2\pi \mathrm{i}}{h_{N-1}}[P_N(x)P_{N-1}(y) - P_{N-1}(x)P_N(y)].
\end{aligned} \tag{2.2.21}$$

Also,

$$\frac{\gamma_N}{\sqrt{h_N h_{N-1}}} = \frac{1}{h_{N-1}}, \tag{2.2.22}$$

hence equation (2.2.19) follows.

2.3. Distribution of eigenvalues and equilibrium measure

2.3.1. Log gas interpretation and heuristics. One basic problem in random matrix theory is to find the limiting global distribution of eigenvalues. In order to deal with this problem, it is useful to think of the eigenvalues of an $N\times N$ matrix as an ensemble of interacting particles on the real line. We can rewrite the distribution of eigenvalues (1.2.13) as

$$\mathrm{d}\mu_N(\lambda) = \widetilde{Z}_N^{-1}\mathrm{e}^{-H_N(\lambda)}\,\mathrm{d}\lambda, \tag{2.3.1}$$

where

$$H_N(\lambda) = -\sum_{j\neq k}\log|\lambda_j - \lambda_k| + N\sum_{j=1}^{N} V(\lambda_j). \tag{2.3.2}$$

This distribution describes a type of *log gas* ensemble; that is, it gives configurations of N particles on the real line which repel each other logarithmically and are confined to the finite space by the potential $NV(x)$. To see the limiting distribution of the particles, we can introduce for each $\lambda = (\lambda_1, \lambda_2, \dots, \lambda_N)$ the normalized counting measure on $\mathbb{R}^1$,

$$\mathrm{d}\nu_\lambda(x) = N^{-1}\sum_{j=1}^{N}\delta(x-\lambda_j)\,\mathrm{d}x. \tag{2.3.3}$$

Then (2.3.2) can be rewritten as

$$H_N(\lambda) = N^2\left[-\iint_{x\neq y}\log|x-y|\,\mathrm{d}\nu_\lambda(x)\,\mathrm{d}\nu_\lambda(y) + \int V(x)\,\mathrm{d}\nu_\lambda(x)\right]. \tag{2.3.4}$$

Let ν be an arbitrary probability measure on $\mathbb{R}^1$. Set

$$I_V(\nu) = -\iint_{x\neq y}\log|x-y|\,\mathrm{d}\nu(x)\,\mathrm{d}\nu(y) + \int V(x)\,\mathrm{d}\nu(x). \tag{2.3.5}$$

Then (2.3.1) reads

$$d\mu_N(\lambda) = \widetilde{Z}_N^{-1} e^{-N^2 I_V(\nu_\lambda)}\, d\lambda. \tag{2.3.6}$$

Because of the factor N^2 in the exponent, we expect that for large N the measure μ_N is concentrated near the minimum of the functional I_V. We call this measure the *equilibrium measure*, and will denote it ν_V. Of course this argument that ν_V is in fact the limiting distribution of eigenvalues is largely heuristic, but it is helpful for our intuition. We will eventually prove that this is indeed the case.

2.3.2. The equilibrium measure. The structure of the equilibrium measure has been well studied, and we can summarize what is known in the following proposition, which collects results from several sources. Consider the minimization problem

$$E_V = \inf_{\nu\in M_1(\mathbb{R})} I_V(\nu), \tag{2.3.7}$$

where

$$M_1(\mathbb{R}) = \left\{\nu : \nu \geq 0, \int_{\mathbb{R}} d\nu = 1\right\}, \tag{2.3.8}$$

is the set of probability measures on the line.

Proposition 2.3.1 (See [28]). *The infimum of $I_V(\nu)$ is attained uniquely at a measure $\nu = \nu_V$, which is called an equilibrium measure. The measure ν_V is absolutely continuous, and it is supported by a finite union of intervals, $J = \bigcup_{j=1}^q [a_j, b_j]$. On each interval of support, its density has the form*

$$\rho_V(x) \equiv \frac{d\nu_V}{dx}(x) = \frac{1}{2\pi i} h(x) R_+^{1/2}(x), \quad R(x) = \prod_{j=1}^q (x-a_j)(x-b_j). \tag{2.3.9}$$

Here $h(x)$ is a function which is real analytic and nonnegative on the interval of support, and $R^{1/2}(x)$ is the branch with cuts on J, which is positive for large positive x, and $R_+^{1/2}(x)$ is the value of $R^{1/2}(x)$ on the upper part of the cut. In the case that $V(x)$ is a polynomial, then the function $h(x)$ is a polynomial, which is the polynomial part of the function $V'(x)/R^{1/2}(x)$ at infinity, i.e.,

$$\frac{V'(x)}{R^{1/2}(x)} = h(x) + \mathcal{O}(x^{-1}). \tag{2.3.10}$$

In particular, $\deg h = \deg V - 1 - q$.

There is a useful formula for the equilibrium density [28]:

$$\frac{d\nu_V(x)}{dx} = \frac{1}{\pi}\sqrt{q(x)}, \tag{2.3.11}$$

where

$$q(x) = -\left(\frac{V'(x)}{2}\right)^2 + \int \frac{V'(x) - V'(y)}{x-y}\, d\nu_V(y). \tag{2.3.12}$$

This in fact is an equation in q, since the right-hand side contains an integration with respect to ν_V. Nevertheless, if V is a polynomial of degree $p = 2p_0$, then (2.3.12) determines uniquely more then a half of the coefficients of the polynomial q,

$$q(x) = -\left(\frac{V'(x)}{2}\right)^2 - \mathcal{O}(x^{p-2}). \tag{2.3.13}$$

Example. If $V(x)$ is *convex* then ν_V is regular (see Section 2.3.3), and the support of ν_V consists of a single interval, see, e.g., [54]. Hence by (2.3.13), for the Gaussian ensemble $V(x) = x^2$, $q(x) = a^2 - x^2$. Since

$$\int_{-a}^{a} \frac{1}{\pi}\sqrt{a^2 - x^2}\,\mathrm{d}x = 1,$$

we find that $a = \sqrt{2}$, hence

$$\rho_V(x) = \frac{1}{\pi}\sqrt{2 - x^2}, \quad |x| \le \sqrt{2}, \tag{2.3.14}$$

which is the well known Wigner semicircle law.

2.3.3. The Euler – Lagrange variational conditions. A nice and important property of minimization problem (2.3.7) is that the minimizer is uniquely determined by the Euler – Lagrange variational conditions, which are obtained simply by taking the variational derivative of the functional (2.3.5) subject to the constraint that ν be a positive probability measure, (see [28]): for some real constant *Lagrange multiplier* l,

$$2\int_{\mathbb{R}} \log|x - y|\,\mathrm{d}\nu_V(y) - V(x) = l, \quad \text{for } x \in J, \tag{2.3.15}$$

$$2\int_{\mathbb{R}} \log|x - y|\,\mathrm{d}\nu_V(y) - V(x) \le l, \quad \text{for } x \in \mathbb{R} \setminus J. \tag{2.3.16}$$

We can classify equilibrium measure as either *regular* or *singular* in the following way.

Definition (See [29]). The equilibrium measure,

$$\mathrm{d}\nu_V(x) = \frac{1}{2\pi\mathrm{i}} h(x) R_+^{1/2}(x)\,\mathrm{d}x, \tag{2.3.17}$$

is called regular (otherwise singular) if

(1) $h(x) \neq 0$ on the (closed) set J.
(2) Inequality (2.3.16) is strict,

$$2\int \log|x - y|\,\mathrm{d}\nu_V(y) - V(x) < l, \quad \text{for } x \in \mathbb{R} \setminus J. \tag{2.3.18}$$

It can be shown that regular equilibrium measures are generic in the sense that the set of polynomials V of a given degree d, which yield a singular equilibrium measure is nowhere dense in the space of all polynomials of the degree d, see [54].

2.3.4. Construction of the equilibrium measure: equations on the end-points. The strategy for constructing the equilibrium measure is the following: first we find the end-points of the support, and then we use equation (2.3.10) to find $h(x)$ and hence the density. The number q of cuts is not in general known, and we try different q's. Consider the resolvent,

$$\omega(z) = \int_J \frac{\mathrm{d}\nu_V(x)}{z - x}, \quad z \in \mathbb{C} \setminus J. \tag{2.3.19}$$

This function can be very useful to construct ρ_V, and has the following properties.

Properties of $\omega(z)$.

(1) $\omega(z)$ is analytic on the set $\mathbb{C} \setminus J$.

(2) On J, the equilibrium condition (2.3.15)) implies that

$$\omega_+(x) + \omega_-(x) = V'(x), \tag{2.3.20}$$

and the Plemelj – Sokhotski formula implies

$$\omega_+(x) - \omega_-(x) = -2\pi \mathrm{i} \rho_V(x). \tag{2.3.21}$$

(3) The Euler – Lagrange variational condition, together with (2.3.21) and (2.3.9) imply that

$$\omega(z) = \frac{V'(z)}{2} - \frac{h(z)R^{1/2}(z)}{2}. \tag{2.3.22}$$

(4) As $z \to \infty$,

$$\omega(z) = \frac{1}{z} + \frac{m_1}{z^2} + \cdots, \qquad m_k = \int_J x^k \rho_V(x)\,\mathrm{d}x. \tag{2.3.23}$$

Quadratic equation on $\omega(z)$. We have that for $x \in J$,

$$\omega_+^2(x) - \omega_-^2(x) = \big(\omega_+(x) - \omega_-(x)\big)\big(\omega_+(x) + \omega_-(x)\big) = -2\pi \mathrm{i} \rho_V(x) V'(x) \tag{2.3.24}$$

hence

$$\omega^2(z) = \int_J \frac{V'(x)\,\mathrm{d}\nu_V(x)}{z-x}, \qquad z \in \mathbb{C} \setminus J, \tag{2.3.25}$$

(recall that $\mathrm{d}\nu_V(x) = \rho_V(x)\,\mathrm{d}x$). Also,

$$\int_J \frac{V'(z)\,\mathrm{d}\nu_V(x)}{z-x} = V'(z)\omega(z), \qquad z \in \mathbb{C} \setminus J. \tag{2.3.26}$$

By adding these equations, we obtain the quadratic equation in $\omega(z)$,

$$\omega^2(z) - V'(z)\omega(z) + \int_J \frac{V'(z) - V'(x)}{z-x}\, d\nu_V(x) = 0. \tag{2.3.27}$$

Solving it we obtain that

$$\omega(z) = \frac{V'(z)}{2} - \sqrt{\left(\frac{V'(z)}{2}\right)^2 - \int_J \frac{V'(z) - V'(x)}{z-x}\,\mathrm{d}\nu_V(x)}, \tag{2.3.28}$$

where the minus sign in front of the square root follows from the asymptotic behavior as $z \to \infty$. Observe that considering the jump of $\omega(z)$ on J, we obtain formulas (2.3.11) – (2.3.12). The equation

$$\frac{V'(z)}{2} - \frac{h(z)R^{1/2}(z)}{2} = \frac{1}{z} + \mathcal{O}(z^{-2}). \tag{2.3.29}$$

gives $q+1$ equations on $a_1, b_1, \ldots, a_q, b_q$, if we substitute formula (2.3.10) for h. The remaining $q-1$ equation are

$$\int_{b_j}^{a_{j+1}} h(x)R^{1/2}(x)\,\mathrm{d}x = 0, \qquad j = 1, \ldots, q-1, \tag{2.3.30}$$

which follow from (2.3.21) and (2.3.22). Observe that the latter equations imply that there is at least one root of $h(x)$ between b_j and a_{j+1}. Hence, when V (and therefore h) is a polynomial, we have $\deg h \geq q-1$. Since $\deg h = \deg V - 1 - q$, we obtain that

$$q \leq \frac{\deg V}{2}. \tag{2.3.31}$$

Example. The even quartic model, $V(M) = (t/2)M^2 + \frac{1}{4}M^4$. By (2.3.31), either $q=1$ or $q=2$. Suppose first $q=1$, so that we have a one-cut equilibrium measure. By the symmetry, the cut is on the symmetric interval $[-a,a]$, and

$$h(z) = kz^2 + c. \tag{2.3.32}$$

To find a, k and c, we look at the asymptotic behavior of different terms in equation (2.3.29) as $z \to \infty$. By equating the coefficients at different powers of z, we obtain that $k=1$ and

$$2c - a^2 = 2t, \quad a^4 + 4ca^2 = 16. \tag{2.3.33}$$

By solving these equations we find that

$$a^2 = \frac{-2t + 2(t^2+12)^{1/2}}{3}, \quad c = \frac{2t + 2(t^2+12)^{1/2}}{3}. \tag{2.3.34}$$

The condition $c \geq 0$ gives that $t \geq -2$. Thus, the one-cut solution exists if $t \geq t_c = -2$, and the density function is equal in this case to

$$\rho_V(x) = \frac{1}{2\pi}(x^2+c)\sqrt{a^2-x^2}. \tag{2.3.35}$$

In particular, for $t=-2$,

$$\rho_V(x) = \frac{1}{2\pi}x^2\sqrt{4-x^2}. \tag{2.3.36}$$

For $t < -2$, the support consists of two intervals, and from the even symmetry the support must be on the two intervals $[-a,-b]$ and $[b,a]$. To calculate the density and to find a and b, we can use the equation (2.3.29). To cancel the z^3 term in the expansion at $z=\infty$, we must have $h(z) \equiv z$, from which it follows that

$$\rho_V(x) = \frac{1}{2\pi}|x|\sqrt{(a^2-x^2)(x^2-b^2)}. \tag{2.3.37}$$

To find a and b, we match the z and z^{-1} terms in (2.3.29) to obtain the equations

$$\frac{t}{2} + \frac{a^2+b^2}{4} = 0 \quad \frac{(a^2+b^2)^2}{16} - \frac{a^2b^2}{4} = 1, \tag{2.3.38}$$

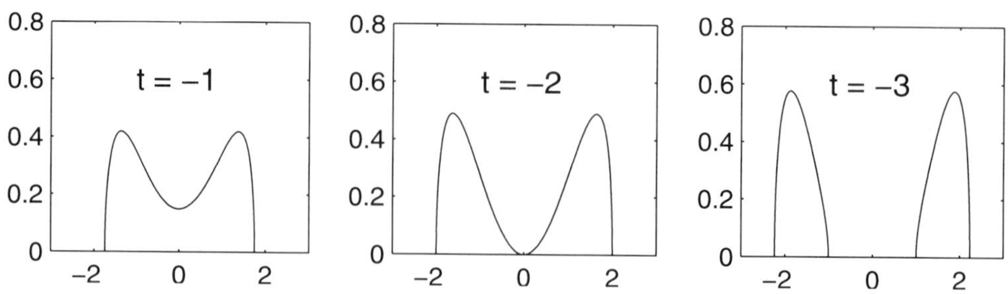

FIGURE 2.1. The density function, $\rho_V(x)$, for the even quartic potential, $V(M) = (t/2)M^2 + \frac{1}{4}M^4$, for $t = -1, -2, -3$.

which can be solved to find

$$a = \sqrt{2-t}, \quad b = \sqrt{-2-t}. \tag{2.3.39}$$

Figure 2.1 shows the density function for the even quartic potential, for $t = -1$, $-2, -3$.

2.4. The Deift – Zhou steepest descent method

We now present the Riemann – Hilbert approach to the large N asymptotics of orthogonal polynomials. The central point of this approach is a construction of an asymptotic solution to the RHP of Section 2.2 as $N \to \infty$. We call such a solution a parametrix. In the original paper of Bleher and Its [10], the RH approach was developed for an even quartic polynomial $V(M)$ via a semiclassical solution of the differential equation for orthogonal polynomials. Then, in a series of papers, Deift, Kriecherbauer, McLaughlin, Venakides, and Zhou (DKMVZ) [28–30] developed the RH approach for a general real analytic V, with some conditions on the growth at infinity. The DKMVZ approach is based on the Deift – Zhou steepest descent method, see [32]. In this section we present the main steps of the DKMVZ approach. For the sake of simplicity, we will assume that V is regular. In this approach a sequence of transformations of the RHP is constructed. The goal of these transformations is to obtain a new RHP whose jump properties and asymptotics at infinity are uniformly close to the identity. This type of RHP has a solution which is generally close to the identity, and can generally be solved by a series of perturbation theory. This sequence of transformations gives the parametrix of the RHP in different regions on complex plane. The motivation for the first transformation comes from the Heine formula for orthogonal polynomials.

2.4.1. Heine's formula for orthogonal polynomials. The Heine formula, see, e.g., [72], gives the Nth orthogonal polynomial as the matrix integral,

$$P_N(z) = \langle \det(z-M) \rangle \equiv Z_N^{-1} \int_{\mathcal{H}_N} \det(z-M) \mathrm{e}^{-N \operatorname{Tr} V(M)} \, \mathrm{d}M. \tag{2.4.1}$$

In the ensemble of eigenvalues,

$$P_N(z) = \left\langle \prod_{j=1}^N (z-\lambda_j) \right\rangle \equiv \widetilde{Z}_N^{-1} \int \prod_{j=1}^N (z-\lambda_j) \prod_{j>k} (\lambda_j - \lambda_k)^2 \prod_{j=1}^N \mathrm{e}^{-NV(\lambda_j)} \, \mathrm{d}\lambda. \tag{2.4.2}$$

Since ν_λ, the normalized counting measure on the set of eigenvalues λ, is close to the equilibrium measure ν_λ for typical λ, we expect that

$$N^{-1} \log \left\langle \prod_{j=1}^N (z-\lambda_j) \right\rangle \approx \int_J \log(z-x) \, \mathrm{d}\nu_V(x),$$

hence by the Heine formula,

$$N^{-1} \log P_N(z) \approx \int_J \log(z-x) \, \mathrm{d}\nu_V(x). \tag{2.4.3}$$

This gives a *heuristic* semiclassical approximation for the orthogonal polynomial,

$$P_N(z) \approx \exp\left[N \int_J \log(z-x) d\nu_V(x) \right], \tag{2.4.4}$$

and it motivates the introduction of the "g-function."

2.4.2. The g-function. Assume the equilibrium measure is supported on the set

$$J = \bigcup_{j=1}^{q} [a_j, b_j], \tag{2.4.5}$$

and define the g-function as

$$g(z) = \int_J \log(z - x)\, \mathrm{d}\nu_V(x), \quad z \in \mathbb{C} \setminus (-\infty, b_q], \tag{2.4.6}$$

where we take the principal branch for the logarithm. This function has the following important properties.

Properties of $g(z)$.

(1) $g(z)$ is analytic in $\mathbb{C} \setminus (-\infty, b_q]$.
(2) As $z \to \infty$

$$g(z) = \log z - \sum_{j=1}^{\infty} \frac{g_j}{z^j}, \quad g_j = \int_J \frac{x^j}{j}\, \mathrm{d}\nu_V(x). \tag{2.4.7}$$

(3) By (2.3.19), (2.3.22),

$$g'(z) = \omega(z) = \frac{V'(z)}{2} - \frac{h(z)R^{1/2}(z)}{2}. \tag{2.4.8}$$

(4) By (2.3.15),

$$g_+(x) + g_-(x) = V(x) + l, \quad x \in J. \tag{2.4.9}$$

(5) By (2.3.18),

$$g_+(x) + g_-(x) < V(x) + l, \quad x \in \mathbb{R} \setminus J. \tag{2.4.10}$$

(6) Equation (2.4.6) implies that the function

$$G(x) \equiv g_+(x) - g_-(x) \tag{2.4.11}$$

is pure imaginary for all real x, and $G(x)$ is constant in each component of $\mathbb{R} \setminus J$,

$$G(x) = \mathrm{i}\Omega_j \quad \text{for } b_j < x < a_{j+1},\ 1 \le j \le q-1, \tag{2.4.12}$$

where

$$\Omega_j = 2\pi \sum_{k=j+1}^{q} \int_{a_k}^{b_k} \rho_V(x)\, \mathrm{d}x, \quad 1 \le j \le q-1. \tag{2.4.13}$$

(7) Also,

$$G(x) = \mathrm{i}\Omega_j - 2\pi\mathrm{i} \int_{b_j}^{x} \rho_V(s)\, \mathrm{d}s \quad \text{for } a_j < x < b_j,\ 1 \le j \le q, \tag{2.4.14}$$

where we set $\Omega_q = 0$.

Observe that from (2.4.14) and (2.3.9) we obtain that $G(x)$ is analytic on (a_j, b_j), and the Cauchy – Riemann equations imply that

$$\left.\frac{\mathrm{d}G(x+\mathrm{i}y)}{\mathrm{d}y}\right|_{y=0} = 2\pi\rho_V(x) > 0, \quad x \in (a_j, b_j),\ 1 \le j \le q. \tag{2.4.15}$$

From (2.4.9) we have also that

$$G(x) = 2g_+(x) - V(x) - l = -[2g_-(x) - V(x) - l], \quad x \in J. \tag{2.4.16}$$

2.4.3. First transformation of the RHP. By (2.4.4) we expect that

$$P_N(z) \approx e^{Ng(z)}, \tag{2.4.17}$$

and we therefore make the following substitution in the RHP:

$$\mathbf{Y}_N(z) = \mathrm{e}^{(Nl/2)\sigma_3} \mathbf{T}_N(z) \mathrm{e}^{N[g(z)-l/2]\sigma_3}, \tag{2.4.18}$$

where

$$\sigma_3 = \begin{pmatrix} 1 & 0 \\ 0 & -1 \end{pmatrix} \tag{2.4.19}$$

is the third Pauli matrix. Then $\mathbf{T}_N(z)$ solves the following RHP:

(1) $\mathbf{T}_N(z)$ is analytic in $\mathbb{C} \setminus \mathbb{R}$.
(2) $\mathbf{T}_{N+}(x) = \mathbf{T}_{N-}(x) j_T(x)$ for $x \in \mathbb{R}$, where

$$j_T(x) = \begin{pmatrix} \mathrm{e}^{-N[g_+(x)-g_-(x)]} & \mathrm{e}^{N[g_+(x)+g_-(x)-V(x)-l]} \\ 0 & \mathrm{e}^{N[g_+(x)-g_-(x)]} \end{pmatrix}. \tag{2.4.20}$$

(3) $\mathbf{T}_N(z) = I + \mathcal{O}(z^{-1})$, as $z \to \infty$.

The above properties of $g(z)$ ensure the following properties of the jump matrix j_T:

(1) $j_T(x)$ is exponentially close to the identity matrix on $(-\infty, a_1) \cup (b_q, \infty)$. Namely,

$$j_T(x) = \begin{pmatrix} 1 & \mathcal{O}\big(\mathrm{e}^{-Nc(x)}\big) \\ 0 & 1 \end{pmatrix}, \quad x \in (-\infty, a_1) \cup (b_q, \infty), \tag{2.4.21}$$

where $c(x) > 0$ is a continuous function such that

$$\lim_{x \to \pm\infty} \frac{c(x)}{\log|x|} = \infty, \quad \lim_{x \to a_1} c(x) = \lim_{x \to b_q} c(x) = 0. \tag{2.4.22}$$

(2) For $1 \le j \le q-1$,

$$j_T(x) = \begin{pmatrix} \mathrm{e}^{-\mathrm{i}N\Omega_j} & \mathcal{O}\big(\mathrm{e}^{-Nc(x)}\big) \\ 0 & \mathrm{e}^{\mathrm{i}N\Omega_j} \end{pmatrix}, \quad x \in (b_j, a_{j+1}), \tag{2.4.23}$$

where $c(x) > 0$ is a continuous function such that

$$\lim_{x \to b_j} c(x) = \lim_{x \to a_{j+1}} c(x) = 0. \tag{2.4.24}$$

(3) On J,

$$j_T(x) = \begin{pmatrix} \mathrm{e}^{-NG(x)} & 1 \\ 0 & \mathrm{e}^{NG(x)} \end{pmatrix}. \tag{2.4.25}$$

The latter matrix can be factorized as follows:

$$\begin{aligned} \begin{pmatrix} \mathrm{e}^{-NG(x)} & 1 \\ 0 & \mathrm{e}^{NG(x)} \end{pmatrix} &= \begin{pmatrix} 1 & 0 \\ \mathrm{e}^{NG(x)} & 1 \end{pmatrix} \begin{pmatrix} 0 & 1 \\ -1 & 0 \end{pmatrix} \begin{pmatrix} 1 & 0 \\ \mathrm{e}^{-NG(x)} & 1 \end{pmatrix} \\ &\equiv j_-(x) j_M j_+(x), \end{aligned} \tag{2.4.26}$$

This leads to the second transformation of the RHP.

2.4.4. Second transformation of the RHP: Opening of lenses. The function $\mathrm{e}^{-NG(x)}$ is analytic on each open interval (a_j, b_j). Observe that $|e^{-NG(x)}|=1$ for real $x \in (a_j, b_j)$, and $\mathrm{e}^{-NG(z)}$ is exponentially decaying for $\operatorname{Im} z > 0$. More precisely, by (2.4.15), there exists $y_0 > 0$ such that $e^{-NG(z)}$ satisfies the estimate,

$$(2.4.27)\quad |\mathrm{e}^{-NG(z)}| \le \mathrm{e}^{-Nc(z)}, \quad z \in R_j^+ = \{z = x + \mathrm{i}y : a_j < x < b_j, 0 < y < y_0\},$$

where $c(z) > 0$ is a continuous function in R_j^+. Observe that $c(z) \to 0$ as $\operatorname{Im} z \to 0$. In addition, $|e^{NG(z)}| = |e^{-NG(\bar{z})}|$, hence

$$(2.4.28)\quad |e^{NG(z)}| \le \mathrm{e}^{-Nc(z)}, \quad z \in R_j^- = \{z = x + \mathrm{i}y : a_j < x < b_j, 0 < -y < y_0\},$$

where $c(z) = c(\bar{z}) > 0$. Consider a C^∞ curve γ_j^+ from a_j to b_j such that

$$(2.4.29)\qquad \gamma_j^+ = \{x + \mathrm{i}y : y = f_j(x)\},$$

where $f_j(x)$ is a C^∞ function on $[a_j, b_j]$ such that

$$(2.4.30)\qquad \begin{gathered} f_j(a_j) = f_j(b_j) = 0; \qquad f_j'(a_j) = -f'(b_j) = \sqrt{3}; \\ 0 < f_j(x) < y_0, \quad a_j < x < b_j. \end{gathered}$$

Consider also the conjugate curve,

$$(2.4.31)\qquad \gamma_j^- = \overline{\gamma_j^+} = \{x - \mathrm{i}y : y = f_j(x)\},$$

see Figure 2.2. The region bounded by the interval $[a_j, b_j]$ and γ_j^+ (resp. γ_j^-) is called the *upper* (resp. *lower*) *lens*, $\mathcal{L}_j^\pm$. Define for $j = 1, \dots, q$,

$$(2.4.32)\qquad \mathbf{S}_N(z) = \begin{cases} \mathbf{T}_N(z) j_+^{-1}(z), & \text{if } z \text{ is in the upper lens, } z \in \mathcal{L}_j^+, \\ \mathbf{T}_N(z) j_-(z), & \text{if } z \text{ is in the lower lens, } z \in \mathcal{L}_j^-, \\ \mathbf{T}_N(z) & \text{otherwise,} \end{cases}$$

where

$$(2.4.33)\qquad j_\pm(z) = \begin{pmatrix} 1 & 0 \\ \mathrm{e}^{\mp NG(z)} & 1 \end{pmatrix}.$$

Then $\mathbf{S}_N(z)$ solves the following RHP:

(1) $\mathbf{S}_N(z)$ is analytic in $\mathbb{C} \setminus (\mathbb{R} \cup \Gamma)$, $\Gamma = \gamma_1^+ \cup \gamma_1^- \cup \dots \cup \gamma_q^+ \cup \gamma_q^-$.

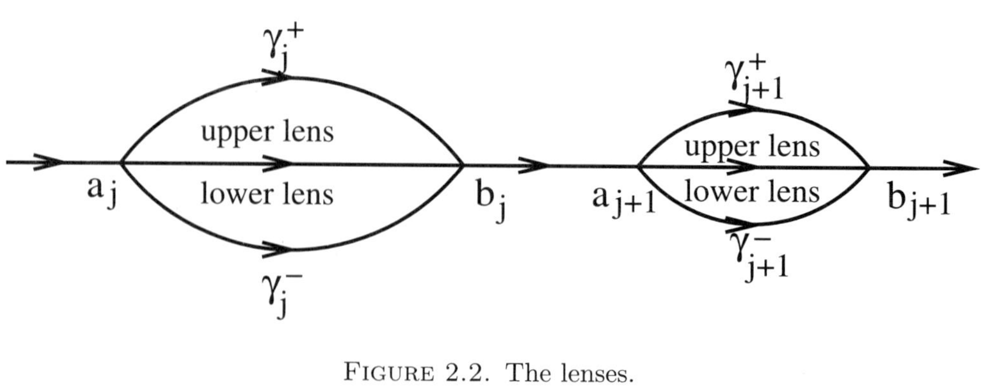

FIGURE 2.2. The lenses.

(2) $\mathbf{S}_N(z)$ satisfies the jump condition

$$\mathbf{S}_{N+}(z) = \mathbf{S}_{N-}(z) j_S(z), \quad z \in \mathbb{R} \cup \Gamma, \tag{2.4.34}$$

where the jump matrix $j_S(z)$ is given by

$$j_S(z) = \begin{cases} \begin{pmatrix} 0 & 1 \\ -1 & 0 \end{pmatrix} & \text{for } z \in J \\ j_T(z) = \begin{pmatrix} \mathrm{e}^{-\mathrm{i}N\Omega_j} & \mathcal{O}(\mathrm{e}^{-Nc(x)}) \\ 0 & \mathrm{e}^{iN\Omega_j} \end{pmatrix} & \text{for } z \in (b_j, a_{j+1}),\ j = 1, \ldots, q-1 \\ j_T(z) = \begin{pmatrix} 1 & \mathcal{O}(\mathrm{e}^{-Nc(x)}) \\ 0 & 1 \end{pmatrix} & \text{for } z \in (-\infty, a_1) \cup (b_q, \infty) \\ j_\pm(z) = \begin{pmatrix} 1 & 0 \\ \mathcal{O}(\mathrm{e}^{-Nc(z)}) & 1 \end{pmatrix} & \text{for } z \in \gamma_j^\pm, \end{cases}$$

where $c(z) > 0$ is a continuous function vanishing only at the endpoints a_j, b_j, $j = 1, \ldots, q$.

(3) $\mathbf{S}_N(z) = I + \mathcal{O}(z^{-1})$, as $z \to \infty$.

We expect, and this will be justified later, that as $N \to \infty$, $\mathbf{S}_N(z)$ converges to a solution of the model RHP, sometimes called a global parametrix, in which we drop the $\mathcal{O}(\mathrm{e}^{-cN})$-terms in the jump matrix $j_S(z)$. Let us consider the model RHP.

2.4.5. The model RHP. We are looking for the 2×2 matrix function $\mathbf{M}(z)$ that solves the following model RHP:

(1) $\mathbf{M}(z)$ is analytic in $\mathbb{C} \setminus [a_1, b_q]$,

(2) On the interval $[a_1, b_q]$, $\mathbf{M}(z)$ satisfies the jump condition

$$\mathbf{M}_+(z) = \mathbf{M}_-(z) j_M(z), \quad z \in [a_1, b_q], \tag{2.4.35}$$

where the jump matrix $j_M(z)$ is given by the following formulas:

$$j_M(z) = \begin{cases} \begin{pmatrix} 0 & 1 \\ -1 & 0 \end{pmatrix} & \text{for } z \in J \\ \begin{pmatrix} \mathrm{e}^{-\mathrm{i}N\Omega_j} & 0 \\ 0 & \mathrm{e}^{\mathrm{i}N\Omega_j} \end{pmatrix} & \text{for } z \in [b_j, a_{j+1}],\ j = 1, \ldots, q-1. \end{cases}$$

(3) $\mathbf{M}(z) = I + \mathcal{O}(z^{-1})$, as $z \to \infty$.

We will construct a solution to the model RHP by following the work [29].

2.4.5.1. *Solution of the model RHP. One-cut case.* Assume that J consists of a single interval $[a, b]$. Then the model RHP reduces to the following:

(1) $\mathbf{M}(z)$ is analytic in $\mathbb{C} \setminus [a, b]$,

(2) $\mathbf{M}_+(z) = \mathbf{M}_-(z) \left(\begin{smallmatrix} 0 & 1 \\ -1 & 0 \end{smallmatrix}\right)$ for $z \in [a, b]$.

(3) $\mathbf{M}(z) = I + \mathcal{O}(z^{-1})$, as $z \to \infty$.

This RHP can be reduced to a pair of scalar RH problems. We have that

$$\begin{pmatrix} 0 & 1 \\ -1 & 0 \end{pmatrix} = \frac{1}{2} \begin{pmatrix} 1 & 1 \\ \mathrm{i} & -\mathrm{i} \end{pmatrix} \begin{pmatrix} \mathrm{i} & 0 \\ 0 & -\mathrm{i} \end{pmatrix} \begin{pmatrix} 1 & -\mathrm{i} \\ 1 & \mathrm{i} \end{pmatrix} \tag{2.4.36}$$

Let

$$\widetilde{\mathbf{M}}(z) = \frac{1}{2} \begin{pmatrix} 1 & -\mathrm{i} \\ 1 & \mathrm{i} \end{pmatrix} \mathbf{M}(z) \begin{pmatrix} 1 & 1 \\ \mathrm{i} & -\mathrm{i} \end{pmatrix}. \tag{2.4.37}$$

Then $\widetilde{\mathbf{M}}(z)$ solves the following RHP:

(1) $\widetilde{\mathbf{M}}(z)$ is analytic in $\mathbb{C} \setminus [a, b]$,
(2) $\widetilde{\mathbf{M}}_+(z) = \widetilde{\mathbf{M}}_-(z)\left(\begin{smallmatrix} \mathrm{i} & 0 \\ 0 & -\mathrm{i} \end{smallmatrix}\right)$ for $z \in [a, b]$.
(3) $\widetilde{\mathbf{M}}(z) = I + \mathcal{O}(z^{-1})$, as $z \to \infty$.

This is a pair of scalar RH problems, which can be solved by the Cauchy integral:

$$
\begin{aligned}
(2.4.38)\qquad \widetilde{\mathbf{M}}(z) &= \begin{pmatrix} \exp\left(\dfrac{1}{2\pi \mathrm{i}} \displaystyle\int_a^b \dfrac{\log \mathrm{i}}{s - z}\, \mathrm{d}s\right) & 0 \\ 0 & \exp\left(\dfrac{1}{2\pi i} \displaystyle\int_a^b \dfrac{\log(-\mathrm{i})}{s - z}\, \mathrm{d}s\right) \end{pmatrix} \\
&= \begin{pmatrix} \exp\left(\dfrac{1}{4} \log \dfrac{z - b}{z - a}\right) & 0 \\ 0 & \exp\left(-\dfrac{1}{4} \log \dfrac{z - b}{z - a}\right) \end{pmatrix} \\
&= \begin{pmatrix} \gamma^{-1}(z) & 0 \\ 0 & \gamma(z) \end{pmatrix},
\end{aligned}
$$

where

$$
(2.4.39)\qquad \gamma(z) = \left(\frac{z - a}{z - b}\right)^{1/4},
$$

with cut on $[a, b]$ and the branch such that $\gamma(\infty) = 1$. Thus,

$$
\begin{aligned}
(2.4.40)\qquad \mathbf{M}(z) &= \frac{1}{2}\begin{pmatrix} 1 & 1 \\ \mathrm{i} & -\mathrm{i} \end{pmatrix} \begin{pmatrix} \gamma^{-1}(z) & 0 \\ 0 & \gamma(z) \end{pmatrix} \begin{pmatrix} 1 & 1 \\ \mathrm{i} & -\mathrm{i} \end{pmatrix}^{-1} \\
&= \begin{pmatrix} \dfrac{\gamma(z) + \gamma^{-1}(z)}{2} & \dfrac{\gamma(z) - \gamma^{-1}(z)}{(-2\mathrm{i})} \\ \dfrac{\gamma(z) - \gamma^{-1}(z)}{2\mathrm{i}} & \dfrac{\gamma(z) + \gamma^{-1}(z)}{2} \end{pmatrix}, \\
\det \mathbf{M}(z) &= 1.
\end{aligned}
$$

At infinity we have

$$
(2.4.41)\qquad \gamma(z) = 1 + \frac{b - a}{4z} + \mathcal{O}(z^{-2}),
$$

hence

$$
(2.4.42)\qquad \mathbf{M}(z) = I + \frac{1}{z}\begin{pmatrix} 0 & \frac{b-a}{(-4\mathrm{i})} \\ \frac{b-a}{4\mathrm{i}} & 0 \end{pmatrix} + \mathcal{O}(z^{-2}).
$$

Let us now generalize to the multicut case.

2.4.5.2. *Solution of the model RHP. Multicut case.* This will be done in three steps.

Step 1. Consider the auxiliary RHP,
(a) $\mathbf{Q}(z)$ is analytic in $\mathbb{C} \setminus J, \quad J = \cup_{j=1}^q [a_j, b_j]$.
(b) $\mathbf{Q}_+(z) = \mathbf{Q}_-(z)\left(\begin{smallmatrix} 0 & 1 \\ -1 & 0 \end{smallmatrix}\right)$ for $z \in J$.
(c) $\mathbf{Q}(z) = I + \mathcal{O}(z^{-1})$, as $z \to \infty$.

Then, similar to the one-cut case, this RHP is reduced to two scalar RHPs, and the solution is

$$\mathbf{Q}(z) = \begin{pmatrix} \dfrac{\gamma(z)+\gamma^{-1}(z)}{2} & \dfrac{\gamma(z)-\gamma^{-1}(z)}{(-2\mathrm{i})} \\ \dfrac{\gamma(z)-\gamma^{-1}(z)}{2\mathrm{i}} & \dfrac{\gamma(z)+\gamma^{-1}(z)}{2} \end{pmatrix}, \tag{2.4.43}$$

where

$$\gamma(z) = \prod_{j=1}^{q} \left(\frac{z-a_j}{z-b_j} \right)^{1/4}, \quad \gamma(\infty) = 1, \tag{2.4.44}$$

with cuts on J. At infinity we have

$$\gamma(z) = 1 + \sum_{j=1}^{q} \frac{b_j - a_j}{4z} + \mathcal{O}(z^{-2}), \tag{2.4.45}$$

hence

$$\mathbf{Q}(z) = I + \frac{1}{z} \begin{pmatrix} 0 & \sum_{j=1}^{q} \frac{b_j-a_j}{(-4\mathrm{i})} \\ \sum_{j=1}^{q} \frac{b_j-a_j}{4\mathrm{i}} & 0 \end{pmatrix} + \mathcal{O}(z^{-2}). \tag{2.4.46}$$

In what follows, we will modify this solution to satisfy part (b) in the jump matrix (2.4.35). This requires the theory of Riemann surfaces and the Riemann theta function.

Step 2. Let X be the two-sheeted Riemann surface of the genus

$$g = q - 1, \tag{2.4.47}$$

associated to $\sqrt{R(z)}$, where

$$R(z) = \prod_{j=1}^{q} (z-a_j)(z-b_j), \tag{2.4.48}$$

with cuts on the intervals (a_j, b_j), $j = 1, \dots, q$, see Figure 2.3. We fix the first sheet of X by the condition that on this sheet,

$$\sqrt{R(x)} > 0, \quad x > b_q. \tag{2.4.49}$$

We would like to introduce $2g$ cycles on X forming a homology basis. To that end, consider, for $j = 1, \dots, g$, a cycle A_j on X, which goes around the interval (b_j, a_{j+1}) in the negative direction, such that the part of A_j in the upper half-plane, $A_j^+ \equiv A_j \cup \{z : \operatorname{Im} z \geq 0\}$, lies on the first sheet of X, and the part in the lower half-plane, $A_j^- \equiv A_j \cup \{z : \operatorname{Im} z \leq 0\}$, lies on the second sheet, $j = 1, \dots, g$. In addition, consider a cycle B_j on X, which goes around the interval (a_1, b_j) on the first sheet in the negative direction, see Figure 2.4. Then the cycles $(A_1, \dots, A_g, B_1, \dots, B_g)$ form a canonical homology basis for X.

Consider the linear space Ω of holomorphic one-forms on X,

$$\Omega = \left\{ \omega = \sum_{j=0}^{q-2} \frac{c_j z^j \, dz}{\sqrt{R(z)}} \right\}. \tag{2.4.50}$$

The dimension of Ω is equal to g. Consider the basis for Ω,

$$\omega = (\omega_1, \dots, \omega_g),$$

with normalization

$$\int_{A_j} \omega_k = \delta_{jk}, \quad j,k = 1,\dots,g. \tag{2.4.51}$$

Such a basis exists and it is unique, see [39]. Observe that the numbers

$$m_{jk} = \int_{A_j} \frac{z^k\,\mathrm{d}z}{\sqrt{R(z)}} = 2\int_{b_j}^{a_{j+1}} \frac{x^k\,\mathrm{d}x}{\sqrt{R(x)}}, \quad 1 \le j \le g,\ 1 \le k \le g-1, \tag{2.4.52}$$

are real. This implies that the basis ω is real, i.e., the one-forms,

$$\omega_j = \sum_{k=1}^{q-1} \frac{c_{jk} z^{k-1}\,\mathrm{d}z}{\sqrt{R(z)}}, \tag{2.4.53}$$

have real coefficients c_{jk}.

Define the associated Riemann matrix of B-periods,

$$\tau = (\tau_{jk}), \quad \tau_{jk} = \int_{B_j} \omega_k, \quad j,k = 1,\dots,g. \tag{2.4.54}$$

Since $\sqrt{R(x)}$ is pure imaginary on (a_j, b_j), the numbers τ_{jk} are pure imaginary. It is known, see, e.g., [39], that the matrix τ is symmetric and $(-i\tau)$ is positive definite.

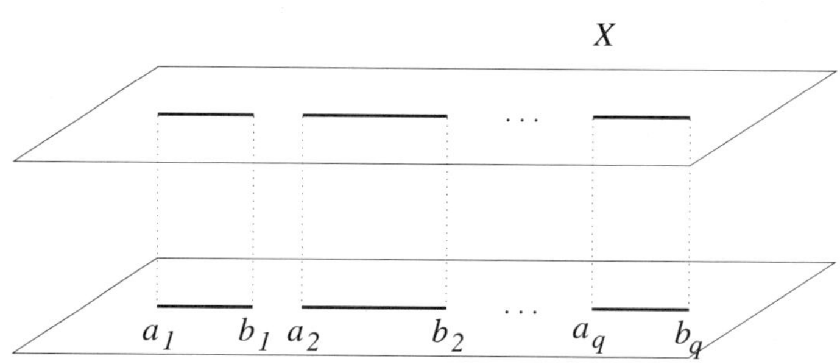

FIGURE 2.3. The Riemann surface associated to $\sqrt{R(z)}$.

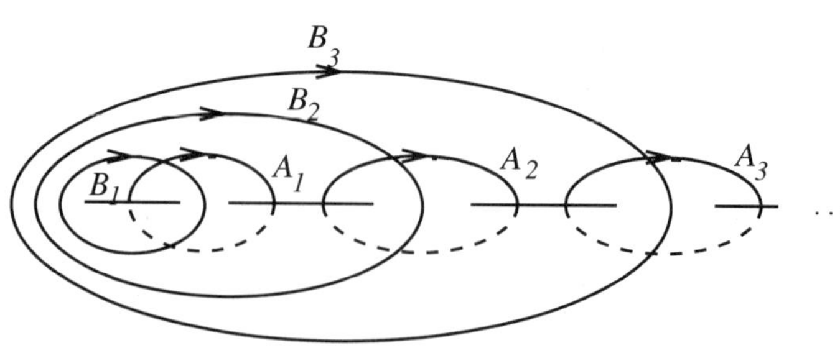

FIGURE 2.4. The basis of cycles on X. The solid paths lie on the first sheet of X, while the dotted paths are on the second sheet. The horizontal line segments are the support of the equilibrium measure, the leftmost segment being the interval (a_1, b_1).

The Riemann theta function with the matrix τ is defined as

$$\theta(s)=\sum_{m\in\mathbb{Z}^g}\mathrm{e}^{2\pi\mathrm{i}(m,s)+\pi\mathrm{i}(m,\tau m)},\quad s\in\mathbb{C}^g;\qquad (m,s)=\sum_{j=1}^{g}m_js_j. \tag{2.4.55}$$

The quadratic form $\mathrm{i}(m,\tau m)$ is negative definite, hence the series is absolutely convergent and $\theta(s)$ is analytic in $\mathbb{C}^g$. The theta function is an even function,

$$\theta(-s)=\theta(s), \tag{2.4.56}$$

and it has the following periodicity properties:

$$\theta(s+e_j)=\theta(s);\quad \theta(s+\tau_j)=\mathrm{e}^{-2\pi\mathrm{i}s_j-\pi\mathrm{i}\tau_{jj}}\theta(s), \tag{2.4.57}$$

where $e_j=(0,\dots,0,1,0,\dots,0)$ is the jth basis vector in $\mathbb{C}^g$, and $\tau_j=\tau e_j$. This implies that the function

$$f(s)=\frac{\theta(s+d+c)}{\theta(s+d)}, \tag{2.4.58}$$

where $c,d\in\mathbb{C}^g$ are arbitrary constant vectors, has the periodicity properties,

$$f(s+e_j)=f(s);\quad f(s+\tau_j)=\mathrm{e}^{-2\pi\mathrm{i}c_j}f(s). \tag{2.4.59}$$

Consider now the theta function associated with the Riemann surface X. It is defined as follows. Introduce the vector function,

$$u(z)=\int_{b_q}^{z}\omega,\quad z\in\mathbb{C}\setminus(a_1,b_q), \tag{2.4.60}$$

where ω is the basis of holomorphic one-forms, determined by equations (2.4.51). The contour of integration in (2.4.60) lies in $\mathbb{C}\setminus(a_1,b_q)$, on the first sheet of X. We will consider $u(z)$ as a function with values in $\mathbb{C}^g/\mathbb{Z}^g$,

$$u\colon\mathbb{C}\setminus(a_1,b_q)\to\mathbb{C}^g/\mathbb{Z}^g. \tag{2.4.61}$$

On $[a_1,b_q]$ the function $u(z)$ is two-valued. From (2.4.54) we have that

$$u_+(x)-u_-(x)=\tau_j,\quad x\in[b_j,a_{j+1}];\qquad 1\le j\le q-1. \tag{2.4.62}$$

Since $\sqrt{R(x)}_-=-\sqrt{R(x)}_+$ on $[a_j,b_j]$, we have that the function $u_+(x)+u_-(x)$ is constant on $[a_j,b_j]$. It follows from (2.4.51) that mod $\mathbb{Z}^g$,

$$u_+(b_j)+u_-(b_j)=u_+(a_{j+1})+u_-(a_{j+1}),\quad 1\le j\le q-1. \tag{2.4.63}$$

Since $u_+(b_q)=u_-(b_q)=0$, we obtain that

$$u_+(x)+u_-(x)=0,\quad x\in J=\bigcup_{j=1}^{q}[a_j,b_j]. \tag{2.4.64}$$

Define

$$f_1(z)=\frac{\theta(u(z)+d+c)}{\theta(u(z)+d)},\quad f_2(z)=\frac{\theta(-u(z)+d+c)}{\theta(-u(z)+d)},\qquad z\in\mathbb{C}\setminus(a_1,b_q), \tag{2.4.65}$$

where $c,d\in\mathbb{C}^g$ are arbitrary constant vectors. Then from (2.4.62) and (2.4.59) we obtain that for $1\le j\le q-1$,

$$f_1(x+\mathrm{i}0)=\mathrm{e}^{-2\pi\mathrm{i}c_j}f_1(x-\mathrm{i}0),\quad f_2(x+\mathrm{i}0)=\mathrm{e}^{2\pi\mathrm{i}c_j}f_2(x-\mathrm{i}0),\qquad x\in(b_j,a_{j+1}), \tag{2.4.66}$$

and from (2.4.64) that

$$f_1(x+\mathrm{i}0)=f_2(x-\mathrm{i}0),\quad f_2(x+\mathrm{i}0)=f_1(x-\mathrm{i}0),\qquad x\in J. \tag{2.4.67}$$

Let us now take

$$c=\frac{N\Omega}{2\pi},\quad \Omega=(\Omega_1,\dots,\Omega_g), \tag{2.4.68}$$

and define the matrix-valued function,

$$\mathbf{F}(z)=\begin{pmatrix}\dfrac{\theta(u(z)+d_1+c)}{\theta(u(z)+d_1)} & \dfrac{\theta(-u(z)+d_1+c)}{\theta(-u(z)+d_1)}\\ \dfrac{\theta(u(z)+d_2+c)}{\theta(u(z)+d_2)} & \dfrac{\theta(-u(z)+d_2+c)}{\theta(-u(z)+d_2)}\end{pmatrix}, \tag{2.4.69}$$

where $d_1,d_2\in\mathbb{C}^g$ are yet undetermined constant vectors. Then from (2.4.67) we obtain that

$$\begin{aligned}\mathbf{F}_+(x)&=\mathbf{F}_-(x)\begin{pmatrix}\mathrm{e}^{-\mathrm{i}N\Omega_j} & 0\\ 0 & \mathrm{e}^{\mathrm{i}N\Omega_j}\end{pmatrix}, && x\in(b_j,a_{j+1});\quad j=1,\dots,q-1,\\ \mathbf{F}_+(x)&=\mathbf{F}_-(x)\begin{pmatrix}0&1\\1&0\end{pmatrix}, && x\in J.\end{aligned} \tag{2.4.70}$$

Step 3. Let us combine formulas (2.4.43) and (2.4.69). We set

$$\begin{aligned}\mathbf{M}(z)&=\mathbf{F}(\infty)^{-1}\\ &\times\begin{pmatrix}\dfrac{\gamma(z)+\gamma^{-1}(z)}{2}\dfrac{\theta(u(z)+d_1+c)}{\theta(u(z)+d_1)} & \dfrac{\gamma(z)-\gamma^{-1}(z)}{(-2\mathrm{i})}\dfrac{\theta(-u(z)+d_1+c)}{\theta(-u(z)+d_1)}\\ \dfrac{\gamma(z)-\gamma^{-1}(z)}{2\mathrm{i}}\dfrac{\theta(u(z)+d_2+c)}{\theta(u(z)+d_2)} & \dfrac{\gamma(z)+\gamma^{-1}(z)}{2}\dfrac{\theta(-u(z)+d_2+c)}{\theta(-u(z)+d_2)}\end{pmatrix},\end{aligned} \tag{2.4.71}$$

where

$$\mathbf{F}(\infty)=\begin{pmatrix}\dfrac{\theta(u(\infty)+d_1+c)}{\theta(u(\infty)+d_1)} & 0\\ 0 & \dfrac{\theta(-u(\infty)+d_2+c)}{\theta(-u(\infty)+d_2)}\end{pmatrix}. \tag{2.4.72}$$

Then $\mathbf{M}(z)$ has the following jumps:

$$\begin{aligned}\mathbf{M}_+(x)&=\mathbf{M}_-(x)\begin{pmatrix}\mathrm{e}^{-\mathrm{i}N\Omega_j} & 0\\ 0 & \mathrm{e}^{\mathrm{i}N\Omega_j}\end{pmatrix}, && x\in(b_j,a_{j+1});\quad j=1,\dots,q-1,\\ \mathbf{M}_+(x)&=\mathbf{M}_-(x)\begin{pmatrix}0&1\\-1&0\end{pmatrix}, && x\in J,\end{aligned} \tag{2.4.73}$$

which fits perfectly to the model RHP, and $\mathbf{M}(\infty)=I$. It remains only to find d_1, d_2 such that $\mathbf{M}(z)$ is analytic at the zeroes of $\theta(\pm u(z)+d_{1,2})$. These zeroes can be cancelled by the zeroes of the functions $\gamma(z)\pm\gamma^{-1}(z)$. Let us consider the latter zeroes.

The zeroes of $\gamma(z)\pm\gamma^{-1}(z)$ are the ones of $\gamma^2(z)\pm 1$, and hence of $\gamma^4(z)-1$. By (2.4.44), the equation $\gamma^4(z)-1=0$ reads

$$p(z)\equiv\prod_{j=1}^{q}\frac{z-a_j}{z-b_j}=1. \tag{2.4.74}$$

It is easy to see that

$$p(b_j + 0) = \infty, \quad p(a_{j+1}) = 0, \qquad 1 \le j \le q-1, \tag{2.4.75}$$

hence equation (2.4.74) has a solution x_j on each interval (b_j, a_{j+1}),

$$p(x_j) = 1, \quad b_j < x_j < a_{j+1}; \quad 1 \le j \le q-1. \tag{2.4.76}$$

Since equation (2.4.74) has $(q-1)$ finite solutions, the numbers $\{x_j, 1 \le j \le q-1\}$ are all the solutions of (2.4.74). The function $\gamma(z)$ defined by equation (2.4.44) with cuts on J is positive on $\mathbb{R} \setminus J$, hence

$$\gamma(x_j) = 1. \tag{2.4.77}$$

Thus, we have $(q-1)$ zeroes of $\gamma(z) - \gamma^{-1}(z)$ and no zeroes of $\gamma(z) + \gamma^{-1}(z)$ on the sheet of $\gamma(z)$ under consideration.

Let us consider the zeroes of the function $\theta(u(z) - d)$. The vector of Riemann constants is given by the formula

$$K = -\sum_{j=1}^{q-1} u(b_j). \tag{2.4.78}$$

Define

$$d := -K + \sum_{j=1}^{q-1} u(x_j). \tag{2.4.79}$$

Then

$$\theta(u(x_j) - d) = 0, \quad 1 \le j \le q-1, \tag{2.4.80}$$

see [29], and $\{x_j, 1 \le j \le q-1\}$ are all the zeroes of the function $\theta(u(z) - d)$. In addition, the function $\theta(u(z) + d)$ has no zeroes at all on the upper sheet of X. In fact, all the zeroes of $\theta(u(z) + d)$ lie on the lower sheet, above the same points $\{x_j, 1 \le j \le q-1\}$. Therefore we set in (2.4.71),

$$d_1 = d, \quad d_2 = -d, \tag{2.4.81}$$

so that

$$\mathbf{M}(z) = \mathbf{F}(\infty)^{-1} \times \begin{pmatrix} \dfrac{\gamma(z) + \gamma^{-1}(z)}{2} \dfrac{\theta(u(z) + d + c)}{\theta(u(z) + d)} & \dfrac{\gamma(z) - \gamma^{-1}(z)}{(-2i)} \dfrac{\theta(-u(z) + d + c)}{\theta(-u(z) + d)} \\ \dfrac{\gamma(z) - \gamma^{-1}(z)}{2i} \dfrac{\theta(u(z) - d + c)}{\theta(u(z) - d)} & \dfrac{\gamma(z) + \gamma^{-1}(z)}{2} \dfrac{\theta(-u(z) - d + c)}{\theta(-u(z) - d)} \end{pmatrix}, \tag{2.4.82}$$

where

$$\mathbf{F}(\infty) = \begin{pmatrix} \dfrac{\theta(u(\infty) + d + c)}{\theta(u(\infty) + d)} & 0 \\ 0 & \dfrac{\theta(-u(\infty) - d + c)}{\theta(-u(\infty) - d)} \end{pmatrix}. \tag{2.4.83}$$

This gives the required solution of the model RHP. As $z \to \infty$,

$$\mathbf{M}(z) = I + \frac{\mathbf{M}_1}{z} + \mathcal{O}(z^{-2}), \tag{2.4.84}$$

where

$$\mathbf{M}_1 = \begin{pmatrix} \left(\frac{\nabla\theta(u+d+c)}{\theta(u+d+c)} - \frac{\nabla\theta(u+d)}{\theta(u(\infty)+d)}, u'(\infty)\right) & \frac{\theta(-u+d+c)\theta(u+d)}{\theta(u+d+c)\theta(-u+d)} \sum_{j=1}^{q} \frac{(b_j-a_j)}{(-4\mathrm{i})} \\ \frac{\theta(u-d+c)\theta(-u-d)}{\theta(-u-d+c)\theta(u-d)} \sum_{j=1}^{q} \frac{(b_j-a_j)}{4\mathrm{i}} & \left(\frac{\nabla\theta(u(\infty)+d-c)}{\theta(u(\infty)+d-c)} - \frac{\nabla\theta(u+d)}{\theta(u+d)}, u'(\infty)\right) \end{pmatrix}, \quad u = u(\infty). \tag{2.4.85}$$

Here $\nabla\theta$ is the gradient of θ,

$$u'(\infty) = (c_{1,q-1}, c_{2,q-1}, \dots, c_{q-1,q-1}), \tag{2.4.86}$$

and the numbers c_{jk} are defined in (2.4.53).

2.4.6. Construction of a parametrix at edge points. Notice that the jumps for $\mathbf{M}$ are uniformly close to the jumps for $\mathbf{S}$, except close to the endpoints of the intervals of support of the equilibrium measure. This indicates that we need a different parametrix close to the endpoints. Consider small disks $D(a_j, r)$, $D(b_j, r)$, $1 \le j \le q$, of radius $r > 0$, centered at the edge points,

$$D(a, r) \equiv \{z : |z - a| \le r\}.$$

We look for a local parametrix $\mathbf{U}_N(z)$ defined on the union of these disks such that

- $\mathbf{U}_N(z)$ is analytic on $D \setminus (\mathbb{R} \cup \Gamma)$, where

$$D = \bigcup_{j=1}^{q} \big(D(a_j, r) \cup D(b_j, r)\big). \tag{2.4.87}$$

- $\mathbf{U}_N$ satisfies the jump conditions,

$$\mathbf{U}_{N+}(z) = \mathbf{U}_{N-}(z) j_S(z), \quad z \in (\mathbb{R} \cup \Gamma) \cap D. \tag{2.4.88}$$

- As $N \to \infty$,

$$\mathbf{U}_N(z) = \left(I + \mathcal{O}\left(\frac{1}{N}\right)\right)\mathbf{M}(z) \quad \text{uniformly for } z \in \partial D. \tag{2.4.89}$$

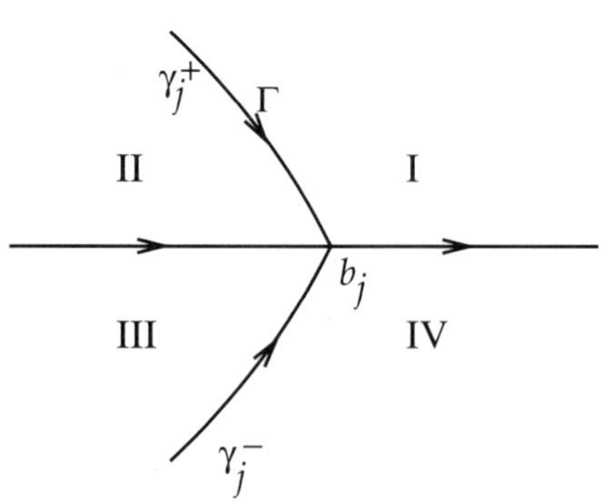

FIGURE 2.5. Partition of a neighborhood of the right edge point.

Consider first the edge point b_j, $1 \le j \le q$, which is the right endpoint of a band of support of the equilibrium measure.

Recall that the jumps j_S near b_j are given as

$$j_S(z) = \begin{cases} \begin{pmatrix} 0 & 1 \\ -1 & 0 \end{pmatrix} & \text{on } [b_j - r, b_j) \\ \begin{pmatrix} 1 & 0 \\ \mathrm{e}^{-NG(z)} & 1 \end{pmatrix} & \text{on } \gamma_j^+ \\ \begin{pmatrix} 1 & 0 \\ \mathrm{e}^{NG(z)} & 1 \end{pmatrix} & \text{on } \gamma_j^- \\ \begin{pmatrix} \mathrm{e}^{-N[g_+(z)-g_-(z)]} & \mathrm{e}^{N(g_+(z)+g_-(z)-V(z)-l)} \\ 0 & \mathrm{e}^{N[g_+(z)-g_-(z)]} \end{pmatrix} & \text{on } (b_j, b_j + r]. \end{cases} \tag{2.4.90}$$

We look for $\mathbf{U}_N(z)$ in the form,

$$\mathbf{U}_N(z) = \mathbf{Q}_N(z) \exp\left(-N \left[g(z) - \frac{V(z)}{2} - \frac{l}{2} \right] \sigma_3 \right). \tag{2.4.91}$$

Then the jump condition (2.4.88) on $\mathbf{U}_N(z)$ is transformed to the jump condition on $\mathbf{Q}_N(z)$,

$$\mathbf{Q}_{N+}(z) = \mathbf{Q}_{N-}(z) j_Q(z), \tag{2.4.92}$$

where

$$\begin{aligned} j_Q(z) = \exp&\left(-N \left[g_-(z) - \frac{V(z)}{2} - \frac{l}{2} \right] \sigma_3 \right) j_S(z) \\ &\times \exp\left(N \left[g_+(z) - \frac{V(z)}{2} - \frac{l}{2} \right] \sigma_3 \right). \end{aligned} \tag{2.4.93}$$

This reduces the problem to one with constant jumps. Indeed, from (2.4.90), (2.4.9), and (2.4.16) we obtain that

$$j_Q = \begin{cases} \begin{pmatrix} 0 & 1 \\ -1 & 0 \end{pmatrix} & \text{on } [b_j - r, b_j) \\ \begin{pmatrix} 1 & 0 \\ 1 & 1 \end{pmatrix} & \text{on } \gamma_j^+ \\ \begin{pmatrix} 1 & 0 \\ 1 & 1 \end{pmatrix} & \text{on } \gamma_j^- \\ \begin{pmatrix} 1 & 1 \\ 0 & 1 \end{pmatrix} & \text{on } (b_j, b_j + r]. \end{cases} \tag{2.4.94}$$

We can construct $\mathbf{Q}_N(z)$ with the help of the Airy function. The Airy function $\mathrm{Ai}(z)$ solves the equation $y'' = zy$ and for any $\varepsilon > 0$, in the sector $-\pi + \varepsilon \le \arg z \le \pi - \varepsilon$, it has the asymptotics as $z \to \infty$,

$$\begin{aligned} \mathrm{Ai}(z) &= \frac{1}{2\sqrt{\pi} z^{1/4}} \mathrm{e}^{-2z^{3/2}/3} \left(1 - \frac{5}{48} z^{-3/2} + \mathcal{O}(z^{-3}) \right) \\ \mathrm{Ai}'(z) &= \frac{1}{2\sqrt{\pi}} z^{1/4} \mathrm{e}^{-2z^{3/2}/3} \left(1 + \frac{7}{48} z^{-3/2} + \mathcal{O}(z^{-3}) \right). \end{aligned} \tag{2.4.95}$$

The functions $\mathrm{Ai}(\omega z)$, $\mathrm{Ai}(\omega^2 z)$, where $\omega = e^{2\pi i/3}$, also solve the equation $y'' = zy$, and we have the linear relation,

$$\mathrm{Ai}(z) + \omega\,\mathrm{Ai}(\omega z) + \omega^2\,\mathrm{Ai}(\omega^2 z) = 0. \tag{2.4.96}$$

Let us write

$$y_0(z) = \mathrm{Ai}(z), \quad y_1(z) = \omega\,\mathrm{Ai}(\omega z), \quad y_2(z) = \omega^2\,\mathrm{Ai}(\omega^2 z), \tag{2.4.97}$$

and we use these functions to define

$$\Phi_r(z) = \begin{cases} \begin{pmatrix} y_0(z) & -y_2(z) \\ y_0'(z) & -y_2'(z) \end{pmatrix}, & \text{for } 0 < \arg z < 2\pi/3 \\ \begin{pmatrix} -y_1(z) & -y_2(z) \\ -y_1'(z) & -y_2'(z) \end{pmatrix}, & \text{for } 2\pi/3 < \arg z < \pi \\ \begin{pmatrix} -y_2(z) & y_1(z) \\ -y_2'(z) & y_1'(z) \end{pmatrix}, & \text{for } -\pi < \arg z < -2\pi/3 \\ \begin{pmatrix} y_0(z) & y_1(z) \\ y_0'(z) & y_1'(z) \end{pmatrix}, & \text{for } -2\pi/3 < \arg z < 0. \end{cases} \tag{2.4.98}$$

Notice that $\det \Phi_r(z)$ has no jumps. In fact, we have (see, e.g., [65])

$$\det \Phi_r(z) \equiv -\frac{1}{2\pi i}. \tag{2.4.99}$$

Observe that equation (2.4.96) reads

$$y_0(z) + y_1(z) + y_2(z) = 0, \tag{2.4.100}$$

and it implies that on the discontinuity rays,

$$\Phi_{r+}(z) = \Phi_{r-}(z) \begin{cases} \begin{pmatrix} 0 & 1 \\ -1 & 0 \end{pmatrix} & \text{for } \arg z = \pi \\ \begin{pmatrix} 1 & 0 \\ 1 & 1 \end{pmatrix} & \text{for } \arg z = 2\pi/3 \\ \begin{pmatrix} 1 & 0 \\ 1 & 1 \end{pmatrix} & \text{for } \arg z = -2\pi/3 \\ \begin{pmatrix} 1 & 1 \\ 0 & 1 \end{pmatrix} & \text{for } \arg z = 0, \end{cases} \tag{2.4.101}$$

where the contours are oriented from left to right on the real axis, and towards the origin on the contours $\arg z = \pm 2\pi/3$. Thus Φ_r satisfies jump conditions similar to (2.4.94), but for jumps on rays emanating from the origin rather than from β_j. We therefore need to map the disk $D(\beta_j, r)$ onto some convex neighborhood of the origin in order to take advantage of the function Φ_r. Our mapping should match the asymptotics of the Airy function in order to have the matching property (2.4.89).

To this end, notice that by (2.3.17) and the regularity condition on $V(x)$, for $t \in [a_j, b_j]$, as $t \to b_j$,

$$\rho_V(t) = C(b_j - t)^{1/2} + \mathcal{O}\big((b_j - t)^{3/2}\big), \quad C > 0. \tag{2.4.102}$$

It follows that for $x \in [a_j, b_j]$, as $x \to b_j$,

$$\int_x^{b_j} \rho_V(t)\,dt = C_0(b_j - x)^{3/2} + \mathcal{O}\big((b_j - x)^{5/2}\big), \quad C_0 = \tfrac{2}{3}C. \tag{2.4.103}$$

Thus the function

$$\psi_{b_j}(z) = -\left\{ \frac{3\pi}{2} \int_z^{b_j} \rho_V(t)\,\mathrm{d}t \right\}^{2/3}, \tag{2.4.104}$$

is analytic at b_j, and thus extends to a conformal map from $D(b_j,r)$ (for small enough r) onto a convex neighborhood of the origin. Furthermore,

$$\psi_{b_j}(b_j) = 0; \quad \psi'_{b_j}(b_j) > 0, \tag{2.4.105}$$

thus ψ_{b_j} is real negative on $(b_j - r, b_j)$, and real positive on $(b_j, b_j + r)$. Also, we can slightly deform the contour Σ_S close to b_j, so that

$$\psi_{b_j}\big(\Gamma \cap D(b_q,r)\big) \subset \{z \mid \arg(z) = \pm 2\pi/3\}.$$

We now set

$$\mathbf{Q}_N(z) = \mathbf{E}_N^{b_j}(z)\Phi_r\big(N^{2/3}\psi_{b_j}(z)\big), \tag{2.4.106}$$

so that

$$\mathbf{U}_N(z) = \mathbf{E}_N^{b_j}(z)\Phi_r\big(N^{2/3}\psi_{b_j}(z)\big)\exp\left[-N\left(g(z) - \frac{V(z)}{2} - \frac{l}{2}\right)\sigma_3\right], \tag{2.4.107}$$

where

$$\begin{aligned} \mathbf{E}_N^{b_j}(z) &= \mathbf{M}(z)\mathrm{e}^{\pm(\mathrm{i}N\Omega_j/2)\sigma_3}\mathbf{L}_N^{b_j}(z)^{-1} \quad \text{for } \pm \operatorname{Im} z > 0, \\ \mathbf{L}_N^{b_j}(z) &= \frac{1}{2\sqrt{\pi}} \begin{pmatrix} N^{-1/6}\psi_{b_j}^{-1/4}(z) & 0 \\ 0 & N^{1/6}\psi_{b_j}^{1/4}(z) \end{pmatrix} \begin{pmatrix} 1 & \mathrm{i} \\ -1 & \mathrm{i} \end{pmatrix}, \end{aligned} \tag{2.4.108}$$

and we take the principal branch of $\psi_{b_j}^{1/4}$, which is positive on $(b_j, b_j + r)$ and has a cut on $(b_j - r, b_j)$. The function $\Phi_r\big(N^{2/3}\psi_{b_j}(z)\big)$ has the jumps j_S, and we claim that the prefactor $\mathbf{E}_N^{\beta_j}$ is analytic in $D(b_j, r)$ and therefore does not change these jumps. This can be seen, as

$$\mathbf{M}_+(z)\mathrm{e}^{(\mathrm{i}N\Omega_j/2)\sigma_3} = \mathbf{M}_-(z)\mathrm{e}^{-(\mathrm{i}N\Omega_j/2)\sigma_3}\mathrm{e}^{(\mathrm{i}N\Omega_j/2)\sigma_3}j_M\mathrm{e}^{(\mathrm{i}N\Omega_j/2)\sigma_3}, \tag{2.4.109}$$

thus the jump for the function $\mathbf{M}(z)\mathrm{e}^{\pm(\mathrm{i}N\Omega_j/2)\sigma_3}$ is

$$\begin{aligned} &\mathrm{e}^{(\mathrm{i}N\Omega_j/2)\sigma_3}j_M\mathrm{e}^{(\mathrm{i}N\Omega_j/2)\sigma_3} \\ &\qquad = \begin{cases} \mathrm{e}^{(\mathrm{i}N\Omega_j/2)\sigma_3}\begin{pmatrix} 0 & 1 \\ -1 & 0 \end{pmatrix}\mathrm{e}^{(\mathrm{i}N\Omega_j/2)\sigma_3} & \text{for } z \in (b_j - r, b_j) \\ \mathrm{e}^{(\mathrm{i}N\Omega_j/2)\sigma_3}\mathrm{e}^{-\mathrm{i}N\Omega_j\sigma_3}\mathrm{e}^{(\mathrm{i}N\Omega_j/2)\sigma_3} & \text{for } z \in (b_j, b_j + r), \end{cases} \end{aligned} \tag{2.4.110}$$

or equivalently,

$$\mathrm{e}^{(\mathrm{i}N\Omega_j/2)\sigma_3}j_M\mathrm{e}^{(\mathrm{i}N\Omega_j/2)\sigma_3} = \begin{cases} \begin{pmatrix} 0 & 1 \\ -1 & 0 \end{pmatrix} & \text{for } z \in (b_j - r, b_j) \\ \begin{pmatrix} 1 & 0 \\ 0 & 1 \end{pmatrix} & \text{for } z \in (b_j, b_j + r), \end{cases} \tag{2.4.111}$$

which is exactly the same as the jump conditions for $\mathbf{L}_N^{b_j}$. It follows that $\mathbf{E}_N^{b_j}(z) = \mathbf{M}(z)\mathrm{e}^{\pm(\mathrm{i}N\Omega_j/2)\sigma_3}\mathbf{L}_N^{b_j}(z)^{-1}$ has no jumps in $D(b_j, r)$. The only other possible singularity for $\mathbf{E}_N^{b_j}$ is at b_j, and this singularity is at most a fourth root singularity, thus removable. Thus, $\mathbf{E}_N^{b_j}$ is analytic in $D(b_j, r)$, and $\mathbf{Q}_N$ has the prescribed jumps. We are left only to prove the matching condition (2.4.89). Using (2.4.95), one can check that, for z in each of the sectors of analyticity, $\Phi_r\big(N^{2/3}\psi_{b_j}(z)\big)$ satisfies the following asymptotics as $N \to \infty$:

$$\begin{aligned}\text{(2.4.112)}\quad \Phi_r\big(N^{2/3}\psi_{b_j}(z)\big) &= \frac{1}{2\sqrt{\pi}} N^{-(1/6)\sigma_3}\psi_{b_j}(z)^{-(1/4)\sigma_3} \\ &\quad\times\left[\begin{pmatrix}1 & \mathrm{i}\\ -1 & \mathrm{i}\end{pmatrix} + \frac{\psi_{b_j}(z)^{-3/2}}{48N}\begin{pmatrix}-5 & 5\mathrm{i}\\ -7 & -7\mathrm{i}\end{pmatrix} + \mathcal{O}(N^{-2})\right] \\ &\quad\times \mathrm{e}^{-(2/3)N\psi_{b_j}(z)^{3/2}\sigma_3},\end{aligned}$$

where we always take the principal branch of $\psi_{b_j}(z)^{3/2}$. As such, $\psi_{b_j}(z)^{3/2}$ is two-valued for $z \in (b_j - r, b_j)$, so that

$$\text{(2.4.113)}\qquad [\tfrac{2}{3}\psi_{b_j}(x)^{3/2}]_\pm = \mp\pi\mathrm{i}\int_x^{b_j}\rho_V(t)\,\mathrm{d}t.$$

Notice that, by (2.4.16),

$$\text{(2.4.114)}\qquad 2g_\pm(x) - V(x) = l \pm 2\pi\mathrm{i}\int_x^{b_j}\rho_V(t)\,\mathrm{d}t = l \pm 2\pi\mathrm{i}\int_x^{b_j}\rho_V(t)\,\mathrm{d}t \pm \mathrm{i}\Omega_j.$$

This implies that for $x \in (b_j - r, b_j)$,

$$\text{(2.4.115)}\qquad \begin{aligned}&[2g_+(b_j) - V(b_j)] - [2g_+(x) - V(x)] = -2\pi\mathrm{i}\int_x^{b_j}\rho_V(t)\,\mathrm{d}t,\\ &[2g_-(b_j) - V(b_j)] - [2g_-(x) - V(x)] = 2\pi\mathrm{i}\int_x^{b_j}\rho_V(t)\,\mathrm{d}t.\end{aligned}$$

Combining these equations with (2.4.113) gives

$$\text{(2.4.116)}\qquad \left[\tfrac{2}{3}\psi_{b_j}(x)^{3/2}\right]_\pm = \tfrac{1}{2}\left[\big(2g_\pm(b_j) - V(b_j)\big) - \big(2g_\pm(x) - V(x)\big)\right].$$

This equation can be extended into the upper and lower planes, respectively, giving

$$\text{(2.4.117)}\quad \tfrac{2}{3}\psi_{b_j}(z)^{3/2} = \tfrac{1}{2}\left[\big(2g_\pm(b_j) - V(b_j)\big) - \big(2g(z) - V(z)\big)\right] \quad \text{for } \pm\operatorname{Im} z > 0.$$

Since by (2.4.114), $2g_\pm(b_j) - V(b_j) = l \pm i\Omega_j$, we get that

$$\text{(2.4.118)}\qquad \tfrac{2}{3}\psi_{b_j}(z)^{3/2} = -g(z) + \frac{V(z)}{2} + \frac{l}{2} \pm \frac{i\Omega_j}{2},$$

for $\pm\operatorname{Im} z > 0$. Plugging (2.4.112) and (2.4.118) into (2.4.107), we get

$$\begin{aligned}\text{(2.4.119)}\quad \mathbf{U}_N(z) &= \mathbf{M}(z)\mathrm{e}^{\pm\mathrm{i}N\Omega_j/2}\mathbf{L}_N^{b_j}(z)^{-1}\frac{1}{2\sqrt{\pi}}N^{-(1/6)\sigma_3}\psi_{b_j}(z)^{-(1/4)\sigma_3} \\ &\quad\times\left[\begin{pmatrix}1 & \mathrm{i}\\ -1 & \mathrm{i}\end{pmatrix} + \frac{\psi_{b_j}(z)^{-3/2}}{48N}\begin{pmatrix}-5 & 5\mathrm{i}\\ -7 & -7\mathrm{i}\end{pmatrix} + \mathcal{O}(N^{-2})\right] \\ &\quad\times \mathrm{e}^{N(g(z)-V(z)/2-l/2\mp\mathrm{i}\Omega_j/2)\sigma_3}\mathrm{e}^{-N(g(z)-V(z)/2-l/2)\sigma_3} \\ &= \mathbf{M}(z)\left[I + \frac{\psi_{b_j}(z)^{-3/2}}{48N}\begin{pmatrix}1 & 6\mathrm{i}\mathrm{e}^{\pm\mathrm{i}N\Omega_j}\\ 6\mathrm{i}\mathrm{e}^{\mp\mathrm{i}N\Omega_j} & -1\end{pmatrix} + \mathcal{O}(N^{-2})\right],\end{aligned}$$

for $\pm\operatorname{Im} z > 0$. Thus we have that $\mathbf{U}_N$ satisfies conditions (2.4.87), (2.4.88), and (2.4.89).

Local analysis near the left endpoint a_j is similar. Namely, we let

$$\psi_{a_j}(z) = -\left\{\frac{3\pi}{2}\int_{a_j}^{z}\rho_V(t)\,\mathrm{d}t\right\}^{2/3}. \tag{2.4.120}$$

Then ψ_{a_j} is analytic throughout $D(a_j, r)$, real valued on the real line, and has negative derivative at a_j. Close to a_j, the jumps j_Q become

$$j_Q(z) = \begin{cases} \begin{pmatrix} 1 & 1 \\ 0 & 1 \end{pmatrix} & \text{for } z \in (a_j - r, a_j) \\ \begin{pmatrix} 1 & 0 \\ 1 & 1 \end{pmatrix} & \text{for } z \in \gamma_j^+ \\ \begin{pmatrix} 1 & 0 \\ 1 & 1 \end{pmatrix} & \text{for } z \in \gamma_j^- \\ \begin{pmatrix} 0 & 1 \\ -1 & 0 \end{pmatrix} & \text{for } z \in (a_j, a_j + r), \end{cases} \tag{2.4.121}$$

After the change of variables ψ_{a_j} (and a slight deformation of the contours $\gamma_j^{\pm}$), these jumps become the following jumps close to the origin:

$$j_Q(\psi_{a_j}(z)) = \begin{cases} \begin{pmatrix} 0 & 1 \\ -1 & 0 \end{pmatrix} & \text{for } \arg\psi_{a_j}(z) = \pi \\ \begin{pmatrix} 1 & 0 \\ 1 & 1 \end{pmatrix} & \text{for } \arg\psi_{a_j}(z) = 2\pi/3 \\ \begin{pmatrix} 1 & 0 \\ 1 & 1 \end{pmatrix} & \text{for } \arg\psi_{a_j}(z) = -2\pi/3 \\ \begin{pmatrix} 1 & 1 \\ 0 & 1 \end{pmatrix} & \text{for } \arg\psi_{a_j}(z) = 0, \end{cases} \tag{2.4.122}$$

where orientation is taken right to left on horizontal contours, and towards the origin on the contours $\arg\psi_{a_j}(z) = \pm 2\pi/3$. These jump conditions are satisfied by the function

$$\Phi_l(z) = \Phi_r(z)\begin{pmatrix} 1 & 0 \\ 0 & -1 \end{pmatrix}. \tag{2.4.123}$$

Then we can take

$$\mathbf{U}_N(z) = \mathbf{E}_N^{a_j}(z)\Phi_l\big(N^{2/3}\psi_{a_j}(z)\big)\mathrm{e}^{-N(g(z)-V(z)/2-l/2)\sigma_3} \tag{2.4.124}$$

for $z \in D(a_j, r)$, where

$$\begin{aligned} \mathbf{E}_N^{a_j}(z) &= \mathbf{M}(z)\mathrm{e}^{\pm(\mathrm{i}N\Omega_{j-1}/2)\sigma_3}\mathbf{L}_N^{a_j}(z)^{-1} \quad \text{for } \pm\operatorname{Im} z > 0, \\ \mathbf{L}_N^{a_j}(z) &= \frac{1}{2\sqrt{\pi}}\begin{pmatrix} N^{-1/6}\psi_{a_j}^{-1/4}(z) & 0 \\ 0 & N^{1/6}\psi_{a_j}^{1/4}(z) \end{pmatrix}\begin{pmatrix} 1 & -\mathrm{i} \\ -1 & -\mathrm{i} \end{pmatrix} \end{aligned} \tag{2.4.125}$$

is an analytic prefactor. Similar to (2.4.112), we have that in each sector of analyticity, $\Phi_l\big(N^{2/3}\psi_{a_j}(z)\big)$ satisfies

$$\begin{aligned}\Phi_l\big(N^{2/3}\psi_{a_j}(z)\big) &= \frac{1}{2\sqrt{\pi}}N^{-(1/6)\sigma_3}\psi_{a_j}(z)^{-(1/4)\sigma_3}\\ &\quad\times\left[\begin{pmatrix}1 & -\mathrm{i}\\ -1 & -\mathrm{i}\end{pmatrix}+\frac{\psi_{a_j}(z)^{-3/2}}{48N}\begin{pmatrix}-5 & -5\mathrm{i}\\ -7 & 7\mathrm{i}\end{pmatrix}+\mathcal{O}(N^{-2})\right]\\ &\quad\times \mathrm{e}^{-(2/3)N\psi_{a_j}(z)^{3/2}\sigma_3}.\end{aligned}\tag{2.4.126}$$

Once again, we have that, for $x\in(a_j,a_j+r)$, $\psi_{a_j}(x)^{3/2}$ takes limiting values from above and below, so that

$$[\tfrac{2}{3}\psi_{a_j}(x)^{3/2}]_\pm = \pm\pi\mathrm{i}\int_{a_j}^{x}\rho_V(t)\,\mathrm{d}t.\tag{2.4.127}$$

In analogue to (2.4.116), we have

$$\tfrac{2}{3}\psi_{a_j}(z)^{3/2} = \tfrac{1}{2}\big[\big(2g_\pm(a_j)-V(a_j)\big)-\big(2g(z)-V(z)\big)\big] \quad \text{for } \pm\operatorname{Im} z>0.\tag{2.4.128}$$

Since by (2.4.114), $2g_\pm(a_j)-V(a_j)=l\pm\pi\mathrm{i}$, we get that

$$\frac{2}{3}\psi_{a_j}(z)^{3/2} = -g(z)+\frac{V(z)}{2}+\frac{l}{2}\pm\frac{\mathrm{i}\Omega_{j-1}}{2} \quad \text{for } \pm\operatorname{Im} z>0.\tag{2.4.129}$$

Plugging (2.4.129) into (2.4.124) and (2.4.126) gives as $N\to\infty$,

$$\begin{aligned}\mathbf{U}_N(z) &= \mathbf{M}(z)\mathrm{e}^{\pm(\mathrm{i}N\Omega_{j-1}/2)\sigma_3}\mathbf{L}_N^{a_j}(z)^{-1}\frac{1}{2\sqrt{\pi}}N^{-(1/6)\sigma_3}\psi_{a_j}(z)^{-(1/4)\sigma_3}\\ &\quad\times\left[\begin{pmatrix}1 & -\mathrm{i}\\ -1 & -\mathrm{i}\end{pmatrix}+\frac{\psi_{a_j}(z)^{-3/2}}{48N}\begin{pmatrix}-5 & -5\mathrm{i}\\ -7 & 7\mathrm{i}\end{pmatrix}+\mathcal{O}(N^{-2})\right]\\ &\quad\times \mathrm{e}^{N(g(z)-V(z)/2-l/2\mp\mathrm{i}\Omega_{j-1}/2)\sigma_3}\mathrm{e}^{-N(g(z)-V(z)/2-l/2)\sigma_3}\\ &= \mathbf{M}(z)\left[I+\frac{\psi_{a_j}(z)^{-3/2}}{48N}\begin{pmatrix}1 & -6\mathrm{i}\mathrm{e}^{\pm\mathrm{i}N\Omega_{j-1}\sigma_3}\\ -6\mathrm{i}\mathrm{e}^{\mp\mathrm{i}N\Omega_{j-1}\sigma_3} & -1\end{pmatrix}+\mathcal{O}(N^{-2})\right],\end{aligned}\tag{2.4.130}$$

for $\pm\operatorname{Im} z>0$.

2.4.7. Third and final transformation of the RHP.

We are now ready to combine $\mathbf{S},\mathbf{M}$ and $\mathbf{U}$ to arrive at a RHP with jumps uniformly close to the identity. In the third and final transformation we put

$$\mathbf{X}_N(z)=\begin{cases}\mathbf{S}_N(z)\mathbf{M}(z)^{-1} & \text{for } z \text{ outside the disks } D(a_j,r), D(b_j,r),\ 1\le j\le q,\\ \mathbf{S}_N(z)\mathbf{U}_N(z)^{-1} & \text{for } z \text{ inside the disks.}\end{cases}\tag{2.4.131}$$

Then $\mathbf{X}_N(z)$ is analytic on $\mathbb{C}\setminus\Gamma_R$, where Γ_R consists of the circles $\partial D(a_j,r)$, $\partial D(b_j,r)$, $1\le j\le q$, the parts of Γ outside of the disks $D(a_j,r)$, $D(b_j,r)$, $1\le j\le q$, and the real intervals $(-\infty,a_1-r),(b_1+r,a_2-r),\dots,(b_{q-1}+r,a_q-r),(b_q+r,\infty)$, see Figure 2.6. There are the jump relations,

$$\mathbf{X}_{N+}(z)=\mathbf{X}_{N-}(z)j_X(z),\tag{2.4.132}$$

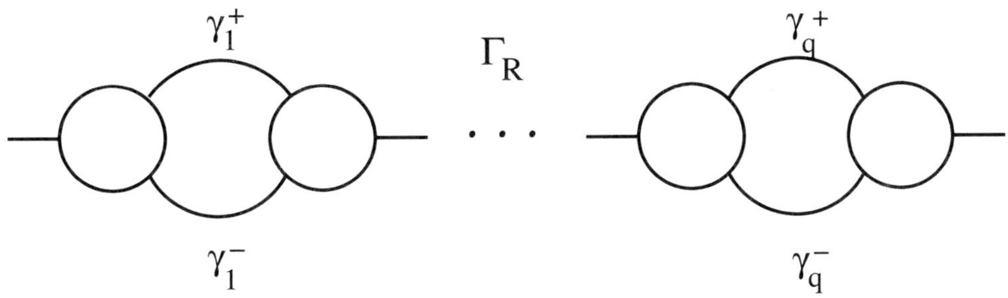

FIGURE 2.6. The contour Γ_R for $\mathbf{X}_N(z)$.

where

$$(2.4.133)\quad j_X(z) = \begin{cases} \mathbf{M}(z)\mathbf{U}_N(z)^{-1} & \text{on the circles, oriented counterclockwise,} \\ \mathbf{M}(z)j_S(z)\mathbf{M}(z)^{-1} & \text{on the remaining parts of } \Gamma_R. \end{cases}$$

We have that

(2.4.134)

$$j_X(z) = \begin{cases} I + \mathcal{O}(N^{-1}) & \text{uniformly on the circles,} \\ I + \mathcal{O}(\mathrm{e}^{-c(z)N}) & \text{for some } c(z) > 0 \text{, on the remaining parts of } \Gamma_R, \end{cases}$$

where $c(x)$ satisfies estimate (2.4.22) as $x \to \pm\infty$. As $z \to \infty$, we have

$$(2.4.135)\qquad \mathbf{X}_N(z) \sim I + \sum_{j=1}^{\infty} \frac{\mathbf{X}_j}{z^j},$$

for some constant matrices $\mathbf{X}_j$. Thus, $\mathbf{X}_N(z)$ solves the following RHP:

(1) $\mathbf{X}_N(z)$ is analytic in $\mathbb{C} \setminus \Gamma_R$. and it is two-valued on Γ_R.
(2) On Γ_R, $\mathbf{X}_N(z)$ satisfies jump condition (2.4.132), where the jump matrix $j_X(z)$ satisfies estimates (2.4.134).
(3) As $z \to \infty$, $\mathbf{X}_N(z)$ has asymptotic expansion (2.4.135).

This RHP can be solved asymptotically as $N \to \infty$ by a perturbation theory series.

2.5. Solution of the RHP for $\mathbf{X}_N(z)$

Set

$$(2.5.1)\qquad j_X^0(z) = j_X(z) - I.$$

Then by (2.4.134),

$$(2.5.2)\quad j_X^0(z) = \begin{cases} \mathcal{O}(N^{-1}) & \text{uniformly on the circles,} \\ \mathcal{O}(e^{-c(z)N}) & \text{for some } c(z) > 0 \text{, on the remaining parts of } \Gamma_R, \end{cases}$$

where $c(x)$ satisfies (2.4.22) as $x \to \infty$. We can apply the following general result.

Proposition 2.5.1. *Assume that $v(z)$, $z \in \Gamma_R$, solves the equation*

$$(2.5.3)\qquad v(z) = I - \frac{1}{2\pi \mathrm{i}} \int_{\Gamma_R} \frac{v(u) j_X^0(u)}{z_- - u}\, du, \qquad z \in \Gamma_R,$$

where z_- means the value of the integral on the minus side of Γ_R. Then

$$\mathbf{X}(z) = I - \frac{1}{2\pi \mathrm{i}} \int_{\Gamma_R} \frac{v(u) j_X^0(u)}{z-u} \, du, \quad z \in \mathbb{C} \setminus \Gamma_R, \tag{2.5.4}$$

solves the following RHP:

(i) $\mathbf{X}(z)$ *is analytic on* $\mathbb{C} \setminus \Gamma_R$.
(ii) $\mathbf{X}_+(z) = \mathbf{X}_-(z) j_X(z)$, $z \in \Gamma_R$.
(iii) $\mathbf{X}(z) = I + \mathcal{O}(z^{-1})$, $z \to \infty$.

PROOF. From (2.5.3), (2.5.4),

$$\mathbf{X}_-(z) = v(z), \quad z \in \Gamma_R. \tag{2.5.5}$$

By the jump property of the Cauchy transform,

$$\mathbf{X}_+(z) - \mathbf{X}_-(z) = v(z) j_X^0(z) = \mathbf{X}_-(z) j_X^0(z), \tag{2.5.6}$$

hence $\mathbf{X}_+(z) = \mathbf{X}_-(z) j_X(z)$. From (2.5.4), $\mathbf{X}(z) = I + \mathcal{O}(z^{-1})$. Proposition 2.5.1 is proved. □

Equation (2.5.3) can be solved by perturbation theory, so that

$$v(z) = I + \sum_{k=1}^{\infty} v_k(z), \tag{2.5.7}$$

where for $k \geq 1$,

$$v_k(z) = -\frac{1}{2\pi \mathrm{i}} \int_{\Gamma_R} \frac{v_{k-1}(u) j_X^0(u)}{z_- - u} \, du, \quad z \in \Gamma_R, \tag{2.5.8}$$

and $v_0(z) = I$. Series (2.5.7) is estimated from above by a convergent geometric series, so it is absolutely convergent. From (2.5.2) we obtain that there exists $C > 0$ such that

$$|v_k(z)| \leq \frac{C^k}{N^k(1+|z|)}. \tag{2.5.9}$$

Observe that

$$v_1(z) = -\frac{1}{2\pi \mathrm{i}} \int_{\Gamma_R} \frac{j_X^0(u)}{z_- - u} \, du, \quad z \in \Gamma_R. \tag{2.5.10}$$

We apply this solution to find $\mathbf{X}_N(z)$. The function $\mathbf{X}_N(z)$ is given then as

$$\mathbf{X}_N(z) = I + \sum_{k=1}^{\infty} \mathbf{X}_{N,k}(z), \tag{2.5.11}$$

where

$$\mathbf{X}_{N,k}(z) = -\frac{1}{2\pi \mathrm{i}} \int_{\Gamma_R} \frac{v_{k-1}(u) j_X^0(u)}{z-u} \, du. \tag{2.5.12}$$

In particular,

$$\mathbf{X}_{N,1}(z) = -\frac{1}{2\pi \mathrm{i}} \int_{\Gamma_R} \frac{j_X^0(u)}{z-u} \, du. \tag{2.5.13}$$

We can write (2.5.8) as

$$v_k(z) = (C_{\Gamma_R}^- J)^k(\mathbf{1}), \tag{2.5.14}$$

where J is the operator of mulitplication by $j_X^0(z)$, and the minus Cauchy operator $C_{\Gamma_R}^-$ is defined as

$$C_{\Gamma_R}^-(f)(z) = -\frac{1}{2\pi i}\int_{\Gamma_R}\frac{f(u)}{z_- - u}\,du. \tag{2.5.15}$$

Consider for a moment J and $C_{\Gamma_R}^-$ as operators on $L^2(\Gamma_R)$. From (2.5.2), we have that the L^2-norm of J is $\mathcal{O}(N^{-1})$. The L^2-norm of $C_{\Gamma_R}^-$ does not depend on N, and if is bounded, then (2.5.14) implies that there exists a constant C such that

$$\|v_k\|_2 < \left(\frac{C}{N}\right)^k, \tag{2.5.16}$$

which implies that, for large enough N, the series (2.5.8) converges in $L^2(\Gamma_R)$.

Now consider $z \in \mathbb{C}$ with $\operatorname{dist}(z, \Gamma_R) > \varepsilon/10$. Applying the Cauchy–Schwartz inequality to (2.5.12) gives

$$|\mathbf{X}_{N,k}(z)| \le \frac{1}{2\pi}\|v_k\|_2 \left\|\frac{j_X^0(u)}{z-u}\right\|_2 \le \frac{1}{2\pi}\left(\frac{C}{N}\right)^{k-1}\left(\frac{C}{N}\right)\left(\frac{C_0}{|z|+1}\right), \tag{2.5.17}$$

for some $C_0 > 0$. In fact, the condition that $\operatorname{dist}(z, \Gamma_R) > \varepsilon/10$ is not necessary, as the contour Γ_R can be deformed. We thus have that there exists $C_1 > 0$ such that for $k \ge 0$,

$$\mathbf{X}_N(z) = I + \sum_{j=1}^{k}\mathbf{X}_{N,j}(z) + \varepsilon_{N,k}(z), \quad |\varepsilon_{N,k}(z)| \le \frac{C_1 C^k}{N^{k+1}(1+|z|)}. \tag{2.5.18}$$

In particular, for $k = 0$,

$$\mathbf{X}_N(z) = I + \varepsilon_N(z), \quad |\varepsilon_N(z)| \le \frac{C_1}{N(1+|z|)}. \tag{2.5.19}$$

2.6. Asymptotics of the recurrence coefficients

Having completed the steepest descent analysis, we can now "unravel" the transformations of the RHP to obtain the large N asymptotics of orthogonal polynomials. From (2.4.131) and (2.5.19) we obtain that

$$\mathbf{S}_N(z) = \begin{cases} \left(I + \mathcal{O}\left(\dfrac{1}{N(|z|+1)}\right)\right)\mathbf{M}(z), & z \in \mathbb{C}\setminus D, \\ \left(I + \mathcal{O}\left(\dfrac{1}{N(|z|+1)}\right)\right)\mathbf{U}_N(z), & z \in D; \end{cases} \tag{2.6.1}$$
$$D = \bigcup_{j=1}^{q}[D(a_j, r)\cup D(b_j, r)].$$

From (2.4.32) we have that

$$\mathbf{T}_N(z) = \begin{cases} \mathbf{S}_N(z)\begin{pmatrix} 1 & 0 \\ e^{-NG(z)} & 1 \end{pmatrix}, & z \in \mathcal{L}^+ = \bigcup_{j=1}^{q} \mathcal{L}_j^+, \\ \mathbf{S}_N(z)\begin{pmatrix} 1 & 0 \\ -e^{NG(z)} & 1 \end{pmatrix}, & z \in \mathcal{L}^- = \bigcup_{j=1}^{q} \mathcal{L}_j^-, \\ \mathbf{S}_N(z), & z \in \mathbb{C} \setminus (\mathcal{L}^+ \cup \mathcal{L}^-). \end{cases} \tag{2.6.2}$$

Finally, from (2.4.18) we obtain that

(2.6.3)

$$\mathbf{Y}_N(z) = \begin{cases} e^{(Nl/2)\sigma_3}\Big(I + \mathcal{O}\big(\frac{1}{N(|z|+1)}\big)\Big)\mathbf{M}(z)\Big(\begin{smallmatrix} 1 & 0 \\ \pm e^{\mp NG(z)} & 1 \end{smallmatrix}\Big)e^{N[g(z)-l/2]\sigma_3}, \\ \qquad\qquad z \in \mathcal{L}^\pm \setminus D, \\ e^{(Nl/2)\sigma_3}\Big(I + \mathcal{O}\big(\frac{1}{N(|z|+1)}\big)\Big)\mathbf{U}_N(z)e^{N[g(z)-l/2]\sigma_3}, \quad z \in D, \\ e^{(Nl/2)\sigma_3}\Big(I + \mathcal{O}\big(\frac{1}{N(|z|+1)}\big)\Big)\mathbf{M}(z)e^{N[g(z)-l/2]\sigma_3}, \quad z \in \mathbb{C}\setminus(D \cup \mathcal{L}^+ \cup \mathcal{L}^-). \end{cases}$$

This gives the large N asymptotics of the orthogonal polynomials and their Cauchy transforms on the complex plane. To get the large N asymptotics of the recurrence coefficients, we use formulas (2.2.11) and (2.2.12). Consider first γ_N^2.

From (2.4.18) we obtain that for large z,

$$\begin{aligned} I + \frac{\mathbf{Y}_1}{z} + \frac{\mathbf{Y}_2}{z^2} + \cdots &= \mathbf{Y}_N(z)z^{-N\sigma_3} \\ &= e^{(Nl/2)\sigma_3}\mathbf{T}(z)e^{N[g(z)-l/2-\log z]\sigma_3} \\ &= e^{(Nl/2)\sigma_3}\left(I + \frac{\mathbf{T}_1}{z} + \frac{\mathbf{T}_2}{z^2} + \cdots\right)e^{N[g(z)-l/2-\log z]\sigma_3}, \end{aligned} \tag{2.6.4}$$

hence

$$[\mathbf{Y}_1]_{12} = e^{Nl}[\mathbf{T}_1]_{12}, \quad [\mathbf{Y}_1]_{21} = e^{-Nl}[\mathbf{T}_1]_{21}, \tag{2.6.5}$$

and

$$\gamma_N^2 = [\mathbf{Y}_1]_{12}[\mathbf{Y}_1]_{21} = [\mathbf{T}_1]_{12}[\mathbf{T}_1]_{21}. \tag{2.6.6}$$

From (2.6.1), (2.6.2) we obtain further that

$$\gamma_N^2 = [\mathbf{M}_1]_{12}[\mathbf{M}_1]_{21} + \mathcal{O}(N^{-1}), \tag{2.6.7}$$

and from (2.4.85),

$$\begin{aligned} [\mathbf{M}_1]_{12} &= \frac{\theta(-u(\infty)+d+c)\theta(u(\infty)+d)}{\theta(u(\infty)+d+c)\theta(-u(\infty)+d)} \sum_{j=1}^{q} \frac{(b_j - a_j)}{(-4\mathrm{i})}, \\ [\mathbf{M}_1]_{21} &= \frac{\theta(u(\infty)-d+c)\theta(-u(\infty)-d)}{\theta(-u(\infty)-d+c)\theta(u(\infty)-d)} \sum_{j=1}^{q} \frac{(b_j - a_j)}{4\mathrm{i}}, \end{aligned} \tag{2.6.8}$$

hence

$$(2.6.9)\quad \gamma_N^2 = \left[\frac{1}{4}\sum_{j=1}^{q}(b_j - a_j)\right]^2 \times \frac{\theta^2(u(\infty)+d)\theta(u(\infty)+N\Omega/(2\pi)-d)\theta(-u(\infty)+N\Omega/(2\pi)+d)}{\theta^2(u(\infty)-d)\theta(-u(\infty)+N\Omega/(2\pi)-d)\theta(u(\infty)+N\Omega/(2\pi)+d)} + \mathcal{O}(N^{-1}),$$

where d is defined in (2.4.79). Also, from (2.2.18) and (2.6.3), the normalizing constants h_N and h_{N-1} are given by

(2.6.10)

$$\begin{aligned} h_N &= -2\pi \mathrm{i} \mathrm{e}^{Nl}[\mathbf{M}_1]_{12}[1+\mathcal{O}(N^{-1})] \\ &= \frac{\pi}{2}\mathrm{e}^{Nl}\left[\sum_{j=1}^{q}(b_j - a_j)\right]\frac{\theta\big(-u(\infty)+d+N\Omega/(2\pi)\big)\theta(u(\infty)+d)}{\theta\big(u(\infty)+d+N\Omega/(2\pi)\big)\theta(-u(\infty)+d)} \\ &\qquad \times [1+\mathcal{O}(N^{-1})], \end{aligned}$$

and

(2.6.11)

$$\begin{aligned} h_{N-1} &= \frac{-2\pi \mathrm{i} \mathrm{e}^{Nl}}{[\mathbf{M}_1]_{12}}[1+\mathcal{O}(N^{-1})] \\ &= 8\pi \mathrm{e}^{Nl}\left[\sum_{j=1}^{q}(b_j - a_j)\right]^{-1}\frac{\theta\big(-u(\infty)-d+N\Omega/(2\pi)\big)\theta(u(\infty)-d)}{\theta\big(u(\infty)-d+N\Omega/(2\pi)\big)\theta(-u(\infty)-d)} \\ &\qquad \times [1+\mathcal{O}(N^{-1})]. \end{aligned}$$

Consider now β_{N-1}.

From (2.6.4) we obtain that

$$(2.6.12)\qquad [\mathbf{Y}_1]_{11} = [\mathbf{T}_1]_{11} + Ng_1, \quad [\mathbf{Y}_2]_{21} = \mathrm{e}^{-Nl}([\mathbf{T}_2]_{21} + [\mathbf{T}_1]_{21}Ng_1),$$

where $g_1 = \int_J x \,\mathrm{d}\nu_V(x)$, see (2.4.7). Hence,

$$(2.6.13)\qquad \beta_{N-1} = \frac{[\mathbf{Y}_2]_{21}}{[\mathbf{Y}_1]_{21}} - [\mathbf{Y}_1]_{11} = \frac{[\mathbf{T}_2]_{21}}{[\mathbf{T}_1]_{21}} - [\mathbf{T}_1]_{11},$$

and by (2.6.1), (2.6.2),

$$(2.6.14)\qquad \beta_{N-1} = \frac{[\mathbf{M}_2]_{21}}{[\mathbf{M}_1]_{21}} - [\mathbf{M}_1]_{11} + \mathcal{O}(N^{-1}).$$

From (2.4.82) we find that

(2.6.15)

$$\begin{aligned}\frac{[\mathbf{M}_2]_{21}}{[\mathbf{M}_1]_{21}} &= \frac{\sum_{j=1}^q (b_j^2 - a_j^2)}{2\sum_{j=1}^q (b_j - a_j)} + \frac{\theta(u(\infty) - d)}{\theta(u(\infty) - d + c)}\left(\nabla_u \frac{\theta(u - d + c)}{\theta(u - d)}\bigg|_{u=u(\infty)}, u'(\infty)\right) \\ &= \frac{\sum_{j=1}^q (b_j^2 - a_j^2)}{2\sum_{j=1}^q (b_j - a_j)} + \left(\frac{\nabla\theta(u(\infty) - d + c)}{\theta(u(\infty) - d + c)} - \frac{\nabla\theta(u(\infty) - d)}{\theta(u(\infty) - d)}, u'(\infty)\right), \\ [\mathbf{M}_1]_{11} &= \frac{\theta(u(\infty) + d)}{\theta(u(\infty) + d + c)}\left(\nabla_u \frac{\theta(u + d + c)}{\theta(u + d)}\bigg|_{u=u(\infty)}, u'(\infty)\right) \\ &= \left(\frac{\nabla\theta(u(\infty) + d + c)}{\theta(u(\infty) + d + c)} - \frac{\nabla\theta(u(\infty) + d)}{\theta(u(\infty) + d)}, u'(\infty)\right).\end{aligned}$$

Hence,

$$\begin{aligned}(2.6.16)\quad \beta_{N-1} = &\frac{\sum_{j=1}^q (b_j^2 - a_j^2)}{2\sum_{j=1}^q (b_j - a_j)} \\ &+ \left(\frac{\nabla\theta(u(\infty) + N\Omega/(2\pi) - d)}{\theta(u(\infty) + N\Omega/(2\pi) - d)} - \frac{\nabla\theta(u(\infty) + N\Omega/(2\pi) + d)}{\theta(u(\infty) + N\Omega/(2\pi) + d)}\right. \\ &\qquad \left. + \frac{\nabla\theta(u(\infty) + d)}{\theta(u(\infty) + d)} - \frac{\nabla\theta(u(\infty) - d)}{\theta(u(\infty) - d)}, u'(\infty)\right) \\ &+ \mathcal{O}(N^{-1}).\end{aligned}$$

This formula can be also written in the shorter form,

$$\begin{aligned}(2.6.17)\quad \beta_{N-1} = &\frac{\sum_{j=1}^q (b_j^2 - a_j^2)}{2\sum_{j=1}^q (b_j - a_j)} \\ &+ \frac{\mathrm{d}}{\mathrm{d}z}\left[\log \frac{\theta(u(z) + N\Omega/(2\pi) - d)\theta(u(z) + d)}{\theta(u(z) + N\Omega/(2\pi) + d)\theta(u(z) - d)}\right]\bigg|_{z=\infty} + \mathcal{O}(N^{-1}).\end{aligned}$$

In the one-cut case, $q = 1$, $a_1 = a$, $b_1 = b$, formulas (2.6.9), (2.6.16) simplify to

$$(2.6.18)\qquad \gamma_N = \frac{b-a}{4} + \mathcal{O}(N^{-1}), \quad \beta_{N-1} = \frac{a+b}{2} + \mathcal{O}(N^{-1}).$$

Formula (2.6.9) is obtained in [29].

2.7. Universality in the random matrix model

By applying asymptotics (2.6.3) to reproducing kernel (2.2.19), we obtain the asymptotics of the eigenvalue correlation functions. First we consider the eigenvalue correlation functions in the bulk of the spectrum. Let us fix a point $x_0 \in \operatorname{Int} J = \bigcup_{j=1}^q (a_j, b_j)$. Then the density $\rho_V(x_0)$ is positive. We have the following universal scaling limit of the reproducing kernel at x_0:

Theorem 2.7.1. *Let x_0 be a point in the interior of the support of the equilibrium measure for V, and let u and v be two bounded real numbers. Then*

$$(2.7.1)\quad \frac{1}{N\rho_V(x_0)} K_N\left(x_0 + \frac{u}{N\rho_V(x_0)}, x_0 + \frac{v}{N\rho_V(x_0)}\right) = \frac{\sin[\pi(u-v)]}{\pi(u-v)} + \mathcal{O}(N^{-1}).$$

PROOF. Assume that for some $1 \le j \le q$ and for some $\varepsilon > 0$, we have $\{x_0, x, y\} \in (a_j + \varepsilon, b_j - \varepsilon)$. By (2.2.19) and (2.4.18),

$$\begin{aligned} K_N(x,y) &= -\frac{\mathrm{e}^{-NV(x)/2}\mathrm{e}^{-NV(y)/2}}{2\pi\mathrm{i}(x-y)} \begin{pmatrix} 0 & 1 \end{pmatrix} \mathbf{Y}_{N+}^{-1}(y)\mathbf{Y}_{N+}(x) \begin{pmatrix} 1 \\ 0 \end{pmatrix} \\ &= -\frac{\mathrm{e}^{-NV(x)/2}\mathrm{e}^{-NV(y)/2}}{2\pi\mathrm{i}(x-y)} \begin{pmatrix} 0 & \mathrm{e}^{N[g_+(y)-l/2]} \end{pmatrix} \mathbf{T}_{N+}^{-1}(y) \\ &\qquad \times \mathbf{T}_{N+}(x) \begin{pmatrix} \mathrm{e}^{N[g_+(x)-l/2]} \\ 0 \end{pmatrix}. \end{aligned} \tag{2.7.2}$$

Now, from (2.4.32) we obtain that

$$\begin{aligned} K_N(x,y) &= -\frac{\mathrm{e}^{-NV(x)/2}\mathrm{e}^{-NV(y)/2}}{2\pi\mathrm{i}(x-y)} \begin{pmatrix} 0 & \mathrm{e}^{N[g_+(y)-l/2]} \end{pmatrix} j_+(y)^{-1}\mathbf{S}_{N+}^{-1}(y) \\ &\qquad \times \mathbf{S}_{N+}(x) j_+(x) \begin{pmatrix} \mathrm{e}^{N[g_+(x)-l/2]} \\ 0 \end{pmatrix} \\ &= -\frac{\mathrm{e}^{-NV(x)/2}\mathrm{e}^{-NV(y)/2}}{2\pi\mathrm{i}(x-y)} \begin{pmatrix} \mathrm{e}^{N[-G(y)+g_+(y)-l/2]} & -\mathrm{e}^{N[g_+(y)-l/2]} \end{pmatrix} \\ &\qquad \times \mathbf{S}_{N+}^{-1}(y)\mathbf{S}_{N+}(x) \begin{pmatrix} \mathrm{e}^{N[g_+(x)-l/2]} \\ \mathrm{e}^{N[-G(x)+g_+(x)-l/2]} \end{pmatrix}. \end{aligned} \tag{2.7.3}$$

By (2.4.16),

$$-\frac{V(x)}{2} + g_+(x) - \frac{l}{2} = \frac{G(x)}{2}, \tag{2.7.4}$$

hence

$$\begin{aligned} &K_N(x,y) \\ &= -\frac{1}{2\pi\mathrm{i}(x-y)} \begin{pmatrix} \mathrm{e}^{-NG(y)/2} & -\mathrm{e}^{NG(y)/2} \end{pmatrix} \mathbf{S}_{N+}^{-1}(y)\mathbf{S}_{N+}(x) \begin{pmatrix} \mathrm{e}^{NG(x)/2} \\ \mathrm{e}^{-NG(x)/2} \end{pmatrix} \end{aligned} \tag{2.7.5}$$

By (2.4.131),

$$\mathbf{S}_{N+}(x) = \mathbf{X}_N(x)\mathbf{M}_+(x). \tag{2.7.6}$$

Observe that $\mathbf{M}_+(x)$ and $\mathbf{X}_N(x)$ are analytic on $(a_j + \varepsilon, b_j - \varepsilon)$ and $\mathbf{X}_N(x)$ satisfies estimate (2.5.19). This implies that as $x - y \to 0$,

$$\mathbf{S}_{N+}^{-1}(y)\mathbf{S}_{N+}(x) = I + \mathcal{O}(x-y), \tag{2.7.7}$$

uniformly in N. Since the function $G(x)$ is pure imaginary for real x, we obtain from (2.7.5) and (2.7.7) that as $x - y \to 0$,

$$K_N(x,y) = \frac{1}{2\pi\mathrm{i}(x-y)}[e^{-N(G(x)-G(y))/2} - e^{N(G(x)-G(y))/2}] + \mathcal{O}(1). \tag{2.7.8}$$

By (2.4.14),

$$-\frac{N[G(x)-G(y)]}{2} = \pi\mathrm{i}N\int_y^x \rho_V(s)\,\mathrm{d}s = \pi\mathrm{i}N\rho_V(\xi)(x-y), \quad \xi \in [x,y], \tag{2.7.9}$$

hence

$$K_N(x,y) = \frac{\sin[\pi N\rho_V(\xi)(x-y)]}{\pi(x-y)} + \mathcal{O}(1). \tag{2.7.10}$$

Let

$$x = x_0 + \frac{u}{N\rho_V(x_0)}, \quad y = x_0 + \frac{v}{N\rho_V(x_0)}, \tag{2.7.11}$$

where u and v are bounded. Then

$$\frac{1}{N\rho_V(x_0)} K_N(x,y) = \frac{\sin[\pi(u-v)]}{\pi(u-v)} + \mathcal{O}(N^{-1}), \tag{2.7.12}$$

which is (2.7.1). The universal limiting kernel $\sin[\pi(u-v)]/[\pi(u-v)]$ is known as the *sine kernel.* □

Consider now the scaling limit at an edge point. Since the density ρ_V is zero at the edge point, we have to expect a different scaling of the eigenvalues. We have the following universal scaling limit of the reproducing kernel at the edge point:

Theorem 2.7.2. *Let $x_0 = b_j$ for some $1 \le j \le q$, and u and v be two bounded real numbers. Then for some $c > 0$,*

$$\begin{aligned} \frac{1}{cN^{2/3}} K_N\left(x_0 + \frac{u}{cN^{2/3}}, x_0 + \frac{v}{cN^{2/3}}\right) \\ = \frac{\operatorname{Ai}(u)\operatorname{Ai}'(v) - \operatorname{Ai}'(u)\operatorname{Ai}(v)}{u-v} + \mathcal{O}(N^{-2/3}). \end{aligned} \tag{2.7.13}$$

Similarly, if $x_0 = a_j$ for some $1 \le j \le q$, then for some $c > 0$, as $N \to \infty$,

$$\begin{aligned} \frac{1}{cN^{2/3}} K_N\left(x_0 - \frac{u}{cN^{2/3}}, x_0 - \frac{v}{cN^{2/3}}\right) \\ = \frac{\operatorname{Ai}(u)\operatorname{Ai}'(v) - \operatorname{Ai}'(u)\operatorname{Ai}(v)}{u-v} + \mathcal{O}(N^{-2/3}). \end{aligned} \tag{2.7.14}$$

Proof. Let us consider the right edge of a band. The left edge is similar. The local solution near b_j divides a neighborhood of b_j into the four regions shown in Figure 2.5. Let us consider a limit as x and y approach the real axis from the region I. In that region, formula (2.7.2) is valid, and we can use that expression along with (2.4.32), (2.4.107), and (2.7.4) to obtain

$$\begin{aligned} K_N(x,y) = -\frac{1}{2\pi i(x-y)} \begin{pmatrix} 0 & 1 \end{pmatrix} \Phi_r\big(N^{2/3}\psi_{b_j}(y)\big)^{-1} \mathbf{E}_N^{b_j}(y)^{-1} \mathbf{X}_N(y)^{-1} \\ \times \mathbf{X}_N(x) \mathbf{E}_N^{b_j}(x) \Phi_r\big(N^{2/3}\psi_{b_j}(x)\big) \begin{pmatrix} 1 \\ 0 \end{pmatrix}, \end{aligned} \tag{2.7.15}$$

where for Φ_r we use the formula (2.4.98) in the region I according to Figure 2.5. Since $\mathbf{E}_N^{b_j}(x)$ and $\mathbf{X}_N(x)$ are analytic in a neighborhood of b_j, and $\mathbf{X}_N(x)$ satisfies estimate (2.5.19), as $x - y \to 0$ we have

$$\mathbf{E}_N^{b_j}(y)^{-1}\mathbf{X}_N(y)^{-1}\mathbf{X}_N(x)\mathbf{E}_N^{b_j}(x) = I + \mathcal{O}(x-y), \tag{2.7.16}$$

From (2.4.98) and (2.4.99),

$$\begin{aligned} \begin{pmatrix} 0 & 1 \end{pmatrix} \Phi_r\big(N^{2/3}\psi_{b_j}(y)\big)^{-1} \Phi_r\big(N^{2/3}\psi_{b_j}(x)\big) \begin{pmatrix} 1 \\ 0 \end{pmatrix} \\ = -2\pi \mathrm{i}\big[-y_0'\big(N^{2/3}\psi_{b_j}(y)\big) y_0\big(N^{2/3}\psi_{b_j}(x)\big) + y_0\big(N^{2/3}\psi_{b_j}(y)\big) y_0'\big(N^{2/3}\psi_{b_j}(x)\big)\big]. \end{aligned} \tag{2.7.17}$$

Recall that $y_0(z) = \mathrm{Ai}(z)$, and that $\psi_{b_j}(b_j) = 0$ and $\psi'_{b_j}(b_j) > 0$. We now make the scaling

$$x = b_j + \frac{u}{cN^{2/3}}, \quad y = b_j + \frac{v}{cN^{2/3}}, \quad c = \psi'_{b_j}(\beta_j). \tag{2.7.18}$$

Then as $N \to \infty$,

$$\begin{aligned}(0 \quad 1)\, \Phi_r\big(N^{2/3}\psi_{b_j}(y)\big)^{-1} \Phi_r\big(N^{2/3}\psi_{b_j}(x)\big) \begin{pmatrix} 1 \\ 0 \end{pmatrix} \\ = -2\pi\mathrm{i}\big(\mathrm{Ai}(u)\,\mathrm{Ai}'(v) - \mathrm{Ai}'(u)\,\mathrm{Ai}(v)\big) + \mathcal{O}(N^{-2/3}).\end{aligned} \tag{2.7.19}$$

Combining (2.7.15), (2.7.16), and (2.7.19), we obtain

$$\begin{aligned} K_N\left(b_j + \frac{u}{cN^{2/3}}, b_j + \frac{v}{cN^{2/3}}\right) \\ = cN^{2/3}\left[\frac{\mathrm{Ai}(u)\,\mathrm{Ai}'(v) - \mathrm{Ai}'(u)\,\mathrm{Ai}(v)}{u - v} + \mathcal{O}(N^{-2/3})\right] + \mathcal{O}(1),\end{aligned} \tag{2.7.20}$$

which implies (2.7.13) in the case $u \geq 0$ and $v \geq 0$. If one or both of u and v are negative then one should consider the limit as x and/or y approach the real line from the region II in Figure 2.5, and the proof is similar. The universal limiting kernel $\big(\mathrm{Ai}(u)\,\mathrm{Ai}'(v) - \mathrm{Ai}'(u)\,\mathrm{Ai}(v)\big)/(u - v)$ is known as the *Airy kernel.* □

CHAPTER 3

Discrete Orthogonal Polynomials on an Infinite Lattice

3.1. The discrete log gas ensemble

Recall the log gas interpretation of the ensemble of eigenvalues of a unitary matrix ensemble. That is, that the eigenvalues of this ensemble may be thought of as particles on the real line which repel each other logarithmically and are confined to the finite space by the potential $NV(x)$. Let us consider a similar problem for particles which sit not on the real line, but on a regular discrete lattice in $\mathbb{R}$. We would like to consider the problem of the distribution of these particles as the number of particles goes to infinity, and therefore we should scale the lattice so that the limiting density of eigenvalues is compact. For a given $N \in \mathbb{N}$, introduce the regular infinite lattice,

$$L_N = \left\{ x_{k,N} = \frac{k}{N},\ k \in \mathbb{Z} \right\}. \tag{3.1.1}$$

We can consider the probability distribution on N-tuples of points $\lambda = (\lambda_1, \dots, \lambda_N) \in L_N^N$,

$$\mathbb{P}_N(\lambda) = Z_N^{-1} \prod_{j>k} (\lambda_j - \lambda_k)^2 \prod_{j=1}^{N} \frac{e^{-NV(\lambda_j)}}{N}, \tag{3.1.2}$$

where

$$Z_N = \sum_{\lambda \in L_N^N} \prod_{j>k} (\lambda_j - \lambda_k)^2 \prod_{j=1}^{N} \frac{e^{-NV(\lambda_j)}}{N}. \tag{3.1.3}$$

Proposition 1.2.2 holds for this discrete ensemble, but now the polynomials are orthogonal with respect to a discrete weight. More specifically, consider the system of monic orthogonal polynomials,

$$P_n(x) = x^n + p_{n,n-1}x^{n-1} + \cdots + p_{n0}, \quad n = 0, 1, \dots,$$

such that

$$\frac{1}{N} \sum_{x \in L_N} P_m(x) P_n(x) e^{-NV(x)} = h_n \delta_{mn}, \tag{3.1.4}$$

for some normalizing coefficients h_n, and the ψ-functions

$$\psi_n(x) = \frac{1}{\sqrt{h_n}} P_n(x) e^{-NV(x)/2}. \tag{3.1.5}$$

The partition function Z_N can again be expressed as

$$Z_N = N! \prod_{j=0}^{N-1} h_j, \tag{3.1.6}$$

and the m-point correlation function $R_{mN}(\lambda_1, \dots \lambda_N)$ (see (1.2.21)) is again given by the formula

$$R_{mN}(\lambda_1, \dots \lambda_m) = \det\bigl(K_N(\lambda_k, \lambda_l)\bigr)_{k,l=1}^{m}, \tag{3.1.7}$$

where $K_N(x, y)$ is the reproducing kernel

$$K_N(x, y) = \sum_{n=0}^{N-1} \psi_n(x)\psi_n(y). \tag{3.1.8}$$

The polynomials $P_n(x)$ once again satisfy the three-term recurrence relation

$$xP_n(x) = P_{n+1}(x) + \beta_n P_n(x) + \gamma_n^2 P_{n-1}(x), \tag{3.1.9}$$

and therefore the Christoffel – Darboux formula

$$K_N(x, y) = \sum_{n=0}^{N-1} \psi_n(x)\psi_n(y) = \gamma_N \frac{\psi_N(x)\psi_{N-1}(y) - \psi_{N-1}(x)\psi_N(y)}{x - y}. \tag{3.1.10}$$

As in the previous chapter, we can thus write K_N in terms of the solution to a Riemann – Hilbert problem. Since the orthogonal polynomials we are considering come from a discrete measure, we must first consider an *interpolation problem.*

3.2. Interpolation problem

We will evaluate the asymptotics of the discrete orthogonal polynomials described above via a steepest descent asymptotic analysis of a Riemann–Hilbert problem. To that end, we first introduce the following Interpolation Problem for discrete orthogonal polynomials (see [5, 13, 14, 19]). In what follows, we employ the notation

$$w_N(x) \equiv \mathrm{e}^{-NV(x)}. \tag{3.2.1}$$

Interpolation Problem. For a given $n = 0, 1, \dots$, find a 2×2 matrix-valued function $\mathbf{P}_n(z) = (\mathbf{P}_n(z)_{ij})_{1 \le i,j \le 2}$ with the following properties:

Analyticity: $\mathbf{P}_n(z)$ is an analytic function of z for $z \in \mathbb{C} \setminus L_N$.

Residues at poles: At each node $x \in L_N$, the elements $\mathbf{P}_n(z)_{11}$ and $\mathbf{P}_n(z)_{21}$ of the matrix $\mathbf{P}_n(z)$ are analytic functions, and the elements $\mathbf{P}_n(z)_{12}$ and $\mathbf{P}_n(z)_{22}$ have a simple pole with the residues,

$$\operatorname*{Res}_{z=x} \mathbf{P}_n(z)_{j2} = \frac{w_N(x)}{N} \mathbf{P}_n(x)_{j1}, \quad j = 1, 2. \tag{3.2.2}$$

Asymptotics at infinity: There exists a function $r(x) > 0$ on L_N such that

$$\lim_{x \to \infty} r(x) = 0, \tag{3.2.3}$$

and such that as $z \to \infty$, $\mathbf{P}_N(z)$ admits the asymptotic expansion,

$$\mathbf{P}_n(z) \sim \left(I + \frac{\mathbf{P}_1}{z} + \frac{\mathbf{P}_2}{z^2} + \cdots\right) \begin{pmatrix} z^n & 0 \\ 0 & z^{-n} \end{pmatrix}, \quad z \in \mathbb{C} \setminus \left[\bigcup_{x \in L_N}^{\infty} D(x, r(x))\right], \tag{3.2.4}$$

where $D\big(x, r(x)\big)$ denotes a disk of radius $r(x) > 0$ centered at x and I is the identity matrix.

It is not difficult to see (see, e.g., [5]) that the IP has a unique solution, which is

$$\mathbf{P}_n(z) = \begin{pmatrix} P_n(z) & C(w_N P_n)(z) \\ (h_{n-1})^{-1} P_{n-1}(z) & (h_{n-1})^{-1} C(w_N P_{n-1})(z) \end{pmatrix}, \tag{3.2.5}$$

where the operator C is the discrete Cauchy transformation defined by the formula,

$$C(f)(z) = \frac{1}{N} \sum_{x \in L_N} \frac{f(x)}{z - x}. \tag{3.2.6}$$

As with the continuous Cauchy transformation, because of the orthogonality condition, as $z \to \infty$,

$$\begin{aligned} C(w_N P_n)(z) &= \frac{1}{N} \sum_{x \in L_N} \frac{w_N(x) P_n(x)}{z - x} \\ &\sim \frac{1}{N} \sum_{x \in L_N} w_N(x) P_n(x) \sum_{j=0}^{\infty} \frac{x^j}{z^{j+1}} = \frac{h_n}{z^{n+1}} + \sum_{j=n+2}^{\infty} \frac{a_j}{z^j}, \end{aligned} \tag{3.2.7}$$

which justifies asymptotic expansion (3.2.4), and we have that

$$h_n = [\mathbf{P}_1]_{12}, \quad h_{n-1}^{-1} = [\mathbf{P}_1]_{21}. \tag{3.2.8}$$

As with the RHP for continuous orthogonal polynomials, the recurrence coefficients in equation (3.1.9) are given by

$$\gamma_n^2 = [\mathbf{P}_1]_{12} [\mathbf{P}_1]_{21}, \quad \beta_{n-1} = \frac{[\mathbf{P}_2]_{21}}{[\mathbf{P}_1]_{21}} - [\mathbf{P}_1]_{11}, \tag{3.2.9}$$

and the reproducing kernel is given by the formula

$$K_n(x, y) = \frac{\mathrm{e}^{-NV(x)/2} e^{-NV(y)/2}}{x - y} \begin{pmatrix} 0 & 1 \end{pmatrix} \mathbf{P}_n^{-1}(x) \mathbf{P}_n(y) \begin{pmatrix} 1 \\ 0 \end{pmatrix}. \tag{3.2.10}$$

We will carry out the asymptotic analysis of this IP for $n = N$, and this will consist of two steps. The first step will be a reduction of the Interpolation Problem to a Riemann – Hilbert problem on a contour on the complex plane, which we accomplish following the general approach introduced in the paper [63] of Miller and in the monograph [48] of Kamvissis, McLaughlin, and Miller. The second step will be an application of the nonlinear steepest descent method of Deift and Zhou [32] to the Riemann – Hilbert problem under consideration. To apply the nonlinear steepest descent method we first need to study the corresponding equilibrium measure.

3.3. Equilibrium measure

Similar to (2.3.1), we can write

$$\mathbb{P}_N(\lambda) = \frac{1}{N^N Z_N} e^{-H_N(\lambda)}, \tag{3.3.1}$$

where

$$H_N(\lambda) = -\sum_{j\neq k} \log|\lambda_j - \lambda_k| + N\sum_{j=1}^{N} V(\lambda_j) \tag{3.3.2}$$
$$= N^2\Big[-\iint_{x\neq y} \log|x-y|\,\mathrm{d}\nu_\lambda(x)\,\mathrm{d}\nu_\lambda(y) + \int V(x)\,\mathrm{d}\nu_\lambda(x)\Big],$$

and

$$\mathrm{d}\nu_\lambda(x) = N^{-1}\sum_{j=1}^{N} \delta(x-\lambda_j)\,\mathrm{d}x. \tag{3.3.3}$$

We again expect that the limiting distribution of particles minimizes the functional

$$I_V(\nu) = -\iint_{x\neq y} \log|x-y|\,\mathrm{d}\nu_\lambda(x)\,\mathrm{d}\nu_\lambda(y) + \int V(x)\,\mathrm{d}\nu_\lambda(x), \tag{3.3.4}$$

but there is a new constraint due to the discreteness of the problem. In particular, notice that the probability of two particles occupying the same site is zero, and thus by (3.3.3) we have that

$$\nu_\lambda(a,b) \leq b - a + \frac{1}{N} \quad \text{for any } -\infty < a < b < \infty. \tag{3.3.5}$$

We should therefore minimize with respect to probability measures ν satifying

$$\nu(a,b) \leq b - a \quad \text{for any } -\infty < a < b < \infty, \tag{3.3.6}$$

which is (3.3.5) in the large N limit. In other words, we should restrict to measures which are less than or equal to Lebesgue measure. With this constraint in mind, we define

$$E_0 = \inf_{\nu\in\mathcal{M}} H(\nu), \tag{3.3.7}$$

where

$$\mathcal{M} = \{\text{Positive probability measures } \nu \text{ on } \mathbb{R} \text{ such that } \nu < \sigma\}, \tag{3.3.8}$$

and σ is Lebesgue measure on $\mathbb{R}$. It is possible to prove that there exists a unique minimizer ν_0, so that

$$E_0 = H(\nu_0), \tag{3.3.9}$$

see, e.g., the works of Saff and Totik [68], Dragnev and Saff [35] and Kuijlaars [53]. We again call this minimizer the *equilibrium measure.*

The Euler – Lagrange variational conditions for the variational problem (3.3.7) are slightly different than those described in (2.3.15) and (2.3.16). Due to the upper constraint, the equilibrium measure ν_0 is uniquely determined by the following Euler – Lagrange variational conditions:

$$2\int \log|x-y|\,\mathrm{d}\nu_0(y) - V(x) \begin{cases} \geq l & \text{for } x \in \operatorname{supp}\nu_0 \\ \leq l & \text{for} x \in \operatorname{supp}(\sigma-\nu_0), \end{cases} \tag{3.3.10}$$

for a constant *Lagrange multiplier* l, see the works [31] of Deift and McLaughlin, and [35]. In particular,

$$2\int \log|x-y|\,\mathrm{d}\nu_0(y) - V(x) = l \quad \text{for } x \in \operatorname{supp}\nu_0 \cap \operatorname{supp}(\sigma-\nu_0). \tag{3.3.11}$$

First, observe that the constraint $\nu_0 \leq \sigma$ implies the existence of the density,

$$\rho(x) = \frac{\mathrm{d}\nu_0}{\mathrm{d}x}. \tag{3.3.12}$$

We can partition $\mathbb{R}$ into the three sets

$$\begin{aligned} I^0 &= \left\{x \in \mathbb{R} : 2\int \log|x-y|\,\mathrm{d}\nu_0(y) - V(x) = l\right\}, \\ I^+ &= \left\{x \in \mathbb{R} : 2\int \log|x-y|\,\mathrm{d}\nu_0(y) - V(x) > l\right\}, \\ I^- &= \left\{x \in \mathbb{R} : 2\int \log|x-y|\,\mathrm{d}\nu_0(y) - V(x) < l\right\}. \end{aligned} \tag{3.3.13}$$

The structure of the equilibrium measure is well described in the following theorem of Kuijlaars, obtained in [53].

Theorem 3.3.1 (Kuijlaars). *For any real analytic potential $V(x)$ satisfying* (1.1.7), *the following hold:*

(1) *The density $\rho(x)$ of the constrained equilibrium measure ν_0 (defined in* (3.3.9)) *is continuous.*

(2) *The sets I^+ and I^- are both finite unions of open intervals.*

(3) *The density ρ is real analytic on the open set $\{x : 0 < \rho(x) < 1\}$.*

(4) *The density ρ has the representation*

$$\rho(x) = \frac{1}{\pi}\sqrt{q_1^+(x)} \quad \text{for } x \in I^0 \cup I^-, \tag{3.3.14}$$

where q_1^+ is the positive part of a function q_1 defined on $I^0 \cup I^-$ which is real analytic on the interior of $I^0 \cup I^-$. The function q_1 is negative on I^-, so that

$$\rho(x) = 0 \quad \text{for } x \in I^-, \tag{3.3.15}$$

and it is nonnegative on I^0, so that

$$\rho(x) = \frac{1}{\pi}\sqrt{q_1(x)} \quad \text{for } x \in I^0. \tag{3.3.16}$$

(5) *The density ρ has the representation*

$$\rho(x) = 1 - \frac{1}{\pi}\sqrt{q_2^+(x)} \quad \text{for } x \in I^0 \cup I^+, \tag{3.3.17}$$

where q_2^+ is the positive part of a function q_2 defined on $I^0 \cup I^+$ which is real analytic on the interior of $I^0 \cup I^+$. The function q_2 is negative on I^+, so that

$$\rho(x) = 1 \quad \text{for } x \in I^+, \tag{3.3.18}$$

and it is nonnegative on I_0, so that

$$\rho(x) = 1 - \frac{1}{\pi}\sqrt{q_2(x)} \quad \text{for } x \in I^0. \tag{3.3.19}$$

Remark. It follows from equations (3.3.16) and (3.3.19) that

$$\frac{1}{\pi}\sqrt{q_1(x)} = 1 - \frac{1}{\pi}\sqrt{q_2(x)} \quad \text{for } x \in I^0, \tag{3.3.20}$$

hence q_1 and q_2 uniquely determine each other.

Notice that, according to point (2) of this theorem, the connected components of I^0 are either closed intervals or isolated points. Since ν_0 has compact support, we can write

$$I^0 = \bigsqcup_{j=1}^{q} [a_j, b_j], \tag{3.3.21}$$

where

$$\begin{aligned} a_j &\le b_j && \text{for } j = 1, \dots, q, \\ b_j &< a_{j+1} && \text{for } j = 1, \dots, q-1. \end{aligned} \tag{3.3.22}$$

Notice that the intervals $(-\infty, a_1)$ and (b_q, ∞) are components of I^-. The interval (b_j, a_{j+1}) for $1 \le j < q$ is a component of either I^+ or I^-. We therefore adopt the notation

$$\begin{aligned} \mathcal{A}_{\mathrm{v}} &= \{j \in \{1, \dots, q-1\} : (b_j, a_{j+1}) \subset I^-\}, \\ \mathcal{A}_{\mathrm{s}} &= \{j \in \{1, \dots, q-1\} : (b_j, a_{j+1}) \subset I^+\}. \end{aligned} \tag{3.3.23}$$

We will call an equilibrium measure ν_0 *regular* if the following hold:

(1) The functions q_1 and q_2 are nonvanishing on the interior of I^0.
(2) I^0 contains no isolated points, so that $a_j < b_j$ for all $j = 1, \dots, q$.
(3) If $j \in \mathcal{A}_{\mathrm{v}}$, then $q_1'(b_j) \neq 0$ and $q_1'(a_{j+1}) \neq 0$.
(4) If $j \in \mathcal{A}_{\mathrm{s}}$, then $q_2'(b_j) \neq 0$ and $q_2'(a_{j+1}) \neq 0$.

For the sake of simplicity we will assume throughout this chapter that our equilibrium measure is regular. In this case, the sets I^0, I^+ and I^- are each finite unions of intervals, so that

$$-\infty < a_1 < b_1 < a_2 < b_2 < \cdots < a_q < b_q < \infty, \tag{3.3.24}$$

and we classify these intervals as follows:

Definition. A *void* is an open subinterval (b_j, a_{j+1}), $j \in \mathcal{A}_{\mathrm{v}}$, or one of the intervals $(-\infty, a_1)$, (b_q, ∞). The union of all voids is I^-.

Definition. A *saturated region* is an open subinterval (b_j, a_{j+1}), $j \in \mathcal{A}_{\mathrm{s}}$. The union of all saturated regions is I^+.

Definition. A *band* is an open subinterval (a_j, b_j), $j = 1, \dots, q$. The union of all bands is the interior of I^0.

Observe that $\rho(x) = 0$ on any void (b_j, a_{j+1}), $\rho(x) = 1$ on any saturated interval (b_j, a_{j+1}), and $0 < \rho(x) < 1$ on any band (a_j, b_j), see Figure 3.1. In addition, at

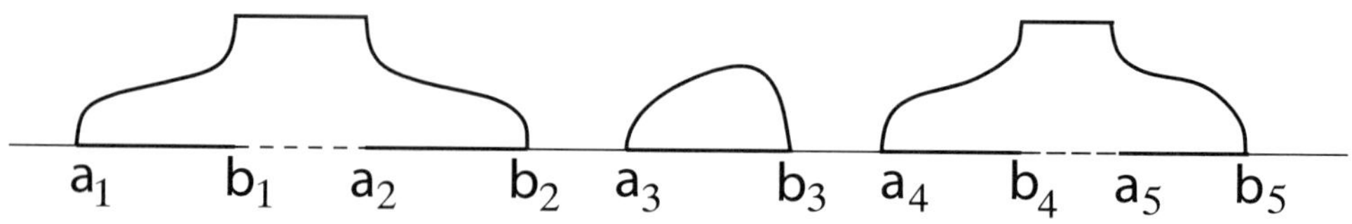

FIGURE 3.1. The graph of the density function for a hypothetical equilibrium measure with $q = 5$. Bands are denoted by bold segments, saturated regions by dashed segments, and voids by thin segments.

the endpoints of any band, $\rho(x)$ has a square-root singularity. Namely, if a_j is a common endpoint of a band and a void then as $x \to +0$,

$$\rho(a_j + x) = C\sqrt{x}\,(1 + \mathcal{O}(x)), \quad C = |q_1'(a_j)|^{1/2} > 0. \tag{3.3.25}$$

and if a_j is a common endpoint of a band and a saturated region then as $x \to +0$,

$$\rho(a_j + x) = 1 - C\sqrt{x}\,(1 + \mathcal{O}(x)), \quad C = |q_2'(a_j)|^{1/2} > 0. \tag{3.3.26}$$

Similarly, if b_j is a common endpoint of a band and a void then as $x \to +0$,

$$\rho(b_j - x) = C\sqrt{x}\,(1 + \mathcal{O}(x)), \quad C = |q_1'(b_j)|^{1/2} > 0. \tag{3.3.27}$$

and if b_j is a common endpoint of a band and a saturated region then as $x \to +0$,

$$\rho(b_j - x) = 1 - C\sqrt{x}\,(1 + \mathcal{O}(x)), \quad C = |q_2'(b_j)|^{1/2} > 0. \tag{3.3.28}$$

3.4. The g-function

As in the previous chapter on continuous orthogonal polynomials, we define the g-function on $\mathbb{C} \setminus (-\infty, b_q]$ as

$$g(z) = \int_{a_1}^{b_q} \log(z - x)\, \mathrm{d}\nu_0(x), \tag{3.4.1}$$

where we take the principal branch for the logarithm. This function satisfies many of the same properties as the g-function of Section 2.4.2, with some small changes to account for the new phenomenon of saturated regions. We summarize these properties below. Introduce the numbers Ω_j for $j = 1, \dots, q-1$ as

$$\Omega_j = \begin{cases} 2\pi \displaystyle\int_{a_{j+1}}^{b_q} \rho(x)\, \mathrm{d}x & \text{for } j \in \mathcal{A}_{\mathrm{v}} \\ 2\pi \displaystyle\int_{a_{j+1}}^{b_q} \rho(x)\, \mathrm{d}x + 2\pi a_{j+1} & \text{for } j \in \mathcal{A}_{\mathrm{s}}. \end{cases} \tag{3.4.2}$$

Properties of $g(z)$.

(1) $g(z)$ is analytic in $\mathbb{C} \setminus (-\infty, b_q]$.

(2) For large z,

$$g(z) = \log z - \sum_{j=1}^{\infty} \frac{g_j}{z^j}, \quad g_j = \int_{a_1}^{b_q} \frac{x^j}{j}\, \mathrm{d}\nu_0(x). \tag{3.4.3}$$

(3) The resolvent of the equilibrium measure is given by

$$g'(z) = \int_{\mathbb{R}} \frac{\rho(x)\, \mathrm{d}x}{z - x}. \tag{3.4.4}$$

(4) From (3.3.13), we have that

$$g_+(x) + g_-(x) \begin{cases} = V(x) + l & \text{for } x \in I^0 \\ > V(x) + l & \text{for } x \in I^+ \\ < V(x) + l & \text{for } x \in I^-, \end{cases} \tag{3.4.5}$$

where g_+ and g_- refer to the limiting values from the upper and lower half-planes, respectively.

(5) Equation (3.4.1) implies that the function

$$G(x) \equiv g_+(x) - g_-(x), \tag{3.4.6}$$

is pure imaginary for all real x, and

$$G(x) = 2\pi \mathrm{i} \int_x^{b_q} \rho(s)\,\mathrm{d}s. \tag{3.4.7}$$

Thus,

$$G(x) = \begin{cases} \mathrm{i}\Omega_j & \text{for } b_j < x < a_{j+1}, \text{ and } j \in \mathcal{A}_\mathrm{v} \\ \mathrm{i}\Omega_j - 2\pi \mathrm{i} x & \text{for } b_j < x < a_{j+1}, \text{ and } j \in \mathcal{A}_\mathrm{s}. \end{cases} \tag{3.4.8}$$

From (3.4.5) and (3.4.7) we obtain that

$$2g_\pm(x) = V(x) + l \pm 2\pi \mathrm{i} \int_x^{b_q} \rho(s)\,\mathrm{d}s \quad \text{for } x \in I^0. \tag{3.4.9}$$

(6) Also, from (3.4.7), we get that $G(x)$ is real analytic on the sets I^+, I^-, and on the interior of I^0. We can therefore extend G into a complex neighborhood of any interval of analyticity for ρ, and the Cauchy–Riemann equations imply that

$$\left.\frac{\mathrm{d}G(x+\mathrm{i}y)}{\mathrm{d}y}\right|_{y=0} = 2\pi\rho(x) \geq 0. \tag{3.4.10}$$

From (3.4.5) we have that

$$G(x) = 2g_+(x) - V(x) - l = -[2g_-(x) - V(x) - l], \quad x \in I^0. \tag{3.4.11}$$

3.5. Reduction of IP to RHP

We now reduce the Interpolation Problem of Section 3.2 to a Riemann–Hilbert Problem. Introduce the function

$$\Pi(z) = \frac{\sin(N\pi z)}{N\pi}. \tag{3.5.1}$$

Notice that

$$\Pi(x_k) = 0, \quad \Pi'(x_k) = \exp(\mathrm{i}N\pi x_k) = (-1)^k, \qquad \text{for } x_k = \frac{k}{N} \in L_N. \tag{3.5.2}$$

Introduce the upper triangular matrices,

$$\mathbf{D}_\pm^u(z) = \begin{pmatrix} 1 & -\dfrac{w_N(z)}{\Pi(z)}\dfrac{\mathrm{e}^{\pm\mathrm{i}N\pi z}}{N} \\ 0 & 1 \end{pmatrix}, \tag{3.5.3}$$

and the lower triangular matrices,

$$\begin{aligned} \mathbf{D}_\pm^l(z) &= \begin{pmatrix} \Pi(z)^{-1} & 0 \\ -\dfrac{N}{w_N(z)}\mathrm{e}^{\pm\mathrm{i}N\pi z} & \Pi(z) \end{pmatrix} \\ &= \begin{pmatrix} \Pi(z)^{-1} & 0 \\ 0 & \Pi(z) \end{pmatrix} \begin{pmatrix} 1 & 0 \\ -\dfrac{N}{\Pi(z)w_N(z)}\mathrm{e}^{\pm\mathrm{i}N\pi z} & 1 \end{pmatrix}. \end{aligned} \tag{3.5.4}$$

Define the matrix-valued functions,

$$\mathbf{R}_N^u(z) = \mathbf{P}_N(z) \times \begin{cases} \mathbf{D}_+^u(z) & \text{when } \operatorname{Im} z \geq 0 \\ \mathbf{D}_-^u(z) & \text{when } \operatorname{Im} z \leq 0, \end{cases} \tag{3.5.5}$$

and

$$\mathbf{R}_N^l(z) = \mathbf{P}_N(z) \times \begin{cases} \mathbf{D}_+^l(z), & \text{when } \operatorname{Im} z \geq 0 \\ \mathbf{D}_-^l(z), & \text{when } \operatorname{Im} z \leq 0. \end{cases} \tag{3.5.6}$$

From (3.2.5) we have that

(3.5.7)

$$\mathbf{R}_N^u(z) = \begin{pmatrix} P_N(z) & -\dfrac{w_N(z)P_N(z)}{\Pi(z)}\dfrac{\mathrm{e}^{\pm \mathrm{i}N\pi z}}{N} + C(w_N P_N)(z) \\ (h_{N-1})^{-1}P_{N-1}(z) & -\dfrac{w_N(z)P_{N-1}(z)}{h_{N-1}\Pi(z)}\dfrac{\mathrm{e}^{\pm \mathrm{i}N\pi z}}{N} + \dfrac{C(w_N P_{N-1})(z)}{h_{N-1}} \end{pmatrix}$$
$$\text{when } \pm \operatorname{Im} z \geq 0,$$

and

(3.5.8)

$$\mathbf{R}_N^l(z) = \begin{pmatrix} \dfrac{P_N(z)}{\Pi(z)} - \dfrac{C(w_N P_N)(z)}{w_N(z)} N\mathrm{e}^{\pm \mathrm{i}N\pi z} & \Pi(z)C(w_N P_N)(z) \\ \dfrac{P_{N-1}(z)}{h_{N-1}\Pi(z)} - \dfrac{C(w_N P_{N-1})(z)}{h_{N-1}w_N(z)} N\mathrm{e}^{\pm \mathrm{i}N\pi z} & \Pi(z)h_{N-1}^{-1}C(w_N P_{N-1})(z) \end{pmatrix}$$
$$\text{when } \pm \operatorname{Im} z \geq 0.$$

Observe that the functions $\mathbf{R}_N^u(z)$, $\mathbf{R}_N^l(z)$ are meromorphic on the closed upper and lower complex planes and they are two-valued on the real axis. Their possible poles are located on the lattice L_N. In fact, due to some cancellations they do not have any poles at all. We have the following proposition.

Proposition 3.5.1. *The matrix-valued functions $\mathbf{R}_N^u(z)$ and $\mathbf{R}_N^l(z)$ have no poles and on the real line they satisfy the following jump conditions at $x \in \mathbb{R}$:*

$$\mathbf{R}_{N+}^u(x) = \mathbf{R}_{N-}^u(x) j_R^u(x), \quad j_R^u(x) = \begin{pmatrix} 1 & -2\pi \mathrm{i}\mathrm{e}^{-NV(x)} \\ 0 & 1 \end{pmatrix}, \tag{3.5.9}$$

and

$$\mathbf{R}_{N+}^l(x) = \mathbf{R}_{N-}^l(x) j_R^l(x), \quad j_R^l(x) = \begin{pmatrix} 1 & 0 \\ -2\pi \mathrm{i}\mathrm{e}^{NV(x)} & 1 \end{pmatrix}. \tag{3.5.10}$$

PROOF. It follows from the definition of $\mathbf{R}_N^u(z)$ that all possible poles of $\mathbf{R}_N^u(z)$ are located on the lattice L_N. Let us show that the residue of all these poles is equal to zero. Consider any $x_k \in L_N$. The residue of the matrix element $[\mathbf{R}_N^u(z)]_{12}$ at x_k is equal to

$$\operatorname*{Res}_{z=x_k} [\mathbf{R}_N^u(z)]_{12} = -\frac{w_N(x_k)P_N(x_k)}{N(-1)^k}(-1)^k + \frac{w_N(x_k)P_N(x_k)}{N} = 0. \tag{3.5.11}$$

Similarly we get that

$$\operatorname*{Res}_{z=x_k} [\mathbf{R}_N(z)]_{22} = 0, \tag{3.5.12}$$

hence $\mathbf{R}_N^u(z)$ has no pole at x_k.

Similarly, the residue of the matrix element $[\mathbf{R}_N^l(z)]_{11}$ at x_k is equal to

$$\operatorname*{Res}_{z=x_k} [\mathbf{R}_N^l(z)]_{11} = \frac{P_N(x_k)}{(-1)^k} - \frac{w_N(x_k)P_N(x_k)(-1)^k}{w_N(x_k)} = 0. \tag{3.5.13}$$

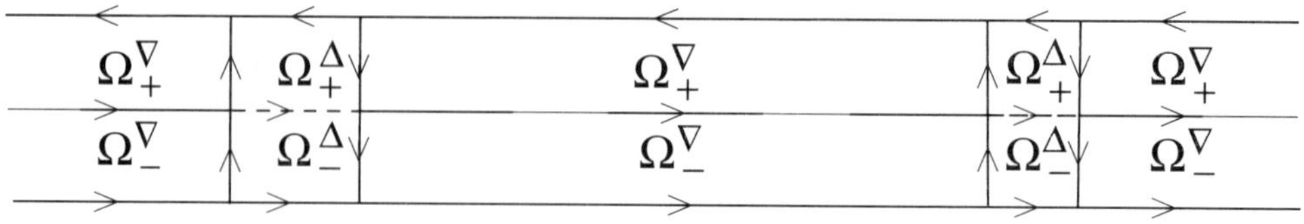

FIGURE 3.2. The contour Σ arising from the hypothetical equilibrium measure in Figure 3.1, dividing an ε-neighborhood of the real line into the regions $\Omega_\pm^\Delta$ and $\Omega_\pm^\nabla$.

In the same way we obtain that

$$\operatorname*{Res}_{z=x_k} [\mathbf{R}_N(z)]_{21} = 0. \tag{3.5.14}$$

In the entry $[\mathbf{R}_N^l(z)]_{21}$, the pole of the function $C(w_N P_N)(z)$ at $z = x_k$ is cancelled by the zero of the function $\Pi(z)$, hence $[\mathbf{R}_N^l(z)]_{21}$ has no pole at x_k. Similarly, $[\mathbf{R}_N^l(z)]_{22}$ has no pole at x_k as well, hence $\mathbf{R}_N^l(z)$ has no pole at x_k.

Let us evaluate the jump matrices for $x \in \mathbb{R}$. From (3.5.5) we have that

$$\begin{aligned} j_R^u(x) &= \mathbf{D}_-^u(x)^{-1}\mathbf{D}_+^u(x) \\ &= \begin{pmatrix} 1 & -\dfrac{w_N(x)}{N\Pi(x)} 2\mathrm{i}\sin(N\pi x) \\ 0 & 1 \end{pmatrix} = \begin{pmatrix} 1 & -2\pi \mathrm{i} e^{-NV(x)} \\ 0 & 1 \end{pmatrix}, \end{aligned} \tag{3.5.15}$$

which proves (3.5.9). Similarly,

$$\begin{aligned} j_R^l(x) &= \mathbf{D}_-^l(x)^{-1}\mathbf{D}_+^l(x) \\ &= \begin{pmatrix} 1 & 0 \\ -\dfrac{N}{\Pi(x) w_N(x)} 2\mathrm{i}\sin(N\pi x) & 1 \end{pmatrix} = \begin{pmatrix} 1 & 0 \\ -2\pi \mathrm{i} e^{NV(x)} & 1 \end{pmatrix}, \end{aligned} \tag{3.5.16}$$

which proves (3.5.10). □

Consider the oriented contour Σ on the complex plane depicted in Figure 3.2, in which the horizontal lines are $\operatorname{Im} z = \varepsilon, 0, -\varepsilon$, where $\varepsilon > 0$ is a small positive constant which will be determined later, and the vertical segments pass through the endpoints of saturated intervals. Consider the regions

$$\begin{aligned} \Omega_\pm^\nabla &= \{I^0 \cup I^-\} \times (0, \pm \mathrm{i}\varepsilon) \\ \Omega_\pm^\Delta &= I^+ \times (0, \pm \mathrm{i}\varepsilon), \end{aligned} \tag{3.5.17}$$

bounded by the contour Σ.

Define

$$\mathbf{R}_N(z) = \begin{cases} \mathbf{K}\mathbf{R}_N^u(z)\mathbf{K}^{-1} & \text{for } z \in \Omega_\pm^\nabla \\ \mathbf{K}\mathbf{R}_N^l(z)\mathbf{K}^{-1} & \text{fo } z \in \Omega_\pm^\Delta \\ \mathbf{K}\mathbf{P}_N(z)\mathbf{K}^{-1} & \text{otherwise,} \end{cases} \tag{3.5.18}$$

where $\mathbf{K} = \begin{pmatrix} 1 & 0 \\ 0 & -2\mathrm{i}\pi \end{pmatrix}$.

Proposition 3.5.2. *The matrix-valued function $\mathbf{R}_N(z)$ has the following jumps on the contour Σ:*

$$\mathbf{R}_{N+}(z) = \mathbf{R}_{N-}(z) j_R(z), \tag{3.5.19}$$

where

(3.5.20)

$$
j_R(z)=\begin{cases}
\begin{pmatrix}1 & \mathrm{e}^{-NV(z)}\\ 0 & 1\end{pmatrix} & \textit{for } z\in I^-\cup I^0\\
\begin{pmatrix}1 & 0\\ -4N^2\pi^2\mathrm{e}^{NV(z)} & 1\end{pmatrix} & \textit{for } z\in I^+\\
\mathbf{K}\mathbf{D}^u_\pm(z)\mathbf{K}^{-1}=\begin{pmatrix}1 & \dfrac{1}{2\mathrm{i}N\pi}\dfrac{\mathrm{e}^{-NV(z)}\mathrm{e}^{\pm\mathrm{i}N\pi z}}{\Pi(z)}\\ 0 & 1\end{pmatrix} & \textit{for } z\in\{I^-\cup I^0\}\pm\mathrm{i}\varepsilon\\
\mathbf{K}\mathbf{D}^l_\pm(z)\mathbf{K}^{-1}=\begin{pmatrix}\Pi(z)^{-1} & 0\\ 2\mathrm{i}N\pi\mathrm{e}^{NV(z)}\mathrm{e}^{\pm\mathrm{i}N\pi z} & \Pi(z)\end{pmatrix} & \textit{for } z\in I^+\pm\mathrm{i}\varepsilon\\
\mathbf{K}\mathbf{D}^l_\pm(z)^{-1}\mathbf{D}^u_\pm(z)\mathbf{K}^{-1}=\begin{pmatrix}\Pi(z) & \dfrac{1}{2\pi\mathrm{i}N}\mathrm{e}^{-NV(z)}\mathrm{e}^{\pm\mathrm{i}N\pi z}\\ -2N\pi\mathrm{i}\mathrm{e}^{NV(z)}\mathrm{e}^{\pm\mathrm{i}N\pi z} & \mp2N\pi\mathrm{i}\mathrm{e}^{\pm\mathrm{i}N\pi z}\end{pmatrix}\\
\qquad\textit{for } z\in(0,\pm i\varepsilon)+b_j \textit{ or } z\in(0,\pm\mathrm{i}\varepsilon)+a_{j+1} \textit{ for } j\in\mathcal{A}_{\mathrm{s}}.
\end{cases}
$$

3.6. First transformation of the RHP

Define the matrix function $\mathbf{T}_N(z)$ as follows from the equation

(3.6.1)
$$\mathbf{R}_N(z)=\mathrm{e}^{(Nl/2)\sigma_3}\mathbf{T}_N(z)\mathrm{e}^{N(g(z)-l/2)\sigma_3},$$

where l is the Lagrange multiplier, the function $g(z)$ is described in Section 3.4, and $\sigma_3=\left(\begin{smallmatrix}1&0\\0&-1\end{smallmatrix}\right)$ is the third Pauli matrix. Then $\mathbf{T}_N(z)$ satisfies the following Riemann–Hilbert Problem:

(1) $\mathbf{T}_N(z)$ is analytic in $\mathbb{C}\setminus\Sigma$.

(2) $\mathbf{T}_{N+}(z)=\mathbf{T}_{N-}(z)j_T(z)$ for $z\in\Sigma$, where

(3.6.2)
$$j_T(z)=\begin{cases}\mathrm{e}^{N(g_-(z)-l/2)\sigma_3}j_R(z)\mathrm{e}^{-N(g_+(z)-l/2)\sigma_3} & \text{for } z\in\mathbb{R}\\ \mathrm{e}^{N(g(z)-l/2)\sigma_3}j_R(z)\mathrm{e}^{-N(g(z)-l/2)\sigma_3} & \text{for } z\in\Sigma\setminus\mathbb{R}.\end{cases}$$

(3) As $z\to\infty$,

(3.6.3)
$$\mathbf{T}_N(z)\sim I+\frac{\mathbf{T}_1}{z}+\frac{\mathbf{T}_2}{z^2}+\cdots.$$

Let's take a closer look at the behavior of the jump matrix j_T described in (3.6.2) on the horizontal segments of Σ. We have that

(3.6.4)

$$
j_T(z)=\begin{cases}
\begin{pmatrix}\mathrm{e}^{-NG(z)} & \mathrm{e}^{N(g_+(z)+g_-(z)-l-V(z))}\\ 0 & \mathrm{e}^{NG(z)}\end{pmatrix} & \text{when } z\in I^0\cup I^-\\
\begin{pmatrix}\mathrm{e}^{-NG(z)} & 0\\ -4N^2\pi^2\mathrm{e}^{-N(g_+(z)+g_-(z)-V(z)-l)} & \mathrm{e}^{NG(z)}\end{pmatrix} & \text{when } z\in I^+\\
\begin{pmatrix}1 & \pm\dfrac{\mathrm{e}^{N(2g(z)-l-V(z))}}{1-\mathrm{e}^{\mp2\mathrm{i}N\pi z}}\\ 0 & 1\end{pmatrix} & \text{when } z\in\{I_0\cup I^-\}\pm\mathrm{i}\varepsilon\\
\begin{pmatrix}\Pi(z)^{-1} & 0\\ 2\mathrm{i}N\pi\mathrm{e}^{\pm\mathrm{i}N\pi z}e^{-N(2g(z)-l-V(z))} & \Pi(z)\end{pmatrix} & \text{when } z\in I^+\pm\mathrm{i}\varepsilon.
\end{cases}
$$

3.7. Second transformation of the RHP

The second transformation is based on two observations. The first is the "opening of the lenses" (see Section 2.4.4) in a neighborhood of the unconstrained support of the equilibrium measure. Namely, notice that for $x \in I^0$, the jump matrix $j_T(x)$ factorizes as

$$(3.7.1)\quad j_T(x) = \begin{pmatrix} e^{-NG(x)} & 1 \\ 0 & e^{NG(x)} \end{pmatrix} = \begin{pmatrix} 1 & 0 \\ e^{NG(x)} & 1 \end{pmatrix} \begin{pmatrix} 0 & 1 \\ -1 & 0 \end{pmatrix} \begin{pmatrix} 1 & 0 \\ e^{-NG(x)} & 1 \end{pmatrix} = j_-(x) j_M j_+(x),$$

which allows us to reduce the jump matrix j_T to the one j_M plus asymptotically small jumps on the lens boundaries. The second observation consists of two facts. Firstly, the jumps on the segments $I^+ \pm i\varepsilon$ behave for large N as $\pm e^{\pm iN\pi z}$. Secondly, note that, for $x \in I^+$, $G(x)$ is a linear function with slope $-2\pi i$. With these facts in mind, we make the second transformation of the RHP. Let

$$(3.7.2)\qquad \mathbf{S}_N(z) = \begin{cases} \mathbf{T}_N(z) j_+(z)^{-1} & \text{for } z \in I^0 \times (0, i\varepsilon) \\ \mathbf{T}_N(z) j_-(z) & \text{for } z \in I^0 \times (0, -i\varepsilon) \\ \mathbf{T}_N(z) \mathbf{A}_+(z) & \text{for } z \in I^+ \times (0, i\varepsilon) \\ \mathbf{T}_N(z) \mathbf{A}_-(z) & \text{for } z \in I^+ \times (0, -i\varepsilon) \\ \mathbf{T}_N(z) & \text{otherwise,} \end{cases}$$

where
(3.7.3)

$$\mathbf{A}_+(z) = \begin{pmatrix} -\frac{1}{2N\pi i} e^{-iN\pi z} & 0 \\ 0 & -2N\pi i e^{iN\pi z} \end{pmatrix}, \quad \mathbf{A}_-(z) = \begin{pmatrix} \frac{1}{2N\pi i} e^{iN\pi z} & 0 \\ 0 & 2N\pi i e^{-iN\pi z} \end{pmatrix},$$

$$j_+(z) = \begin{pmatrix} 1 & 0 \\ e^{-NG(z)} & 1 \end{pmatrix}, \qquad j_-(z) = \begin{pmatrix} 1 & 0 \\ e^{NG(z)} & 1 \end{pmatrix}.$$

This function satisfies a similar RHP to $\mathbf{T}$, but jumps now occur on a new contour Σ_S, which is obtained from Σ by adding the segments $(a_1 - i\varepsilon, a_1 + i\varepsilon)$, $(b_q - i\varepsilon, b_q + i\varepsilon)$, $(a_{j+1} - i\varepsilon, a_{j+1} + i\varepsilon)$, $(b_j - i\varepsilon, b_j + i\varepsilon)$ for $j \in \mathcal{A}_v$, see Figure 3.3.

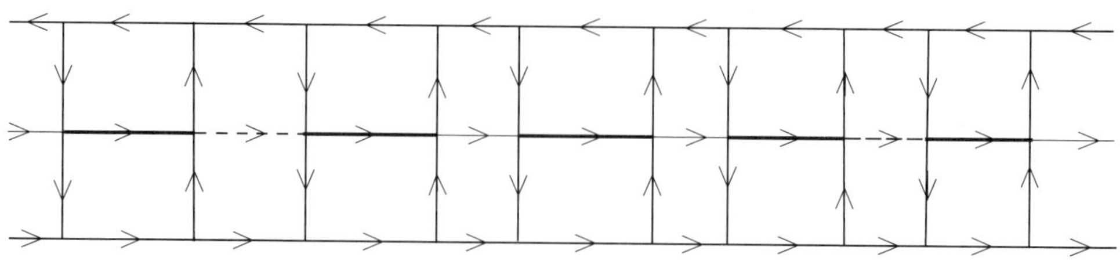

FIGURE 3.3. The contour Σ_S arising from the hypothetical equilibrium measure shown in Figure 3.1.

On horizontal segments, we have that

(3.7.4)

$$j_S(z) = \begin{cases} \begin{pmatrix} 0 & 1 \\ -1 & 0 \end{pmatrix} & \text{for } z \in I^0 \\ \begin{pmatrix} (1-\mathrm{e}^{\pm 2\pi \mathrm{i} N z})^{-1} & \pm\dfrac{\mathrm{e}^{\pm N G(z)}}{1-\mathrm{e}^{\mp 2\pi \mathrm{i} N z}} \\ \mp \mathrm{e}^{\mp N G(z)} & 1 \end{pmatrix} & \text{for } z \in I^0 \pm \mathrm{i}\varepsilon \\ \begin{pmatrix} -\mathrm{e}^{-\mathrm{i} N \Omega_j} & 0 \\ -\mathrm{e}^{-N(g_+(z)+g_-(z)-l-V(z))} & -\mathrm{e}^{\mathrm{i} N \Omega_j} \end{pmatrix} & \text{for } z \in (b_j, a_{j+1}),\ j \in \mathcal{A}_{\mathrm{s}} \\ \begin{pmatrix} (1-\mathrm{e}^{\pm 2\pi \mathrm{i} N z})^{-1} & 0 \\ \mp \mathrm{e}^{-N(2g(z)-V(z)-l)} & 1-\mathrm{e}^{\pm 2\pi \mathrm{i} N z} \end{pmatrix} & \text{for } z \in I^+ \pm \mathrm{i}\varepsilon \\ \begin{pmatrix} \mathrm{e}^{-\mathrm{i} N \Omega_j} & \mathrm{e}^{N(g_+(z)+g_-(z)-l-V(z))} \\ 0 & \mathrm{e}^{\mathrm{i} N \Omega_j} \end{pmatrix} & \text{for } z \in (b_j, a_{j+1}),\ j \in \mathcal{A}_{\mathrm{v}} \\ \begin{pmatrix} 1 & \pm\dfrac{\mathrm{e}^{N(2g(z)-l-V(z))}}{1-\mathrm{e}^{\mp 2\pi \mathrm{i} N z}} \\ 0 & 1 \end{pmatrix} & \text{for } z \in I^- \pm \mathrm{i}\varepsilon. \end{cases}$$

By formula (3.4.5) for the G-function and the upper constraint on the density ρ, we obtain that, for sufficiently small $\varepsilon > 0$ and $x \in (a_j, b_j)$,

$$0 < \pm \operatorname{Re} G(x \pm \mathrm{i}\varepsilon) = 2\pi\varepsilon\rho(x) + \mathcal{O}(\varepsilon^2) < 2\pi\varepsilon + \mathcal{O}(\varepsilon^2). \tag{3.7.5}$$

This, combined with property (3.4.7) of the g-function, implies that all jumps on horizontal segments are exponentially close to the identity matrix, provided that they are bounded away from the segment (a_1, b_q).

3.8. Model RHP

The model RHP appears when we drop in the jump matrix $j_S(z)$ the terms that vanish as $N \to \infty$. Introduce the numbers $\Omega_{j,N}$ for $j = 0, \ldots, q$ as

$$\Omega_{j,N} = \begin{cases} N\Omega_j & \text{for } j \in \mathcal{A}_{\mathrm{v}} \\ \pi + N\Omega_j & \text{for } j \in \mathcal{A}_{\mathrm{s}} \\ 2\pi N & \text{for } j = 0 \\ 0 & \text{for } j = q. \end{cases} \tag{3.8.1}$$

The model RHP is then

(1) $\mathbf{M}(z)$ is analytic in $\mathbb{C} \setminus [a_1, b_q]$.

(2) $\mathbf{M}_+(z) = \mathbf{M}_-(z) j_M(z)$ for $z \in [a_1, b_q]$, where

$$j_M(z) = \begin{cases} \left(\begin{smallmatrix} 0 & 1 \\ -1 & 0 \end{smallmatrix}\right) & \text{for } z \in I^0 \\ \mathrm{e}^{-\mathrm{i}\Omega_{j,N}\sigma_3} & \text{for } z \subset (b_j, a_{j+1}), \end{cases} \tag{3.8.2}$$

(3) As $z \to \infty$,

$$\mathbf{M}(z) \sim I + \frac{\mathbf{M}_1}{z} + \frac{\mathbf{M}_2}{z^2} + \cdots. \tag{3.8.3}$$

This RHP is identical to the model problem for continuous orthogonal polynomials (see Section 2.4.5). We let $\theta(s)$ and $u(z)$ be as in Section 2.4.5, so that they define a Riemann theta function associated with $\sqrt{R(z)}$, where

$$R(z) = \prod_{j=1}^{q}(z-a_q)(z-b_q). \tag{3.8.4}$$

Also, we let

$$\gamma(z) = \prod_{j=1}^{q}\left(\frac{z-a_j}{z-b_j}\right)^{1/4}, \quad \gamma(\infty) = 1, \tag{3.8.5}$$

with cuts on I_0. Define the vector

$$\Omega_N = (\Omega_{1,N}, \dots, \Omega_{q-1,N}). \tag{3.8.6}$$

Then the solution is

$$\begin{aligned}\mathbf{M}(z) &= \mathbf{F}(\infty)^{-1} \\ &\times \begin{pmatrix} \frac{\gamma(z)+\gamma^{-1}(z)}{2}\frac{\theta(u(z)+\Omega_N/(2\pi)+d)}{\theta(u(z)+d)} & \frac{\gamma(z)-\gamma^{-1}(z)}{-2\mathrm{i}}\frac{\theta(u(z)-\Omega_N/(2\pi)-d)}{\theta(u(z)-d)} \\ \frac{\gamma(z)-\gamma^{-1}(z)}{2\mathrm{i}}\frac{\theta(u(z)+\Omega_N/(2\pi)-d)}{\theta(u(z)-d)} & \frac{\gamma(z)+\gamma^{-1}(z)}{2}\frac{\theta(u(z)-\Omega_N/(2\pi)+d)}{\theta(u(z)+d)} \end{pmatrix},\end{aligned} \tag{3.8.7}$$

where

$$\mathbf{F}(\infty) = \begin{pmatrix} \frac{\theta(u(\infty)+\Omega_N/(2\pi)+d)}{\theta(u(\infty)+d)} & 0 \\ 0 & \frac{\theta(u(\infty)-\Omega_N/(2\pi)+d)}{\theta(u(\infty)+d)} \end{pmatrix}, \tag{3.8.8}$$

and the vector d is defined from the vector of Riemann constants as in (2.4.79).

3.9. Parametrix at band-void edge points

We now consider small disks $D(a_j,\varepsilon)$ for $j-1 \in \mathcal{A}_{\mathrm{v}} \cup \{0\}$, and $D(b_j,\varepsilon)$ for $j \in \mathcal{A}_{\mathrm{v}} \cup \{q\}$, centered at the endpoints of bands which are adjacent to a void. Denote

$$D = \left(\bigcup_{j-1\in\mathcal{A}_{\mathrm{v}}\cup\{0\}} D(a_j,\varepsilon)\right) \cup \left(\bigcup_{j\in\mathcal{A}_{\mathrm{v}}\cup\{q\}} D(b_j,\varepsilon)\right), \tag{3.9.1}$$

and we seek a local parametrix $\mathbf{U}_N(z)$ defined on D such that

(1) $\mathbf{U}_N(z)$ is analytic on $D \setminus \Sigma_S$.
(2) $\mathbf{U}_{N+}(z) = \mathbf{U}_{N-}(z) j_S(z)$ for $z \in D \cap \Sigma_S$.
(3) $\mathbf{U}_N(z) = \mathbf{M}(z)\big(I + \mathcal{O}(N^{-1})\big)$ uniformly for $z \in \partial D$.

This RHP is identical to the one encountered in Section 2.4.6. If we let

$$y_0(z) = \mathrm{Ai}(z), \quad y_1(z) = \omega\,\mathrm{Ai}(\omega z), \quad y_2(z) = \omega^2\,\mathrm{Ai}(\omega^2 z), \tag{3.9.2}$$

where $\omega = \mathrm{e}^{2\pi \mathrm{i}/3}$, and

$$\Phi_{\mathrm{rv}}(z) = \begin{cases} \begin{pmatrix} y_0(z) & -y_2(z) \\ y_0'(z) & -y_2'(z) \end{pmatrix} & \text{for } \arg z \in (0, \pi/2) \\ \begin{pmatrix} -y_1(z) & -y_2(z) \\ -y_1'(z) & -y_2'(z) \end{pmatrix} & \text{for } \arg z \in (\pi/2, \pi) \\ \begin{pmatrix} -y_2(z) & y_1(z) \\ -y_2'(z) & y_1'(z) \end{pmatrix} & \text{for } \arg z \in (-\pi, -\pi/2) \\ \begin{pmatrix} y_0(z) & y_1(z) \\ y_0'(z) & y_1'(z) \end{pmatrix} & \text{for } \arg z \in (-\pi/2, 0), \end{cases} \tag{3.9.3}$$

and

$$\psi_{b_j}(z) = -\left\{ \frac{3\pi}{2} \int_z^{b_j} \rho(t)\,\mathrm{d}t \right\}^{2/3}, \tag{3.9.4}$$

then the local parametrix near b_j for $j \in \mathcal{A}_{\mathrm{v}}$ is given by

$$\mathbf{U}_N(z) = \mathbf{E}_N^{b_j}(z)\Phi_{\mathrm{rv}}\big(N^{2/3}\psi_{b_j}(z)\big)\mathrm{e}^{-N(g(z)-V(z)/2-l/2)\sigma_3}, \tag{3.9.5}$$

where

$$\begin{aligned} \mathbf{E}_N^{b_j}(z) &= \mathbf{M}(z)\mathrm{e}^{\pm(\mathrm{i}\Omega_{j,N}/2)\sigma_3}\mathbf{L}_N^{b_j}(z)^{-1} \quad \text{for } \pm \operatorname{Im} z > 0, \\ \mathbf{L}_N^{b_j}(z) &= \frac{1}{2\sqrt{\pi}} \begin{pmatrix} N^{-1/6}\psi_{b_j}^{-1/4}(z) & 0 \\ 0 & N^{1/6}\psi_{b_j}^{1/4}(z) \end{pmatrix} \begin{pmatrix} 1 & \mathrm{i} \\ -1 & \mathrm{i} \end{pmatrix}, \end{aligned} \tag{3.9.6}$$

and we take the principal branch of $\psi_{b_j}^{1/4}$, which is positive on $(b_j, b_j + \varepsilon)$ and has a cut on $(b_j - \varepsilon, b_j)$.

Similarly, the parametrix at a_j for $j-1 \in \mathcal{A}_{\mathrm{v}}$ is given by

$$\mathbf{U}_N(z) = \mathbf{E}_N^{a_j}(z)\Phi_{\mathrm{lv}}\big(N^{2/3}\psi_{a_j}(z)\big)e^{-N(g(z)-\frac{V(z)}{2}-\frac{l}{2})\sigma_3}, \tag{3.9.7}$$

for $z \in D(a_j, \varepsilon)$, where

$$\begin{aligned} \mathbf{E}_N^{a_j}(z) &= \mathbf{M}(z)\mathrm{e}^{\pm(\mathrm{i}\Omega_{j-1,N}/2)\sigma_3}\mathbf{L}_N^{a_j}(z)^{-1} \quad \text{for } \pm \operatorname{Im} z > 0, \\ \mathbf{L}_N^{a_j}(z) &= \frac{1}{2\sqrt{\pi}} \begin{pmatrix} N^{-1/6}\psi_{a_j}^{-1/4}(z) & 0 \\ 0 & N^{1/6}\psi_{a_j}^{1/4}(z) \end{pmatrix} \begin{pmatrix} 1 & -\mathrm{i} \\ -1 & -\mathrm{i} \end{pmatrix}, \end{aligned} \tag{3.9.8}$$

where

$$\psi_{a_j}(z) = -\left\{ \frac{3\pi}{2} \int_{a_j}^{z} \rho(t)\,\mathrm{d}t \right\}^{2/3}, \tag{3.9.9}$$

and

$$\Phi_{\mathrm{lv}}(z) = \Phi_{\mathrm{rv}}(z) \begin{pmatrix} 1 & 0 \\ 0 & -1 \end{pmatrix}. \tag{3.9.10}$$

The subscripts lv and rv in the Φ-functions above stand for "left void" and "right void," respectively. The construction of the local parametrix at the endpoints which separate bands and saturated regions is slightly different, and is considered in the next section.

3.10. Parametrix at the band-saturated region end points

We now consider small disks $D(a_j,\varepsilon)$ for $j-1\in\mathcal{A}_{\rm s}$, and $D(b_j,\varepsilon)$ for $j\in\mathcal{A}_{\rm s}$, centered at the endpoints of bands which are adjacent to a saturated region. Denote

$$\widetilde{D}=\Big(\bigcup_{j-1\in\mathcal{A}_{\rm s}} D(a_j,\varepsilon)\Big)\cup\Big(\bigcup_{j\in\mathcal{A}_{\rm s}} D(b_j,\varepsilon)\Big). \tag{3.10.1}$$

We will seek a local parametrix $\mathbf{U}_N(z)$ defined on $\widetilde{D}$ such that

(1) $\mathbf{U}_N(z)$ is analytic on $\widetilde{D}\setminus\Sigma_S$.
(2) $\mathbf{U}_{N+}(z)=\mathbf{U}_{N-}(z)j_S(z)$ for $z\in\widetilde{D}\cap\Sigma_S$.
(3) For $z\in\partial\widetilde{D}$, we have the uniform estimate,

$$\mathbf{U}_N(z)=\mathbf{M}(z)\big(I+\mathcal{O}(N^{-1})\big). \tag{3.10.2}$$

The local analysis here is slightly different than in the continuous case, due to the presence of the saturated region. Nonetheless, the RHP is still solved using Airy functions. We first construct the parametrix near b_j for $j\in\mathcal{A}_{\rm s}$. Let

$$\mathbf{U}_N(z)=\widetilde{\mathbf{Q}}_N(z)\mathrm{e}^{\mp \mathrm{i}N\pi z\sigma_3}\mathrm{e}^{-N(g(z)-V(z)/2-l/2)\sigma_3}\quad \text{for } \pm\operatorname{Im} z>0. \tag{3.10.3}$$

Then the jumps for $\widetilde{\mathbf{Q}}_N$ are

$$j_{\widetilde{Q}}(z)=\begin{cases}\begin{pmatrix}0 & 1\\ -1 & 0\end{pmatrix} & \text{for } z\in(b_j-\varepsilon,b_j)\\ \begin{pmatrix}-1 & 0\\ -1 & -1\end{pmatrix} & \text{for } z\in(b_j,b_j+\varepsilon)\\ \begin{pmatrix}1 & -1\\ 0 & 1\end{pmatrix} & \text{for } z\in(b_j,b_j+i\varepsilon)\\ \begin{pmatrix}1 & 1\\ 0 & 1\end{pmatrix} & \text{for } z\in(b_j,b_j-i\varepsilon),\end{cases} \tag{3.10.4}$$

where orientation is taken from left to right on horizontal contours, and down to up on vertical contours according to Figure 3.3. We now take

$$\Phi_{\rm rs}(z)=\begin{cases}\begin{pmatrix}y_2(z) & -y_0(z)\\ y_2'(z) & -y_0'(z)\end{pmatrix} & \text{for } \arg z\in(0,\pi/2)\\ \begin{pmatrix}y_2(z) & y_1(z)\\ y_2'(z) & y_1'(z)\end{pmatrix} & \text{for } \arg z\in(\pi/2,\pi)\\ \begin{pmatrix}y_1(z) & -y_2(z)\\ y_1'(z) & -y_2'(z)\end{pmatrix} & \text{for } \arg z\in(-\pi,-\pi/2)\\ \begin{pmatrix}y_1(z) & y_0(z)\\ y_1'(z) & y_0'(z)\end{pmatrix} & \text{for } \arg z\in(-\pi/2,0).\end{cases} \tag{3.10.5}$$

Then $\Phi_{\rm rs}(z)$ solves a RHP similar to that of $\widetilde{\mathbf{Q}}_N$, but for jumps emanating from the origin rather than from b_j. Notice that

$$\det\Phi_{\rm rs}(z)\equiv\frac{1}{2\pi \mathrm{i}}. \tag{3.10.6}$$

Once again,

$$\psi_{b_j}(z)=-\left\{\frac{3\pi}{2}\int_z^{b_j}\big(1-\rho(t)\big)\,\mathrm{d}t\right\}^{2/3}, \tag{3.10.7}$$

extends to a conformal map from $D(b_j, \varepsilon)$ onto a convex neighborhood of the origin, with

$$\psi_{b_j}(b_j) = 0, \quad \psi'_{b_j}(b_j) > 0. \tag{3.10.8}$$

Again, we can slightly deform the vertical pieces of the contour Σ_S close to b_j, so that

$$\psi_{b_j}\{D(b_j,\varepsilon) \cap \Sigma_S\} = (-\varepsilon, \varepsilon) \cup (-\mathrm{i}\varepsilon, \mathrm{i}\varepsilon). \tag{3.10.9}$$

We now take

$$\widetilde{\mathbf{Q}}_N(z) = \mathbf{E}_N^{b_j}(z)\Phi_{\mathrm{rs}}\big(N^{2/3}\psi_{b_j}(z)\big), \tag{3.10.10}$$

where

$$\begin{aligned} \mathbf{E}_N^{b_j}(z) &= \mathbf{M}(z)\mathrm{e}^{\pm(\mathrm{i}\Omega_{j,N}/2)\sigma_3}\mathbf{L}_N^{b_j}(z)^{-1} \quad \text{for } \pm \operatorname{Im} z \geq 0, \\ \mathbf{L}_N^{b_j}(z) &= \frac{1}{2\sqrt{\pi}}\begin{pmatrix} N^{-1/6}\psi_{b_j}^{-1/4}(z) & 0 \\ 0 & N^{1/6}\psi_{b_j}^{1/4}(z)\end{pmatrix}\begin{pmatrix} 1 & \mathrm{i} \\ 1 & -\mathrm{i}\end{pmatrix}, \end{aligned} \tag{3.10.11}$$

and we take the principal branch of $\psi_{b_j}^{1/4}$. The function $\Phi_{\mathrm{rs}}\big(N^{2/3}\psi_{b_j}(z)\big)$ has the jumps j_S. Similar to the prefactor $\mathbf{E}_N^{b_j}$ at band-void endpoints, the prefactor $\mathbf{E}_N^{b_j}$ is analytic in $D(b_j, \varepsilon)$ and thus does not change these jumps.

We now check that $\mathbf{U}_N$ satisfies the matching condition (3.10.2). The large N asymptotics of $\Phi_{\mathrm{rs}}\big(N^{2/3}\psi_{b_j}(z)\big)$ are given in each of the different regions of analyticity as

$$\begin{aligned} \Phi_{\mathrm{rs}}\big(N^{2/3}\psi_{b_j}(z)\big) &= \frac{1}{2\sqrt{\pi}}N^{-(1/6)\sigma_3}\psi_{b_j}(z)^{-(1/4)\sigma_3} \\ &\times\left[\pm\begin{pmatrix} -\mathrm{i} & -1 \\ -\mathrm{i} & 1\end{pmatrix} \pm \frac{\psi_{b_j}(z)^{-3/2}}{48N}\begin{pmatrix} -5\mathrm{i} & 5 \\ 7\mathrm{i} & 7\end{pmatrix} + \mathcal{O}(N^{-2})\right] \\ &\times \mathrm{e}^{(2/3)N\psi_{b_j}(z)^{3/2}\sigma_3} \quad \text{for } \pm \operatorname{Im} z > 0, \end{aligned} \tag{3.10.12}$$

where we always take the principal branch of $\psi_{b_j}(z)^{3/2}$. As such, $\psi_{b_j}(z)^{3/2}$ is two-valued for $x \in (b_j - \varepsilon, b_j)$, so that

$$[\tfrac{2}{3}\psi_{b_j}(x)^{3/2}]_\pm = \mp\pi\mathrm{i}\int_x^{b_j}\big(1-\rho(t)\big)\,\mathrm{d}t = \mp\pi\mathrm{i}(b_j - x) \pm \pi\mathrm{i}\int_x^{b_j}\rho(t)\,\mathrm{d}t. \tag{3.10.13}$$

From (3.4.9) we have that

$$2g_\pm(x) - V(x) = l \pm 2\pi\mathrm{i}\int_x^{b_q}\rho(t)\,\mathrm{d}t = l \pm 2\pi\mathrm{i}\int_x^{b_j}\rho(t)\,\mathrm{d}t \pm \mathrm{i}\Omega_j \mp 2\pi\mathrm{i}b_j \tag{3.10.14}$$

for $x \in (b_j - \varepsilon, b_j)$. These equations imply that

$$\big(2g_\pm(x) - V(x)\big) - \big(2g_\pm(b_j) - V(b_j)\big) = \pm 2\pi\mathrm{i}\int_x^{b_j}\rho(t)\,\mathrm{d}t. \tag{3.10.15}$$

We can therefore write (3.10.13) as

$$[\tfrac{2}{3}\psi_{b_j}(x)^{3/2}]_\pm = \mp\pi\mathrm{i}(b_j - x) + \tfrac{1}{2}\big[\big(2g_\pm(x) - V(x)\big) - \big(2g_\pm(b_j) - V(b_j)\big)\big]. \tag{3.10.16}$$

We can extend these equations into the upper and lower half-planes, respectively, obtaining

$$\tag{3.10.17} \tfrac{2}{3}\psi_{b_j}(z)^{3/2} = \mp\pi\mathrm{i}(b_j - z) + \tfrac{1}{2}\big[\big(2g(z) - V(z)\big) - \big(2g_\pm(b_j) - V(b_j)\big)\big] \quad \text{for } \pm\operatorname{Im} z > 0.$$

Using (3.10.14) at $x = b_j$, we can write

$$\tag{3.10.18} \frac{2}{3}\psi_{b_j}(z)^{3/2} = g(z) - \frac{V(z)}{2} - \frac{l}{2} \pm \pi\mathrm{i}z \mp \frac{\mathrm{i}(\Omega_{j,N} - \pi)}{2N} \quad \text{for } \pm\operatorname{Im} z > 0.$$

Plugging (3.10.12) and (3.10.18) into (3.10.10) gives as $N \to \infty$,

$$\tag{3.10.19} \begin{aligned} \mathbf{U}_N(z) &= \mathbf{M}(z)\mathrm{e}^{(\mathrm{i}\Omega_{j,N}/2)\sigma_3}\mathbf{L}_N^{b_j}(z)^{-1}\frac{1}{2\sqrt{\pi}}N^{-(1/6)\sigma_3}\psi_{b_j}(z)^{-(1/4)\sigma_3} \\ &\quad\times\left[\pm\begin{pmatrix} -\mathrm{i} & -1 \\ -\mathrm{i} & 1\end{pmatrix} \pm \frac{\psi_{b_j}(z)^{-3/2}}{48N}\begin{pmatrix} -5\mathrm{i} & 5 \\ 7\mathrm{i} & 7\end{pmatrix} + \mathcal{O}(N^{-2})\right] \\ &\quad\times \mathrm{e}^{N(g(z)-l/2-V(z)/2)\sigma_3}\mathrm{e}^{\pm\mathrm{i}N\pi z\sigma_3}\mathrm{e}^{\mp(\mathrm{i}\Omega_{j,N}/2)\sigma_3} \\ &\quad\times \mathrm{e}^{\pm(\mathrm{i}\pi/2)\sigma_3}\mathrm{e}^{\mp\mathrm{i}N\pi z\sigma_3}\mathrm{e}^{-N(g(z)-V(z)/2-l/2)\sigma_3} \\ &= \mathbf{M}(z)\left[I + \frac{\psi_{b_j}(z)^{-3/2}}{48N}\begin{pmatrix} -1 & -6\mathrm{i}\mathrm{e}^{\pm\mathrm{i}\Omega_{j,N}} \\ -6\mathrm{i}\mathrm{e}^{\mp\mathrm{i}\Omega_{j,N}} & 1\end{pmatrix} + \mathcal{O}(N^{-2})\right] \\ &\qquad \text{for } \pm\operatorname{Im}(z) > 0. \end{aligned}$$

We can make a similar construction near a_j for $j - 1 \in \mathcal{A}_{\mathrm{s}}$. Let

$$\tag{3.10.20} \psi_{a_j}(z) = -\left\{\frac{3\pi}{2}\int_{a_j}^{z}\big(1 - \rho(t)\big)\,\mathrm{d}t\right\}^{2/3}.$$

This function is analytic in $D(a_j,\varepsilon)$ and has negative derivative at a_j, thus $\operatorname{Im} z$ and $\operatorname{Im}\psi_{a_j}(z)$ have opposite signs for $z \in D(a_j,\varepsilon)$. Then the jumps for $\widetilde{\mathbf{Q}}_N$ are

$$\tag{3.10.21} j_{\widetilde{Q}}(z) = \begin{cases} \begin{pmatrix} 0 & 1 \\ -1 & 0\end{pmatrix} & \text{for } z \in (a_j, a_j + \varepsilon) \\ \begin{pmatrix} -1 & 0 \\ -1 & -1\end{pmatrix} & \text{for } z \in (a_j - \varepsilon, a_j) \\ \begin{pmatrix} 1 & -1 \\ 0 & 1\end{pmatrix} & \text{for } z \in (a_j, a_j + i\varepsilon) \\ \begin{pmatrix} 1 & 1 \\ 0 & 1\end{pmatrix} & \text{for } z \in (a_j, a_j - i\varepsilon), \end{cases}$$

where the contour is oriented from left to right on horizontal segments and up to down on vertical segments according to Figure 3.3. After a slight deformation of the vertical contours and the change of variables ψ_{a_j}, these jumps become the following jumps close to the origin:

$$j_{\widetilde{Q}}(\psi_{a_j}(z)) = \begin{cases} \begin{pmatrix} 0 & 1 \\ -1 & 0 \end{pmatrix} & \text{for } \psi_{a_j}(z) \in (-\varepsilon, 0) \\ \begin{pmatrix} -1 & 0 \\ -1 & -1 \end{pmatrix} & \text{for } \psi_{a_j}(z) \in (0, \varepsilon) \\ \begin{pmatrix} 1 & -1 \\ 0 & 1 \end{pmatrix} & \text{for } \psi_{a_j}(z) \in (-i\varepsilon, 0) \\ \begin{pmatrix} 1 & 1 \\ 0 & 1 \end{pmatrix} & \text{for } \psi_{a_j}(z) \in (0, i\varepsilon), \end{cases} \tag{3.10.22}$$

where the contour is oriented from right to left on horizontal segments and down to up on vertical segments. These jump conditions are satisfied by the function

$$\Phi_{\text{ls}}(z) = \Phi_{\text{rs}}(z) \begin{pmatrix} 1 & 0 \\ 0 & -1 \end{pmatrix}. \tag{3.10.23}$$

Then we can take for $z \in D(a_j, \varepsilon)$,

$$\begin{aligned} \mathbf{U}_N(z) = \mathbf{M}(z) \mathrm{e}^{\pm(\mathrm{i}\Omega_{j-1,N}/2)\sigma_3} \mathbf{L}_N^{a_j}(z)^{-1} \Phi_{\text{ls}}\big(N^{2/3}\psi_{a_j}(z)\big) \\ \times \mathrm{e}^{\mp \mathrm{i} N\pi z\sigma_3} \mathrm{e}^{-N(g(z)-V(z)/2-l/2)\sigma_3} \quad \text{for } \pm \operatorname{Im} z > 0, \end{aligned} \tag{3.10.24}$$

where

$$\mathbf{L}_N^{a_j}(z) = \frac{1}{2\sqrt{\pi}} \begin{pmatrix} N^{-1/6}\psi_{a_j}^{-1/4}(z) & 0 \\ 0 & N^{1/6}\psi_{a_j}^{1/4}(z) \end{pmatrix} \begin{pmatrix} -1 & \mathrm{i} \\ -1 & -\mathrm{i} \end{pmatrix}. \tag{3.10.25}$$

We once again have as $N \to \infty$,

$$\begin{aligned} \Phi_{\text{ls}}\big(N^{2/3}\psi_{a_j}(z)\big) &= \frac{1}{2\sqrt{\pi}} N^{-(1/6)\sigma_3} \psi_{a_j}(z)^{-(1/4)\sigma_3} \\ &\times \left[\pm \begin{pmatrix} -\mathrm{i} & 1 \\ -\mathrm{i} & -1 \end{pmatrix} \pm \frac{\psi_{a_j}(z)^{-3/2}}{48N} \begin{pmatrix} -5\mathrm{i} & -5 \\ 7\mathrm{i} & -7 \end{pmatrix} + \mathcal{O}(N^{-2}) \right] \\ &\times \mathrm{e}^{(2/3)N\psi_{a_j}(z)^{3/2}\sigma_3} \\ &\qquad \text{for } \pm \operatorname{Im} \psi_{a_j}(z) > 0 \text{ (so } \mp \operatorname{Im} z > 0), \end{aligned} \tag{3.10.26}$$

and for $z \in D(a_j, \varepsilon)$,

$$\frac{2}{3}\psi_{a_j}^{3/2}(z) = \pm \mathrm{i}\pi z + g(z) - \frac{V(z)}{2} - \frac{l}{2} \mp \frac{\mathrm{i}(\Omega_{j-1,N} - \pi)}{2N} \quad \text{for } \pm \operatorname{Im} z > 0. \tag{3.10.27}$$

Combining (3.10.24), (3.10.26), and (3.10.27) gives as $N \to \infty$,

(3.10.28)

$$\begin{aligned} \mathbf{U}_N(z) = \mathbf{M}(z) \mathrm{e}^{\pm(\mathrm{i}\Omega_{j-1,N}/2)\sigma_3} \mathbf{L}_N^{a_j}(z)^{-1} \frac{1}{2\sqrt{\pi}} N^{-(1/6)\sigma_3} \psi_{a_j}(z)^{-(1/4)\sigma_3} \\ \times \left[\pm \begin{pmatrix} \mathrm{i} & -1 \\ \mathrm{i} & 1 \end{pmatrix} \pm \frac{\psi_{a_j}(z)^{-3/2}}{48N} \begin{pmatrix} 5\mathrm{i} & 5 \\ -7\mathrm{i} & 7 \end{pmatrix} + \mathcal{O}(N^{-2}) \right] \\ \times \mathrm{e}^{\pm \mathrm{i} N\pi z\sigma_3} \mathrm{e}^{N(g(z)-V(z)/2-l/2)\sigma_3} \mathrm{e}^{\mp(\mathrm{i}\Omega_{j-1,N}/2)\sigma_3} \\ \times \mathrm{e}^{\pm(\mathrm{i}\pi/2)\sigma_3} \mathrm{e}^{\mp \mathrm{i} N\pi z\sigma_3} \mathrm{e}^{-N(g(z)-V(z)/2-l/2)\sigma_3} \end{aligned}$$

$$= \mathbf{M}(z)\mathrm{e}^{\pm(\mathrm{i}\Omega_{j-1,N}/2)\sigma_3}\left[I + \frac{\psi_{a_j}(z)^{-3/2}}{48N}\begin{pmatrix} -1 & 6\mathrm{i} \\ 6\mathrm{i} & 1 \end{pmatrix} + \mathcal{O}(N^{-2})\right] \times \mathrm{e}^{\mp(\mathrm{i}\Omega_{j-1,N}/2)\sigma_3}$$

$$= \mathbf{M}(z)\left[I + \frac{\psi_{a_j}(z)^{-3/2}}{48N}\begin{pmatrix} -1 & 6\mathrm{i}\mathrm{e}^{\pm\mathrm{i}\Omega_{j-1,N}} \\ 6\mathrm{i}\mathrm{e}^{\mp\mathrm{i}\Omega_{j-1,N}} & 1 \end{pmatrix} + \mathcal{O}(N^{-2})\right] \quad \text{for } \pm\operatorname{Im} z > 0.$$

3.11. The third and final transformation of the RHP

We now consider the contour Σ_X, which consists of the circles $\partial D(a_j, \varepsilon)$, and $\partial D(b_j, \varepsilon)$, for $j = 1, \dots q$, all oriented counterclockwise, together with the parts of $\Sigma_S \setminus (\bigcup_j [a_j, b_j])$ which lie outside of the disks $D(a, \varepsilon)$, $D(a', \varepsilon)$, $D(\beta', \varepsilon)$, and $D(\beta, \varepsilon)$, see Figure 3.4.

We let

$$\mathbf{X}_N(z) = \begin{cases} \mathbf{S}_N(z)\mathbf{M}(z)^{-1} & \text{for } z \text{ outside the disks } D(a_j, \varepsilon), D(b_j, \varepsilon) \\ \mathbf{S}_N(z)\mathbf{U}_N(z)^{-1} & \text{for } z \text{ inside the disks } D(a_j, \varepsilon), D(b_j, \varepsilon). \end{cases} \tag{3.11.1}$$

Then $\mathbf{X}_N(z)$ solves the following RHP:

(1) $\mathbf{X}_N(z)$ is analytic on $\mathbb{C} \setminus \Sigma_X$.

(2) $\mathbf{X}_N(z)$ has the jump properties

$$\mathbf{X}_{N+}(x) = \mathbf{X}_{N-}(z) j_X(z) \tag{3.11.2}$$

where

$$j_X(z) = \begin{cases} \mathbf{M}(z)\mathbf{U}_N(z)^{-1} & \text{for } z \text{ on the circles} \\ \mathbf{M}(z) j_S(z)\mathbf{M}(z)^{-1} & \text{otherwise.} \end{cases} \tag{3.11.3}$$

(3) As $z \to \infty$,

$$\mathbf{X}_N(z) \sim I + \frac{\mathbf{X}_1}{z} + \frac{\mathbf{X}_2}{z^2} + \cdots. \tag{3.11.4}$$

Additionally, we have that $j_X(z)$ is uniformly close to the identity in the following sense:

$$j_X(z) = \begin{cases} I + \mathcal{O}(N^{-1}) & \text{uniformly on the circles} \\ I + \mathcal{O}(e^{-c(z)N}) & \text{on the rest of } \Sigma_X, \end{cases} \tag{3.11.5}$$

where $c(z)$ is a positive, continuous function satisfying (2.4.22). If we set

$$j_X^0(z) = j_X(z) - I, \tag{3.11.6}$$

then (3.11.5) becomes

$$j_X^0(z) = \begin{cases} \mathcal{O}(N^{-1}) & \text{uniformly on the circles} \\ \mathcal{O}(\mathrm{e}^{-c(z)N}) & \text{on the rest of } \Sigma_X. \end{cases} \tag{3.11.7}$$

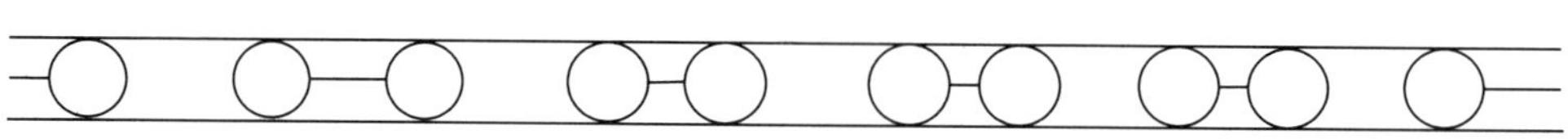

FIGURE 3.4. The contour Σ_X arising from the hypothetical equilibrium measure shown in Figure 3.1.

The solution to the RHP for $\mathbf{X}_N$ follows from Proposition 2.5.1 and the arguments following it, which give that

$$\mathbf{X}_N(z) = I + \sum_{k=1}^{\infty} \mathbf{X}_{N,k}(z), \tag{3.11.8}$$

where

$$\mathbf{X}_{N,k}(z) = -\frac{1}{2\pi \mathrm{i}} \int_{\Sigma_X} \frac{v_{k-1}(u) j_X^0(u)}{z-u} \mathrm{d}u. \tag{3.11.9}$$

In particular, this implies that

$$\mathbf{X}_N(z) \sim I + \mathcal{O}\left(\frac{1}{N(|z|+1)}\right), \quad \text{as } N \to \infty, \tag{3.11.10}$$

uniformly for $z \in \mathbb{C} \setminus \Sigma_X$.

3.12. Asymptotics of recurrence coefficients

The transformations (3.5.18), (3.6.1), (3.7.2), (3.11.1) give that, for z bounded away from the real line,

$$\mathbf{P}_N(z) = \mathbf{K}^{-1} \mathrm{e}^{(Nl/2)\sigma_3} \mathbf{X}_N(z) \mathbf{M}(z) e^{N(g(z)-l/2)\sigma_3} \mathbf{K}, \tag{3.12.1}$$

and for z close to the real line but bounded away from the support of the equilibrium measure,

$$\mathbf{P}_N(z) = \mathbf{K}^{-1} \mathrm{e}^{(Nl/2)\sigma_3} \mathbf{X}_N(z) \mathbf{M}(z) e^{N(g(z)-l/2)\sigma_3} \mathbf{K} \mathbf{D}_{\pm}^u(z)^{-1} \quad \text{for } \pm \operatorname{Im} z \geq 0. \tag{3.12.2}$$

Using (3.2.4), (3.2.8), and (3.2.9), and a straightforward large z expansion of equation (3.12.1), we can derive asymptotics for the recurrence coefficients, which we present in the following theorem. The details of the proof are left to the reader.

Theorem 3.12.1 (Asymptotics of recurrence coefficients). *Let $V(x)$ be a real analytic function satisfying* (1.1.7) *which yields a regular equilibrium measure* (3.3.9), *and let $\{P_n\}_{n=0}^{\infty}$ be the system of orthogonal polynomials defined according to* (3.1.4). *Then as $N \to \infty$, the normalizing constants in* (3.1.4) *and recurrence coefficients in* (3.1.9) *admit the following asymptotic expansions.*

$$h_N = \frac{\pi}{2} \mathrm{e}^{Nl} \left(\sum_{j=1}^{q} (b_j - a_j) \right) \frac{\theta(u(\infty)+d)\theta(u(\infty) - \Omega_N/(2\pi) - d)}{\theta(u(\infty)-d)\theta(u(\infty) + (\Omega_N/(2\pi) + d)} \times \left[1 + \mathcal{O}\left(\frac{1}{N}\right)\right], \tag{3.12.3}$$

$$h_{N-1} = 8\pi \mathrm{e}^{Nl} \left(\sum_{j=1}^{q} (b_j - a_j) \right)^{-1} \frac{\theta(u(\infty)-d)\theta(u(\infty) - \Omega_N/(2\pi) + d)}{\theta(u(\infty)+d)\theta(u(\infty) + \Omega_N/(2\pi) - d)} \times \left[1 + \mathcal{O}\left(\frac{1}{N}\right)\right], \tag{3.12.4}$$

$$\begin{aligned}(3.12.5)\qquad \gamma_N^2 =& \left(\frac{\sum_{j=1}^q (b_j - a_j)}{4}\right)^2\\ &\times \frac{\theta(u(\infty)+d)^2\theta(u(\infty)-\Omega_N/(2\pi)-d)\theta(u(\infty)+\Omega_N/(2\pi)-d)}{\theta(u(\infty)-d)^2\theta(u(\infty)+\Omega_N/(2\pi)+d)\theta(u(\infty)-\Omega_N/(2\pi)+d)}\\ &+\mathcal{O}\left(\frac{1}{N}\right),\end{aligned}$$

$$\begin{aligned}(3.12.6)\quad \beta_{N-1} =& \frac{\sum_{j=1}^q (b_j^2 - a_j^2)}{2\sum_{j=1}^q (b_j - a_j)}\\ &+\Bigg(\frac{\nabla\theta(u(\infty)+\Omega_N/(2\pi)-d)}{\theta(u(\infty)+\Omega_N/(2\pi)-d)} - \frac{\nabla\theta(u(\infty)+\Omega_N/(2\pi)+d)}{\theta(u(\infty)+\Omega_N/(2\pi)+d)}\\ &\qquad + \frac{\nabla\theta(u(\infty)+b)}{\theta(u(\infty)+b)} - \frac{\nabla\theta(u(\infty)-d)}{\theta(u(\infty)-d)}, u'(\infty)\Bigg)\\ &+\mathcal{O}\left(\frac{1}{N}\right).\end{aligned}$$

where $\nabla\theta$ is the gradient of θ,

$$(3.12.7)\qquad u'(\infty) = (c_{1,q-1}, c_{2,q-1}, \dots, c_{q-1,q-1}),$$

and the numbers $c_{j,q-1}$ are the leading coefficients of the basic holomorphic one forms on the Riemann surface associated with $\sqrt{R(z)}$, see (2.4.53).

Notice that these asymptotics are very similar to the results obtained in the previous chapter for continuous orthogonal polynomials.

3.13. Universality in the discrete log gas ensemble

We now turn to the question of scaling limits of the correlation kernel $K_N(x, y)$ given in (3.2.10). Inside a band or at the edge of a band, the scaling limits are very similar to the continuous case. We have the following theorems.

Theorem 3.13.1. *Let x_0 be a point in the interior I_0, and let u_N and v_N be two bounded sequences of real numbers chosen such that*

$$(3.13.1)\qquad x_0 + \frac{u_N}{N\rho(x_0)} \in L_N, \quad x_0 + \frac{v_N}{N\rho(x_0)} \in L_N,$$

for each $N \in \mathbb{N}$. In particular this implies

$$(3.13.2)\qquad u_N - v_N \in \rho(x_0)\mathbb{Z}.$$

Then

$$(3.13.3)\quad \frac{1}{N\rho(x_0)} K_N\left(x_0 + \frac{u_N}{N\rho(x_0)}, x_0 + \frac{v_N}{N\rho(x_0)}\right) = \frac{\sin[\pi(u_N - v_N)]}{\pi(u_N - v_N)} + \mathcal{O}(N^{-1}).$$

The limiting kernel $\sin[\pi(u_N - v_N)]/\pi(u_N - v_N)$ is called the *discrete sine kernel,* since u_N and v_N must be chosen from a discrete set. The proof is similar to the proof of Theorem 2.7.1, and we omit it here. The scaling limit for the correlation kernel at the edge of a void and band is identical to that of the continuous kernel (see Theorem 2.7.2).

Theorem 3.13.2. *Let $x_0 = b_j$ for some $q \in \mathcal{A}_{\mathrm{v}}$, and u and v be two bounded real numbers. Then for some $c > 0$,*

$$\frac{1}{cN^{2/3}} K_N\left(x_0 + \frac{u}{cN^{2/3}}, x_0 + \frac{v}{cN^{2/3}}\right) = \frac{\mathrm{Ai}(u)\,\mathrm{Ai}'(v) - \mathrm{Ai}'(u)\,\mathrm{Ai}(v)}{u-v} + \mathcal{O}(N^{-2/3}). \tag{3.13.4}$$

Similarly, if $x_0 = a_{j+1}$ for some $q \in \mathcal{A}_{\mathrm{v}}$, then for some $c > 0$, as $N \to \infty$,

$$\frac{1}{cN^{2/3}} K_N\left(x_0 - \frac{u}{cN^{2/3}}, x_0 - \frac{v}{cN^{2/3}}\right) = \frac{\mathrm{Ai}(u)\,\mathrm{Ai}'(v) - \mathrm{Ai}'(u)\,\mathrm{Ai}(v)}{u-v} + \mathcal{O}(N^{-2/3}). \tag{3.13.5}$$

The proof of this theorem is similar to that of Theorem 2.7.2, and we omit it here.

Let us now consider the scaling limit of the kernel $K_N(x,y)$ at the edge of a band and a saturated region. This situation is unique to the discrete orthogonal polynomials, and we have the following theorem.

Theorem 3.13.3. *Let $x_0 = b_j$ for some $j \in \mathcal{A}_{\mathrm{s}}$, and let u_N and v_N be two bounded sequences of real numbers chosen such that*

$$x \equiv x_0 + \frac{u_N}{cN^{2/3}} \in L_N, \quad y \equiv x_0 + \frac{v_N}{cN^{2/3}} \in L_N. \tag{3.13.6}$$

Notice that this means that

$$x = \frac{k_N}{N}, \quad y = \frac{m_N}{N}, \tag{3.13.7}$$

for some sequences of integers k_N and m_N. There is a constant c such that for $x \neq y$, as $N \to \infty$,

$$-\frac{(-1)^{k_N - m_N}}{cN^{2/3}} K_N(x,y) = \frac{\mathrm{Ai}(u_N)\,\mathrm{Ai}'(v_N) - \mathrm{Ai}'(u_N)\,\mathrm{Ai}(v_N)}{u_N - v_N} + \mathcal{O}(N^{-2/3}), \tag{3.13.8}$$

and for $x = y$,

$$\frac{N^{1/3}}{c}\left(1 - \frac{1}{N} K_N(x,x)\right) = \mathrm{Ai}'(u_N)^2 - u_N\,\mathrm{Ai}(u_N)^2 + \mathcal{O}(N^{-2/3}). \tag{3.13.9}$$

Similarly, if $x_0 = a_{j+1}$ for some $j \in \mathcal{A}_{\mathrm{s}}$, and u_N and v_N are two bounded sequences of real numbers chosen such that

$$x \equiv x_0 - \frac{u_N}{N\rho(x_0)} = \frac{k_N}{N} \in L_N, \quad y \equiv x_0 - \frac{v_N}{N\rho(x_0)} = \frac{m_N}{N} \in L_N, \tag{3.13.10}$$

then there is a constant c such that for $x \neq y$, as $N \to \infty$,

$$\frac{(-1)^{k_N - m_N}}{cN^{2/3}} K_N(x,y) = \frac{\mathrm{Ai}(u_N)\,\mathrm{Ai}'(v_N) - \mathrm{Ai}'(u_N)\,\mathrm{Ai}(v_N)}{u_N - v_N} + \mathcal{O}(N^{-2/3}), \tag{3.13.11}$$

and for $x = y$,

$$\frac{N^{1/3}}{c}\left(1 - \frac{1}{N} K_N(x,x)\right) = \mathrm{Ai}'(u_N)^2 - u_N\,\mathrm{Ai}(u_N)^2 + \mathcal{O}(N^{-2/3}). \tag{3.13.12}$$

PROOF. Consider $x \neq y$ in a neighborhood of a right endpoint of a band, b_j for $j \in \mathcal{A}_s$. The analysis at the left endpoint is similar. The local solution near b_j divides a neighborhood of b_j into four regions. Let us consider a limit as x and y approach the real line from the part of the upper half plane lying above the band. In that region (3.2.10) and (3.5.18) imply

$$K_N(x,y) = \frac{\mathrm{e}^{-NV(x)/2}\mathrm{e}^{-NV(y)/2}}{x-y}\begin{pmatrix}0 & 1\end{pmatrix}\mathbf{K}^{-1}\mathbf{R}_N(x)^{-1}\mathbf{R}_N(y)\mathbf{K}\begin{pmatrix}1\\0\end{pmatrix}. \tag{3.13.13}$$

Further unwrapping the RHP using (3.6.1), (3.7.2), and (3.11.1) as well as the identity (3.4.11), we obtain

$$\begin{aligned} K_N(x,y) &= -\frac{1}{2\pi\mathrm{i}(x-y)}\begin{pmatrix}0 & \mathrm{e}^{NG(x)/2}\end{pmatrix} j_+(x)^{-1}\mathbf{U}_N(x)^{-1}\mathbf{X}_N(x)^{-1} \\ &\qquad\qquad \times \mathbf{X}_N(y)\mathbf{U}_N(y)j_+(y)\begin{pmatrix}\mathrm{e}^{NG(y)/2}\\0\end{pmatrix} \\ &= \frac{1}{2\pi\mathrm{i}(x-y)}\begin{pmatrix}\mathrm{e}^{-NG(x)/2} & -\mathrm{e}^{NG(x)/2}\end{pmatrix}\mathbf{U}_N(x)^{-1}\mathbf{X}_N(x)^{-1} \\ &\qquad\qquad \times \mathbf{X}_N(y)\mathbf{U}_N(y)\begin{pmatrix}\mathrm{e}^{NG(y)/2}\\ \mathrm{e}^{-NG(y)/2}\end{pmatrix}. \end{aligned} \tag{3.13.14}$$

Using the formulas (3.10.3) and (3.10.10) for $\mathbf{U}_N$ and noting that

$$\mathbf{E}_N^{b_j}(x)^{-1}\mathbf{X}_N(x)^{-1}\mathbf{X}_N(y)\mathbf{E}_N^{b_j}(y) = I + \mathcal{O}(x-y), \tag{3.13.15}$$

we obtain

$$\begin{aligned} K_N(x,y) &= \frac{1}{x-y}\begin{pmatrix}1 & -1\end{pmatrix}\mathrm{e}^{\mathrm{i}N\pi x\sigma_3}\Phi_{\mathrm{rs}}\big(N^{2/3}\psi_{b_j}(x)\big)^{-1} \\ &\qquad \times \Phi_{\mathrm{rs}}\big(N^{2/3}\psi_{b_j}(y)\big)\mathrm{e}^{-\mathrm{i}N\pi y\sigma_3}\begin{pmatrix}1\\1\end{pmatrix} + \mathcal{O}(1), \end{aligned} \tag{3.13.16}$$

where for Φ_{rs} we use the formula (3.10.5) for z in the second quadrant. Multiplying out (3.13.16) and using (3.10.6) gives

$$K_N(x,y) = \frac{f_2(x)f_1(y) - f_1(x)f_2(y)}{x-y} + \mathcal{O}(1), \tag{3.13.17}$$

where

$$\begin{aligned} f_1(x) &:= y_1\big(N^{2/3}\psi_{b_j}(x)\big)\mathrm{e}^{\mathrm{i}N\pi x} + y_2\big(N^{2/3}\psi_{b_j}(x)\big)\mathrm{e}^{-\mathrm{i}N\pi x}, \\ f_2(x) &:= y_1'\big(N^{2/3}\psi_{b_j}(x)\big)\mathrm{e}^{\mathrm{i}N\pi x} + y_2'\big(N^{2/3}\psi_{b_j}(x)\big)\mathrm{e}^{-\mathrm{i}N\pi x}. \end{aligned} \tag{3.13.18}$$

Using the linear relation (2.4.100), for any integer k,

$$\begin{aligned} f_1\left(\frac{k}{N}\right) &= (-1)^k \operatorname{Ai}\left(N^{2/3}\psi_{b_j}\left(\frac{k}{N}\right)\right), \\ f_2\left(\frac{k}{N}\right) &= (-1)^k \operatorname{Ai}'\left(N^{2/3}\psi_{b_j}\left(\frac{k}{N}\right)\right). \end{aligned} \tag{3.13.19}$$

If x and y are both on the lattice L_N, then for some $k_N, m_N \in \mathbb{Z}$,

$$x = \frac{k_N}{N}, \quad y = \frac{m_N}{N}, \tag{3.13.20}$$

and so

$$(3.13.21)\quad \begin{aligned} &f_2(x)f_1(y)-f_1(x)f_2(y)\\ &\quad=(-1)^{k_N-m_N+1}\left[\operatorname{Ai}\left(N^{2/3}\psi_{b_j}\left(\frac{k_N}{N}\right)\right)\operatorname{Ai}'\left(N^{2/3}\psi_{b_j}\left(\frac{m_N}{N}\right)\right)\right.\\ &\qquad\left.-\operatorname{Ai}'\left(N^{2/3}\psi_{b_j}\left(\frac{k_N}{N}\right)\right)\operatorname{Ai}\left(N^{2/3}\psi_{b_j}\left(\frac{m_N}{N}\right)\right)\right].\end{aligned}$$

In the scaling (3.13.6), the equation (3.13.17) then becomes

$$(3.13.22)\quad K_N(x,y)=\frac{cN^{2/3}(-1)^{k_N-m_N+1}}{u-v}(\operatorname{Ai}(u)\operatorname{Ai}'(v)-\operatorname{Ai}'(u)\operatorname{Ai}(v))+\mathcal{O}(1),$$

where

$$(3.13.23)\quad c=\psi'_{b_j}(b_j)>0.$$

This proves (3.13.8) in the case $u\le 0$ and $v\le 0$. If one or both of u and v are positive, one must consider slightly different formulas for the local solution in the saturated region, but the proof is similar.

For $x=y$, we can use L'Hospital's rule for the expression (3.13.17), obtaining

$$(3.13.24)\quad K_N(x,x)=f_2'(x)f_1(x)-f_1'(x)f_2(x).$$

For $x\in L_N$, we can use the fact that $y''_{1,2}(z)=zy_{1,2}(z)$ and the formulas (3.13.18) and (3.13.24) to obtain

$$(3.13.25)\quad \begin{aligned} K_N(x,x)&=N^{4/3}\psi'_{b_j}(x)\psi_{b_j}(x)f_1(x)^2-N^{2/3}\psi_{b_j}(x)f_2(x)^2\\ &\quad+2\pi\mathrm{i}N\Big(y_1'\big(N^{2/3}\psi_{b_j}(x)\big)y_2\big(N^{2/3}\psi_{b_j}(x)\big)\\ &\qquad-y_1\big(N^{2/3}\psi_{b_j}(x)\big)y_2'\big(N^{2/3}\psi_{b_j}(x)\big)\Big).\end{aligned}$$

From (3.10.5) and (3.10.6), $y_1'(z)y_2(z)-y_1(z)y_2'(z)\equiv 1/(2\pi\mathrm{i})$, and so we in fact have

$$(3.13.26)\quad K_N(x,x)=N+N^{4/3}\psi'_{b_j}(x)\psi_{b_j}(x)f_1(x)^2-N^{2/3}\psi_{b_j}(x)f_2(x)^2.$$

In the scaling (3.13.6), this is

$$(3.13.27)\quad \begin{aligned} K_N(x,x)=N-N^{2/3}c(\operatorname{Ai}'(u_N)^2-u_N\operatorname{Ai}(u_N)^2)+\mathcal{O}(1),\\ c=\psi'_{b_j}(b_j)>0,\end{aligned}$$

which immediately gives (3.13.9) in the case $u_N\le 0$. Once again, the proof is similar for $u_N\ge 0$. □

The oscillating factor $(-1)^{k_N-m_N}$ in (3.13.8) and (3.13.12) is somewhat odd from the physical point of view of the discrete orthogonal polynomial ensemble. Indeed, in the determinantal formula (3.1.7) for the correlation functions, this oscillating factor plays no role due to the symmetry of the matrix $\big(K_N(\lambda_k,\lambda_l)\big)_{k,l=1}^m$. From the point of view of the orthogonal polynomials, the oscillating factor can be explained by the phenomenon of saturation. In the saturated region, for large N there is exactly one zero of the polynomial $P_N(x)$ between adjacent nodes of the lattice L_N. In fact, the zeroes of the orthogonal polynomials are exponentially close to the lattice points, a phenomenon known as *exponential confinement of zeroes*, see [5]. It is clear then that, in formula (3.1.10) for the Christoffel–Darboux kernel, when one of x or y moves to an adjacent lattice point a zero is crossed and the sign of $K_N(x, y)$ changes.

CHAPTER 4

Introduction to the Six-Vertex Model

4.1. Definition of the model

The six-vertex model, or the model of two-dimensional ice, is a statistical mechanical model stated on a square $n \times n$ lattice with arrows on edges. The arrows obey the rule that at every vertex there are two arrows pointing in and two arrows pointing out. This rule is sometimes called the *ice-rule.* There are only six possible configurations of arrows at each vertex, hence the name of the model, see Figure 4.1, and the states of the system are the possible configurations of arrows which obey this rule at every vertex.

The six-vertex model is the prototypical "ice-type" model, meaning that the states of the model can be identified with two-dimensional H_2O crystals. This is achieved by placing an oxygen atom at each vertex of the graph and a hydrogen atom at each edge, and specifying that there is a bond between a hydrogen and an adjacent oxygen if the arrow on that edge points towards the oxygen atom. Other "ice-type" ensembles include the ensemble of alternating sign matrices (ASM's), the three color model, and the eight vertex solid-on-solid model.

The Gibbs measure for this ensemble, which gives the probability of a particular state σ, is defined in the following way. For each vertex type in Figure 4.1, we assign a weight $w_i > 0$, $i = 1, \dots, 6$, and define the partition function as a sum over all possible arrow configurations of the product of the vertex weights,

$$(4.1.1) \qquad Z_n = \sum_{\text{arrow configurations } \sigma} w(\sigma), \quad w(\sigma) = \prod_{x \in V_n} w_{t(x;\sigma)} = \prod_{i=1}^{6} w_i^{N_i(\sigma)},$$

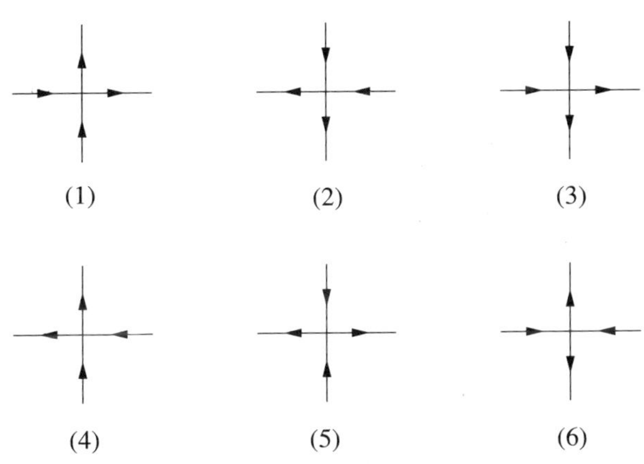

FIGURE 4.1. The six arrow configurations allowed at a vertex.

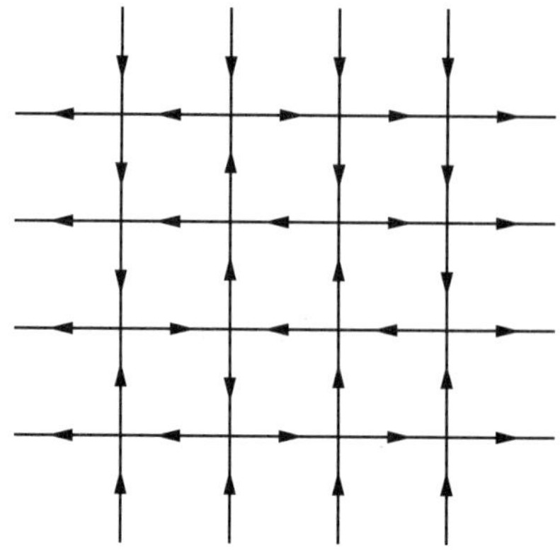

FIGURE 4.2. An example of 4×4 configuration with DWBC.

where V_n is the $n \times n$ set of vertices, $t(x;\sigma) \in \{1, \dots, 6\}$ is the type of configuration σ at vertex x according to Figure 4.1, and $N_i(\sigma)$ is the number of vertices of type i in the configuration σ. The sum is taken over all possible configurations obeying the given boundary condition. The Gibbs measure is then defined as

$$\mu_n(\sigma) = \frac{w(\sigma)}{Z_n}. \tag{4.1.2}$$

The six-vertex model was first introduced by J. C. Slater in [69], and solved exactly in the thermodynamic limit by Lieb, [56–59], and Sutherland [71], for periodic boundary conditions. The six-vertex model differs from other standard models of statistical mechanics, such as the Ising model, in that the limiting free energy is sensitive to boundary conditions. The model has been studied with free boundary conditions, anti-periodic boundary conditions, boundary loop conditions, etc. In this book, we consider *domain wall boundary conditions* (DWBC), in which the arrows on the upper and lower boundaries point into the square, and the ones on the left and right boundaries point out. One possible configuration with DWBC on the 4×4 lattice is shown on Figure 4.2. Our main goal is to obtain the large n asymptotics of the partition function Z_n for the six-vertex model with DWBC.

One consequence of domain wall boundary conditions is that they imply some conservation laws in the system which allow us to reduce the number of parameters.

4.2. Height function and reduction of parameters

The six-vertex model has six parameters: the weights w_i. By using some conservation laws it can be reduced to only two parameters. It is convenient to derive the conservation laws from the *height function*. Let E be the set of edges in the $n \times n$ integer lattice, and consider the dual lattice,

$$V' = \{x = (i + \tfrac{1}{2}, j + \tfrac{1}{2}), 0 \le i, j \le n\}. \tag{4.2.1}$$

Given a configuration σ on E, an integer-valued function $h = h_\sigma$ on V' is called a *height function* of σ if for any two neighboring points $x, y \in V'$, $|x - y| = 1$, we have that

$$h(y) - h(x) = (-1)^s, \tag{4.2.2}$$

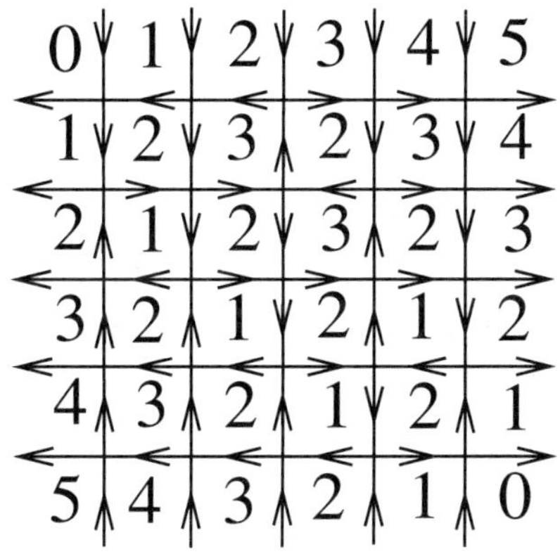

FIGURE 4.3. A 5×5 configuration with a height function.

where $s = 0$ if the arrow σ_e on the edge $e \in E$, crossing the segment $[x, y]$, is oriented in such a way that it points from left to right with respect to the vector $\vec{xy}$, and $s = 1$ if σ_e is oriented from right to left with respect to $\vec{xy}$. The ice-rule ensures that the height function $h = h_\sigma$ exists for any configuration σ. It is unique up to addition of a constant. An example of a configuration and its corresponding height function is given in Figure 4.3.

To derive conservation laws, consider the height function $h = h_\sigma$ on a diagonal sequence of points defined by the formula,

$$x_j = x_0 + (j, j), \quad 0 \le j \le k, \tag{4.2.3}$$

where both x_0 and x_k lie on the boundary B' of the dual lattice V',

$$\begin{aligned} B' = &\{x = (i + \tfrac{1}{2}, \tfrac{1}{2}), 0 \le i \le n\} \cup \{x = (m + \tfrac{1}{2}, j + \tfrac{1}{2}), 0 \le j \le n\} \\ &\cup \{x = (i + \tfrac{1}{2}, n + \tfrac{1}{2}), 0 \le i \le n\} \cup \{x = (\tfrac{1}{2}, j + \tfrac{1}{2}), 0 \le j \le n\}. \end{aligned} \tag{4.2.4}$$

Then it follows from the definition of the height function that

$$h(x_j) - h(x_{j-1}) = \begin{cases} 2 & \text{if } t(x; \sigma) = 3 \\ -2 & \text{if } t(x; \sigma) = 4 \\ 0 & \text{if } t(x; \sigma) = 1, 2, 5, 6, \end{cases} \tag{4.2.5}$$

where

$$x = \frac{x_j + x_{j-1}}{2}. \tag{4.2.6}$$

Hence

$$0 = h(x_k) - h(x_0) = 2N_3(\sigma; L) - 2N_4(\sigma; L), \tag{4.2.7}$$

where $N_i(\sigma; L)$ is the number of vertex states of type i in σ on the line

$$L = \{x = x_0 + (t, t), t \in \mathbb{R}\}. \tag{4.2.8}$$

The line L is parallel to the diagonal $y = x$. By summing up over all possible lines L, we obtain that

$$N_3(\sigma) - N_4(\sigma) = 0, \tag{4.2.9}$$

where $N_i(\sigma)$ is the total number of vertex states of the type i in the configuration σ.

Similarly, by considering lines L parallel to the diagonal $y = -x$, we obtain that

$$N_1(\sigma) - N_2(\sigma) = 0. \tag{4.2.10}$$

Also,

$$N_5(\sigma) - N_6(\sigma) = n, \tag{4.2.11}$$

which follows if we consider lines L parallel to the x-axis, see Section 4.3.1 below. These conservation laws are also obtained in the paper [40] of Ferrari and Spohn as a corollary of a path representation of the six-vertex model.

The conservation laws allow us to reduce the weights $w_1, \dots, w_6$ to 3 parameters. Namely, we have that

$$w_1^{N_1} w_2^{N_2} w_3^{N_3} w_4^{N_4} w_5^{N_5} w_6^{N_6} = C(n) a^{N_1} a^{N_2} b^{N_3} b^{N_4} c^{N_5} c^{N_6}, \tag{4.2.12}$$

where

$$a = \sqrt{w_1 w_2}, \quad b = \sqrt{w_3 w_4}, \quad c = \sqrt{w_5 w_6}, \tag{4.2.13}$$

and the constant

$$C(n) = \left(\frac{w_5}{w_6}\right)^{n/2}. \tag{4.2.14}$$

This implies the relation between the partition functions,

$$Z_n(w_1, w_2, w_3, w_4, w_5, w_6) = C(n) Z_n(a, a, b, b, c, c), \tag{4.2.15}$$

and between the Gibbs measures,

$$\mu_n(\sigma; w_1, w_2, w_3, w_4, w_5, w_6) = \mu_n(\sigma; a, a, b, b, c, c). \tag{4.2.16}$$

Therefore, for fixed boundary conditions like DWBC, the general weights are reduced to the case when

$$w_1 = w_2 = a, \quad w_3 = w_4 = b, \quad w_5 = w_6 = c. \tag{4.2.17}$$

Furthermore,

$$Z_n(a, a, b, b, c, c) = c^{n^2} Z_n\left(\frac{a}{c}, \frac{a}{c}, \frac{b}{c}, \frac{b}{c}, 1, 1\right) \tag{4.2.18}$$

and

$$\mu_n(\sigma; a, a, b, b, c, c) = \mu_n\left(\sigma; \frac{a}{c}, \frac{a}{c}, \frac{b}{c}, \frac{b}{c}, 1, 1\right), \tag{4.2.19}$$

so that a general weight reduces to the two parameters, $a/c, b/c$.

4.3. Mappings of the six-vertex model onto other ensembles

4.3.1. Alternating sign matrices. According to DWBC, in any row the far left arrow points to the left and the far right arrow points to the right. It follows that there must be some vertex in the row for which the arrow to the left of this vertex points to the left while the arrow to the right points to the right. In other words, there is at least one vertex of type (5). Furthermore, between any two type (5) vertices in any row, there must be exactly one type (6) vertex, for which the arrow to the left of this vertex points to the right while the arrow to the right points to the left. The same is true in any column. This is the basis for the mapping of the six-vertex model with DWBC onto the ensemble of alternating sign matrices.

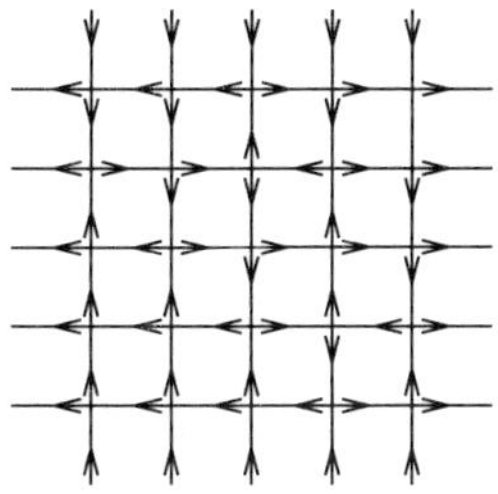

$$\begin{pmatrix} 0 & 0 & 1 & 0 & 0 \\ 1 & 0 & -1 & 1 & 0 \\ 0 & 1 & 0 & 0 & 0 \\ 0 & 0 & 1 & -1 & 1 \\ 0 & 0 & 0 & 1 & 0 \end{pmatrix}$$

FIGURE 4.4. A 5×5 state of the six-vertex model with DWBC and the corresponding 5×5 alternating sign matrix.

Given a state of the $n \times n$ six-vertex model with DWBC, replace all vertices of type (5) with a 1, all vertices of type (6) with a -1, and all other vertices with a 0. The result will be an $n \times n$ matrix A with the following properties:

(1) The entries of A are all 1, (-1), or 0.
(2) The sum along any row or column of A is 1.
(3) Along any row or column of A, the 1's and (-1)'s alternate.

Such a matrix is called an *alternating sign matrix* (ASM), see Figure 4.4.

These matrices are a natural generalization of permutation matrices, which satisfy the above 3 properties, but do not permit entries of (-1). It is clear from the above discussion that there is a bijection between states of the $n \times n$ six-vertex model and ASM's of size $n \times n$. It was conjectured in 1983 by Mills, Robbins, and Rumsey [62] that the number of alternating sign matrices of size $n \times n$, denoted A_n, is given by the formula

$$A_n = \prod_{j=0}^{n-1} \frac{(3j+1)!}{(n+j)!}. \tag{4.3.1}$$

This number should thus also give the partition function of the six-vertex model when the weights are all equal to 1. That is, we have

$$A_n = Z_n(1,1,1), \tag{4.3.2}$$

where $Z_n(a,b,c) \equiv Z_n(a,a,b,b,c,c)$ as in (4.2.18). This conjecture was subsequently proven by Zeilberger [77] and later Kuperberg [55]. In the book [21] of Bressoud the author gives a thorough and engaging exposition of the story of the Alternating Sign Matrix Conjecture, as (4.3.1) came to be known.

In constructing an ensemble of ASM's, we clearly have less parameters than the six-vertex model, since we can no longer distinguish between vertices of weight a and b. Therefore any ensemble of ASM's is equivalent to the six-vertex model with DWBC when $a = b$, and there is therefore only one possible parameter. This parameter is usually described as the *q-enumeration* of ASM's. Introduce the parameter $q > 0$, let A be an alternating sign matrix of size $n \times n$, and let $N(-1,A)$ be the number of (-1)'s in the matrix A. Consider the probability measure on the set of ASM's of size $n \times n$,

$$\mu(A) = \frac{1}{A_n(q)} q^{N(-1,A)}, \quad A_n(q) = \sum_{n \times n \text{ ASM's}} q^{N(-1,A)}. \tag{4.3.3}$$

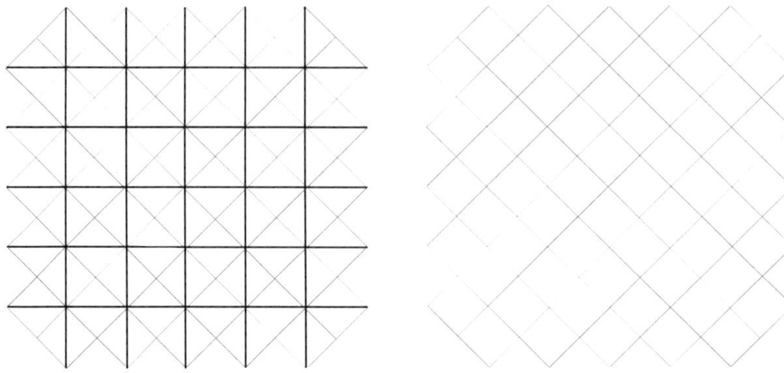

FIGURE 4.5. The square lattice on which the six-vertex model is stated is in dark lines in the left picture. The Aztec diamond is overlaid in light lines on the left, and shown without the original six-vertex lattice on the right.

Notice that $N(-1, A) = N_6(\sigma)$ where σ is the corresponding state of the six-vertex model. It follows that q-enumeration ensemble of ASM's is equivalent to a six-vertex model with DWBC with $a = b = 1$ and

$$q^{N_6(\sigma)} \sim c^{N_5(\sigma)+N_6(\sigma)} = c^{2N_6(\sigma)+n}, \tag{4.3.4}$$

where $\sim$ means that these quantities are equal up to some factor which does not depend on σ (it may depend on q, c, and n). This is clearly satisfied exactly when $q = c^2$. Following the structure of (4.2.19), we can say that the q-enumerative ensemble of ASM's corresponds to the six-vertex model with DWBC and $a/c = b/c = 1/\sqrt{q}$. See Figure 4.8, which is the phase diagram for the six-vertex model with DWBC, and for which the dotted line corresponds to the ASM ensembles. The points corresponding to 1-, 2-, and 3-enumeration are marked specifically.

4.3.2. Domino tilings of the Aztec diamond. Consider the lattice formed by taking the diagonals of an $n \times n$ square lattice, as shown by the light lines in Figure 4.5. This lattice is still a sublattice of $\mathbb{Z}^2$ (rotated 45 degrees), but is no longer a square. It has a particular shape known as an *Aztec diamond.* Notice that in Figure 4.5, every other vertex of the light lattice is on a vertex of the dark lattice.

Now consider domino tilings of the Aztec diamond. A domino tiling of the Aztec diamond can be realized by darkening some of the light edges in Figure 4.5 such that:

(1) Each face of the lattice is enclosed by dark edges.
(2) Each connected component enclosed by dark edges consists of exactly two faces of the lattice.

Clearly then, all of the boundary edges need to be darkened. We would like to find a map between six-vertex configurations with DWBC on the horizontal/vertical (dark) lattice on the left in Figure 4.5, henceforth referred to as the six-vertex lattice, and domino tilings on the diagonal (light) lattice, henceforth referred to as the Aztec diamond. It is not difficult to check that replacing each vertex on the six-vertex lattice with a configuration of dark lines on the Aztec diamond gives such a map. This correspondence is shown in Figure 4.6. Furthermore, one may check

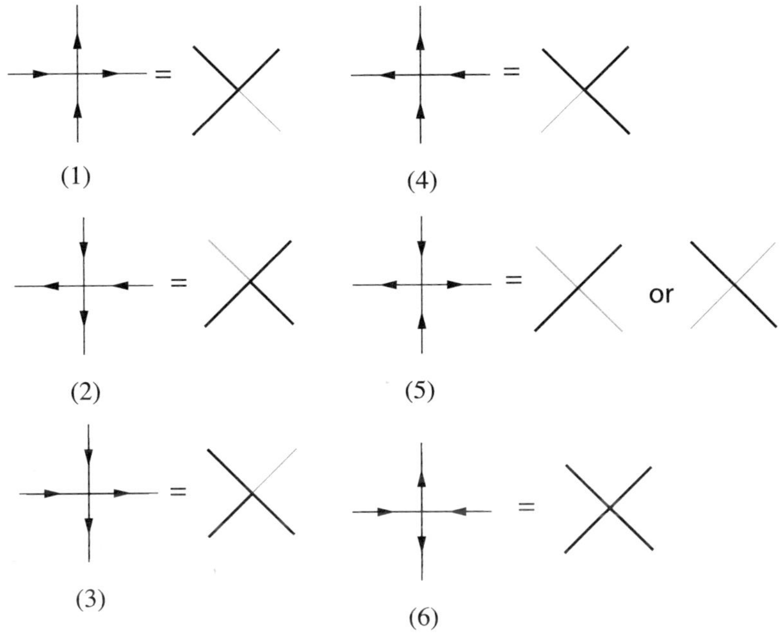

FIGURE 4.6

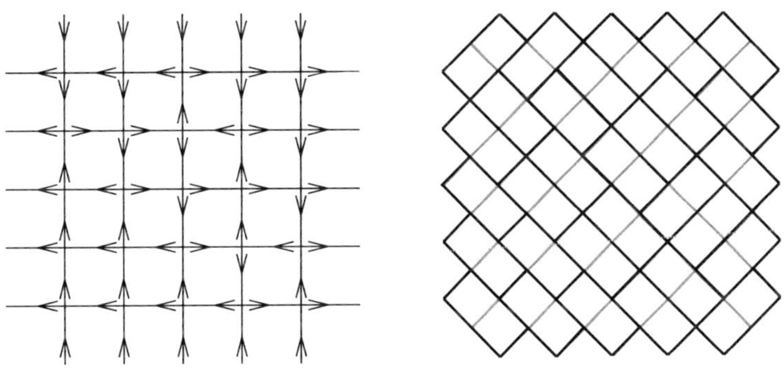

FIGURE 4.7

that the Aztec diamond shape corresponds exactly to the Domain Wall Boundary conditions.

Notice that this map is not a bijection. For each six-vertex configuration σ, there are $2^{N_5(\sigma)}$ corresponding domino tilings of the Aztec diamond, where $N_5(\sigma)$ is the number of vertices of type 5. In Figure 4.7, there is a six-vertex configuration with seven type-5 vertices, and one of the $2^7 = 128$ corresponding domino tilings of the Aztec diamond.

Now let us consider the Boltzmann weights of the domino tilings. There are two species of dominos in Figure 4.7: those which are aligned in the north-east direction, and those aligned in the south-east direction. Let us weight the north-east dominos with the weight $p > 0$, and the south-east dominoes with $q > 0$. More specifically, we consider the probability distribution on domino tilings T

$$\mathbb{P}(T) = \frac{1}{Z} p^{N_{\rm ne}(T)} q^{N_{\rm se}(T)}, \tag{4.3.5}$$

where $N_{\rm ne}(T)$ is the number of north-east dominos in the tiling T, $N_{\rm se}(T)$ is the number of south-east dominos in the tiling T, and Z is the normalizing partition function. Notice that the total number of dominos is

$$N_{\rm ne}(T) + N_{\rm se}(T) = n(n+1). \tag{4.3.6}$$

Notice also that each type-1 vertex and type-2 vertex yields a north-east domino, while each type-3 and type-4 vertex yields a south-east domino. Each type-5 vertex yields either a pair of north-east dominos or a pair of south-east dominos. A type-6 vertex yields no domino. Thus, in relating the Boltzmann weights of the six-vertex model with those of the tilings of the Aztec diamond, we let $p = a$, $q = b$.

For a given tiling T corresponding to a six-vertex configuration σ, let $N_{ne,5}(T)$ be the number of pairs of north-east dominos in T which correspond to a type-5 vertex in the six-vertex model. Then the number of pairs of south-east dominos in T which correspond to a type-5 vertex in the six-vertex model is $N_5(\sigma) - N_{\rm ne,5}(T)$. The total number of north-east dominos is $N_1(\sigma)+N_2(\sigma)+2N_{\rm ne,5}(T)$, and the total number of south-east dominos is $N_3(\sigma) + N_4(\sigma) + 2N_5(\sigma) - 2N_{\rm ne,5}(T)$. We seek to find the weight c in the six-vertex model such that the weight of the six-vertex configuration σ is proportional to the sum of the weights of all 2^{N_5} corresponding domino tilings. In other words, so that

$$\begin{aligned} a^{N_1(\sigma)+N_2(\sigma)} b^{N_3(\sigma)+N_4(\sigma)} c^{N_5(\sigma)+N_6(\sigma)} \\ \sim \sum a^{N_1(\sigma)+N_2(\sigma)+2N_{\rm ne,5}(T)} b^{N_3(\sigma)+N_4(\sigma)+2N_5(\sigma)-2N_{\rm ne,5}(T)}, \end{aligned} \tag{4.3.7}$$

where the sum is over all domino tilings T corresponding to the six-vertex-configuration σ. In light of conservation law (4.2.11), this reduces to

$$c^{2N_5(\sigma)} \sim \sum a^{2N_{\rm ne,5}(T)} b^{2N_5(\sigma)-2N_{\rm ne,5}(T)}. \tag{4.3.8}$$

The sum on the right can be written as

$$\sum_{N_{\rm ne,5}=0}^{N_5} \binom{N_5}{N_{\rm ne,5}} a^{2N_{\rm ne,5}} b^{2N_5(\sigma)-2N_{\rm ne,5}} = (a^2+b^2)^{N_5(\sigma)}. \tag{4.3.9}$$

Thus, we need to have $c^2 = a^2 + b^2$. This condition on the weights of the six-vertex model is called the *free fermion condition.* For these weights, the six-vertex model is equivalent to the Aztec diamond tilings, which is a dimer model. As we shall see in Section 5.4, in this case the partition function of the six-vertex model takes on a particularly simple form. The *free fermion line* in the phase diagram for the six-vertex model is the quarter-circle in Figure 4.8.

4.4. Exact solution of the six-vertex model for a finite n

Introduce the parameter

$$\Delta = \frac{a^2+b^2-c^2}{2ab}. \tag{4.4.1}$$

There are three physical phases in the six-vertex model: the ferroelectric phase, $\Delta > 1$; the antiferroelectric phase, $\Delta < -1$; and the disordered phase, $-1 < \Delta < 1$. Notice that $|a-b| > c$ in the ferroelectric phase region and $c > a + b$ in the

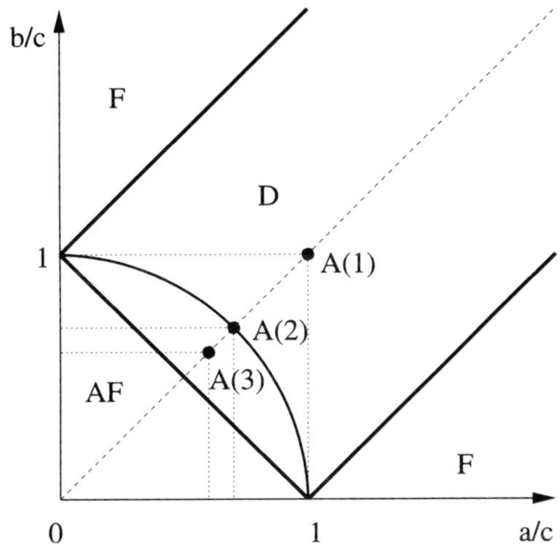

FIGURE 4.8. The phase diagram of the model, where **F**, **AF** and **D** mark ferroelectric, antiferroelectric, and disordered phases, respectively. The circular arc corresponds to the so-called "free fermion" line, when $\Delta = 0$, and the three dots correspond to 1-, 2-, and 3-enumeration of alternating sign matrices.

antiferroelectric phase region, while in the disordered phase region a, b, c satisfy the triangle inequalities. Let us parametrize the weights in the following ways: for the ferroelectric phase,

$$a = \sinh(t - \gamma), \quad b = \sinh(t + \gamma), \quad c = \sinh(2|\gamma|), \qquad 0 < |\gamma| < t; \tag{4.4.2}$$

for the antiferroelectric phase,

$$a = \sinh(\gamma - t), \quad b = \sinh(\gamma + t), \quad c = \sinh(2\gamma), \qquad |t| < \gamma; \tag{4.4.3}$$

and for the disordered phase

$$a = \sin(\gamma - t), \quad b = \sin(\gamma + t), \quad c = \sin(2\gamma), \qquad |t| < \gamma. \tag{4.4.4}$$

The phase diagram of the six-vertex model is shown on Figure 4.8.

The phase diagram and the Bethe Ansatz solution of the six-vertex model for periodic and anti-periodic boundary conditions are thoroughly discussed in the works of Lieb [56–59], Lieb, Wu [60], Sutherland [71], Baxter [7], Batchelor, Baxter, O'Rourke, Yung [6]. See also the work of Wu, Lin [75], in which the Pfaffian solution for the six-vertex model with periodic boundary conditions is obtained on the free fermion line, $\Delta = 0$.

In the paper [50], Korepin derived an important recursion relation for the partition function of the six-vertex model with DWBC. This led to a beautiful determinantal formula of Izergin [46] for the partition function, commonly called the Izergin – Korepin formula. A detailed proof of this formula and its generalizations are given in the paper of Izergin, Coker, and Korepin [47]. When the weights are parameterized according to (4.4.2) – (4.4.4), the Izergin – Korepin formula is

$$Z_n = \frac{(ab)^{n^2}}{\left(\prod_{j=0}^{n-1} j!\right)^2} \tau_n, \tag{4.4.5}$$

where τ_n is the Hankel determinant,

$$\tau_n = \det\left(\frac{\mathrm{d}^{j+k-2}\phi}{\mathrm{d}t^{j+k-2}}\right)_{j,k=1}^{n}, \tag{4.4.6}$$

and

$$\phi(t) = \frac{c}{ab} = \begin{cases} \dfrac{\sinh(2|\gamma|)}{\sinh(t-\gamma)\sinh(t+\gamma)} & \text{in the ferroelectric phase} \\ \dfrac{\sinh(2\gamma)}{\sinh(\gamma-t)\sinh(\gamma+t)} & \text{in the antiferroelectric phase} \\ \dfrac{\sin(2\gamma)}{\sin(\gamma-t)\sin(\gamma+t)} & \text{in the disordered phase.} \end{cases} \tag{4.4.7}$$

An elegant derivation of the Izergin – Korepin formula from the Yang – Baxter equation is given in the papers of Korepin and Zinn-Justin [51] and of Kuperberg [55] (see also the book of Bressoud [21]).

It was noticed by Zinn-Justin that τ_n has the structure of the partition function of a unitary matrix ensemble. Indeed, suppose the function ϕ can be written as a Laplace-type transform of some measure μ on the real line, so that

$$\phi(t) = \int_{\mathbb{R}} \mathrm{e}^{tx}\,\mathrm{d}\mu(x). \tag{4.4.8}$$

Then

$$\phi^{(k)}(t) = x^k \int_{\mathbb{R}} \mathrm{e}^{tx}\,\mathrm{d}\mu(x), \tag{4.4.9}$$

and the determinant τ_n becomes

$$\begin{aligned} \tau_n &= \int\cdots\int_{\mathbb{R}^n} \det\begin{pmatrix} \mathrm{e}^{tx_1} & x_1\mathrm{e}^{tx_1} & \dots & x_1^{n-1}\mathrm{e}^{tx_1} \\ x_2\mathrm{e}^{tx_2} & x_2^2\mathrm{e}^{tx_2} & \dots & x_2^{n}\mathrm{e}^{tx_2} \\ x_3^2\mathrm{e}^{tx_3} & x_3^3\mathrm{e}^{tx_3} & \dots & x_3^{n+1}\mathrm{e}^{tx_3} \\ \vdots & \vdots & \ddots & \vdots \\ x_n^{n-1}\mathrm{e}^{tx_n} & x_n^{n}\mathrm{e}^{tx_n} & \dots & x_n^{2n-2}\mathrm{e}^{tx_n} \end{pmatrix} \prod_{k=1}^{n}\mathrm{d}\mu(x_k) \\ &= \int\cdots\int_{\mathbb{R}^n} \Delta(x)\prod_{k=1}^{n} x_k^{k-1}\prod_{k=1}^{n}\mathrm{e}^{tx_k}\,\mathrm{d}\mu(x_k) \\ &= \int\cdots\int_{\mathbb{R}^n} \Delta(x)\prod_{k=1}^{n} x_k^{k-1}\prod_{k=1}^{n}\mathrm{e}^{tx_k}\,\mathrm{d}\mu(x_k), \end{aligned} \tag{4.4.10}$$

where $\Delta(x) = \prod_{i<j}(x_i - x_j)$ is the Vandermonde determinant. Up to sign, this expression for τ_n is invariant with respect to any permutation of x_k's. Multiplying by their signs and then summing over all permutations, we get

$$n!\tau_n = \int\cdots\int_{\mathbb{R}^n} \Delta(x)\sum_{\pi\in S_n}(-1)^{\pi}\prod_{k=1}^{n}(x_k)^{\pi(k)-1}\prod_{k=1}^{n}\mathrm{e}^{tx_k}\,\mathrm{d}\mu(x_k). \tag{4.4.11}$$

We therefore have

$$\tau_n = \frac{1}{n!}\int\cdots\int_{\mathbb{R}^n} \Delta(x)^2\prod_{k=1}^{n}\mathrm{e}^{tx_k}\,\mathrm{d}\mu(x_k), \tag{4.4.12}$$

which is the eigenvalue partition function for an ensemble of random matrices. If we assume that there exists a system of monic orthogonal polynomials $\{P_k(x)\}_{k=0}^{\infty}$ with respect to the measure $e^{tx}\mu(x)$, so that

$$\int_{\mathbb{R}} P_j(x)P_k(x)e^{tx}\,d\mu(x) = h_k\delta_{jk}, \tag{4.4.13}$$

for some normalizing constants $\{h_k\}_{k=0}^{\infty}$, then it follows from Proposition 1.2.2 that

$$\tau_n = \prod_{k=0}^{n-1} h_k. \tag{4.4.14}$$

From Theorem 1.4.2, we see that τ_n satisfies the Toda equation

$$\tau_n\tau_n'' - \tau_n'^2 = \tau_{n+1}\tau_{n-1}, \quad n \geq 1, \qquad (') = \frac{\partial}{\partial t}, \tag{4.4.15}$$

(see also the work of Sogo [70]). The Toda equation was used by Korepin and Zinn-Justin [51] to derive the free energy of the six-vertex model with DWBC, assuming some Ansatz on the behavior of subdominant terms in the large n asymptotics of the free energy.

It follows that evaluation of the partition function Z_n in the thermodynamic limit reduces to finding a large n asymptotic formula for the normalizing coefficients of a system of orthogonal polynomials. This can generally be done via the Riemann – Hilbert approach described in Chapters 2 and 3 of this book. In Chapters 6–8, we will carry out this asymptotic analysis in the various regions of the phase diagram, but first we give a proof of the Izergin – Korepin formula.

CHAPTER 5

The Izergin–Korepin Formula

In order to prove the Izergin – Korepin formula, we must first consider a slighlty more general six-vertex model, in which weights at vertices depend not only on the type of vertex configuration but also on the location of the vertex in the lattice. Such a model is called the *inhomogeneous six-vertex model.* More specifically, we emunerate the rows of the lattice from top to bottom, and the columns from left to right. To the jth row we assign the parameter λ_j, and to the jth column we assign the parameter μ_j. In general, these parameters may be any complex numbers, although at some point we will make the proper restrictions to them which correspond to positive weights. We now consider the vertex weights which depend on both the type of vertex and the location of the vertex in the lattice, so that for any configuration σ,

$$w_{jk}(\sigma)=\begin{cases} a(\lambda_j,\mu_k) & \text{if vertex } (j,k) \text{ of } \sigma \text{ is of type 1 or 2} \\ b(\lambda_j,\mu_k) & \text{if vertex } (j,k) \text{ of } \sigma \text{ is of type 3 or 4} \\ c(\lambda_j,\mu_k) & \text{if vertex } (j,k) \text{ of } \sigma \text{ is of type 5 or 6,} \end{cases} \tag{5.0.1}$$

where

$$a(\lambda,\mu)=\sinh(\lambda-\mu-\gamma),\quad b(\lambda,\mu)=\sinh(\lambda-\mu+\gamma),\quad c(\lambda,\mu)=\sinh(2\gamma). \tag{5.0.2}$$

The weight of any particular configuration is then given as

$$w(\sigma)=\prod_{j,k=1}^{n} w_{jk}(\sigma), \tag{5.0.3}$$

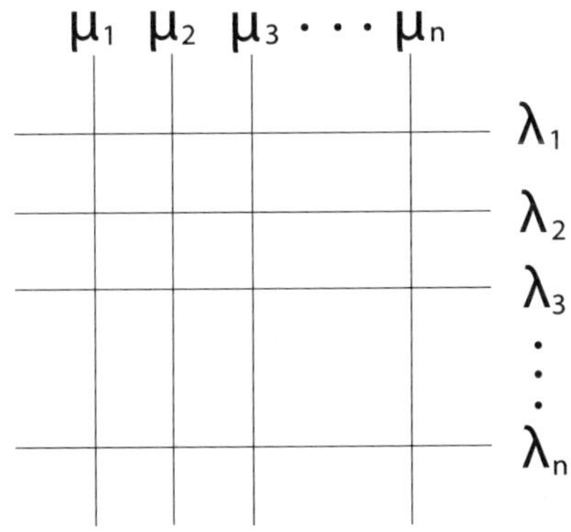

FIGURE 5.1. The inhomogeneous model with spectral parameters λ_j and μ_k.

and the partition function as

$$Z_n(\lambda_1,\dots,\lambda_n,\mu_1,\dots,\mu_n)=\sum_\sigma w(\sigma)=\sum_\sigma\prod_{j,k=1}^n w_{jk}(\sigma),\tag{5.0.4}$$

where the sum is taken over all six-vertex configurations σ satisfying DWBC. Of course this partition function depends on the parameter γ as well, but we supress it from the notation here, as it is assumed to remain fixed throughout this chapter. Also for brevity we will write $\lambda=(\lambda_1,\dots,\lambda_n)$, $\mu=(\mu_1,\dots,\mu_n)$. The Izergin–Korepin determinantal formula is then given in the following theorem.

Theorem 5.0.1 (Izergin). *The partition function for the inhomogeneous six-vertex model with domain wall boundary conditions with weights parametrized by* (5.0.1) *and* (5.0.2) *is*

$$Z_n(\lambda,\mu)=\frac{\prod_{j,k=1}^n[\sinh(\lambda_j-\mu_k+\gamma)\sinh(\lambda_j-\mu_k-\gamma)]}{\prod_{1\le j<k\le n}[\sinh(\lambda_k-\lambda_j)\sinh(\mu_j-\mu_k)]}\tau_n(\lambda,\mu),\tag{5.0.5}$$

where

$$\tau_n=\det\big(\phi(\lambda_j-\mu_k)\big)_{j,k=1}^n,\quad \phi(x)=\frac{\sinh(2\gamma)}{\sinh(x+\gamma)\sinh(x-\gamma)}.\tag{5.0.6}$$

The above formula is valid for any complex values of the parameters λ_j, μ_k, and γ, even those which give nonphysical weights. In the parametrization (5.0.2), the weights are positive if and only if $0<\gamma<\lambda_j-\mu_k$. This restriction corresponds to the region $b>a+c$ in the phase diagram, which is part of the ferroelectric phase (see Figure 4.8). By changing the parametrization (5.0.2) slightly to $c=\sinh(2|\gamma|)$ and considering $0<|\gamma|<\lambda_j-\mu_k$, we also obtain the region $a>b+c$ of the ferroelectric phase. For the Izergin–Korepin formula in the rest of the phase diagram, it is convenient to consider slightly different parametrizations of the weights, which can be obtained by analytically continuing the parametrizations (5.0.2). In the antiferroelectric phase region, the parameterization has the form,

$$w_{jk}(\sigma_x)=\begin{cases}a(\lambda_j,\mu_k)=\sinh(\gamma-\lambda_j+\mu_k)\\ b(\lambda_j,\mu_k)=\sinh(\gamma+\lambda_j-\mu_k)\\ c(\lambda_j,\mu_k)=\sinh(2\gamma),\end{cases}\tag{5.0.7}$$

and formula (5.0.5) becomes

$$Z_n(\lambda,\mu)=\frac{\prod_{j,k=1}^n[\sinh(\gamma-\lambda_j+\mu_k)\sinh(\gamma+\lambda_j-\mu_k)]}{\prod_{1\le j<k\le n}\sinh(\lambda_j-\lambda_k)\prod_{1\le j<k\le n}\sinh(\mu_j-\mu_k)}\times\det\left(\frac{\sinh(2\gamma)}{\sinh(\gamma-\lambda_j+\mu_k)\sinh(\gamma+\lambda_j-\mu_k)}\right)_{j,k=1}^n.\tag{5.0.8}$$

In the disordered phase region, the parameterization has the form,

$$w_{jk}(\sigma_x)=\begin{cases}a(\lambda_j,\mu_k)=\sin(\gamma-\lambda_j+\mu_k)\\ b(\lambda_j,\mu_k)=\sin(\gamma+\lambda_j-\mu_k)\\ c(\lambda_j,\mu_k)=\sin(2\gamma),\end{cases}\tag{5.0.9}$$

and (5.0.5) becomes,

$$Z_n(\lambda,\mu) = \frac{\prod_{j,k=1}^{n}[\sin(\gamma-\lambda_j+\mu_k)\sin(\gamma+\lambda_j-\mu_k)]}{\prod_{1\le j<k\le n}\sin(\lambda_j-\lambda_k)\prod_{1\le j<k\le n}\sin(\mu_j-\mu_k)} \times \det\left(\frac{\sin(2\gamma)}{\sin(\gamma-\lambda_j+\mu_k)\sin(\gamma+\lambda_j-\mu_k)}\right)_{j,k=1}^{n}. \tag{5.0.10}$$

Notice that, although the denominator of (5.0.5) (as well as (5.0.8) and (5.0.10)) vanishes when two of the spectral parameters coincide ($\lambda_j = \lambda_k$ or $\mu_j = \mu_k$ for $j \neq k$), the determinant τ_n also vanishes in this case since two rows (or columns) will then coincide, and therefore is is possible to take the homogeneous limit of this expression. The proof of Theorem 5.0.1 follows from the following three conditions satisfied by $Z_n(\lambda,\mu)$:

(1) $Z_1(\lambda,\mu) = \sinh(2\gamma)$.

(2) $Z_n(\lambda,\mu)$ satisfies the recursion relation

$$Z_n(\lambda,\mu)\big|_{\lambda_j=\mu_k-\gamma} = \sinh(2\gamma)Z_{n-1}(\hat{\lambda}^j,\hat{\mu}^k)\prod_{i\neq j}\sinh(\lambda_i-\mu_k-\gamma)\prod_{i\neq k}\sinh(\lambda_j-\mu_i-\gamma), \tag{5.0.11}$$

where $\hat{\lambda}^j = (\lambda_1,\dots,\lambda_{j-1},\lambda_{j+1},\dots,\lambda_n)$, and $\hat{\mu}^k = (\mu_1,\dots,\mu_{k-1},\mu_{k+1},\dots,\mu_n)$.

(3) For any (j,k), we have

$$Z_n(\lambda,\mu) = \mathrm{e}^{-(n-1)\lambda_j}p_{n-1}(\mathrm{e}^{2\lambda_j}) = \mathrm{e}^{-(n-1)\mu_k}q_{n-1}(\mathrm{e}^{2\mu_k}),$$

where p_{n-1} and q_{n-1} are polynomials of degree exactly $n-1$.

We claim that these conditions determine $Z_n(\lambda,\mu)$ uniquely, which is easily seen by induction on n. Indeed, assume that Z_{n-1} is uniquely determined by the above conditions. Then by condition (2), for any fixed j, we know the value $Z_n(\lambda_j)$ at the n points $\lambda_j = \mu_k + \gamma$, $k = 1,2,\dots,n$. Since by condition (3), $Z_n(\lambda_j) = \mathrm{e}^{-(n-1)\lambda_j}p_{n-1}(\mathrm{e}^{2\lambda_j})$, the recursion relation (2) determines the polynomial p_{n-1} completely. This holds for any j, and thus $Z_n(\lambda,\mu)$ is uniquely determined by conditions (2) and (3) and Z_{n-1}. To complete the induction the condition (1) explicitly gives Z_1.

Thus in order to prove Theorem 5.0.1, we need to show that both sides of equation (5.0.5) satisfy conditions (1)–(3). In the case of condition (1), this is trivial. It is also fairly easy to see that the partition function $Z_n(\lambda,\mu)$ satisfies condition (3). Indeed, consider a particular $n\times n$ six-vertex configuration σ, and fix $1 \le j \le n$. The dependence of $w(\sigma)$ on λ_j comes only from vertices in the jth row. The weight of type-c vertices does not depend on λ_j, while

$$\begin{aligned} a(\lambda_j,\mu_k) &= \sinh(\lambda_j-\mu_k-\gamma) = \mathrm{e}^{-\lambda_j}(A\mathrm{e}^{2\lambda_j}+B),\\ b(\lambda,\mu) &= \sinh(\lambda-\mu+\gamma) = \mathrm{e}^{-\lambda_j}(C\mathrm{e}^{2\lambda_j}+D), \end{aligned} \tag{5.0.12}$$

where A, B, C and D do not depend on λ_j. It follows that, as a function of λ_j,

$$\begin{aligned} w(\sigma) &= \mathrm{e}^{-\lambda_j(N_a+N_b)}(A\mathrm{e}^{2\lambda_j}+B)^{N_a}(C\mathrm{e}^{2\lambda_j}+D)^{N_b}c^{N_c}\\ &= \mathrm{e}^{-\lambda_j(n-1)}\mathrm{e}^{\lambda_j(N_c-1)}(A\mathrm{e}^{2\lambda_j}+B)^{N_a}(C\mathrm{e}^{2\lambda_j}+D)^{N_b}c^{N_c}, \end{aligned} \tag{5.0.13}$$

where N_a, N_b and N_c are the number of vertices of type a, b, and c, respectively, in row j. Clearly then, $N_a + N_b + N_c = n$. Furthermore, the domain wall boundary conditions ensure that there are an odd number of vertices of type c in the jth row of σ. Therefore, $N_c - 1$ is even, and the quantity

$$e^{\lambda_j(N_c-1)}(Ae^{2\lambda_j} + B)^{N_a}(Ce^{2\lambda_j} + D)^{N_b}c^{N_c} \tag{5.0.14}$$

is a polynomial in $e^{2\lambda_j}$ of degree $N_a + N_b + (N_c - 1)/2 \le n - 1$. Thus

$$w(\sigma) = e^{-\lambda_j(n-1)}p_\sigma(e^{2\lambda_j}), \tag{5.0.15}$$

where p_σ is a polynomial of degree at most $n - 1$. The partition function is then similarly given by

$$Z_n(\lambda_j) = \sum_\sigma w(\sigma) = e^{-\lambda_j(n-1)}p(e^{2\lambda_j}), \tag{5.0.16}$$

where p is a polynomial whose degree is the maximum of the degrees of p_σ over all configurations σ. This maximum is achieved when $N_c = 1$, and there clearly exist such configurations, therefore the degree of p is exactly $n - 1$. This holds for any row j, and indeed for any column k, thus the partition function Z_n satisfies condition (3). To see that the right-hand side of (5.0.5) also satisfies condition (3), one need only use expressions (5.0.12) to highlight the type of dependence on λ_j (or μ_k), and expand the determinant as a polynomial in its entries.

We are left to prove that both sides of (5.0.5) satisfy the recursion condition (2). This is indeed the most difficult piece in the proof of the Izergin – Korepin formula, and the proof is based on the Yang – Baxter equations. We present this proof in the next section.

5.1. The Yang – Baxter equation

In order to prove that the partition function Z_n satisfies recursion relation (5.0.11) we must first prove the following fact, which is important enough that we present it as a proposition.

Proposition 5.1.1. *The partition function $Z_n(\lambda, \mu)$ is a symmetric function of the row spectral parameters $\{\lambda_j\}$ and of the column spectral parameters $\{\mu_k\}$.*

The proof of Proposition 5.1.1 is based on the Yang – Baxter equation, which is expressed graphically in Figure 5.2.

This figure shows graphs consisting of three vertices, three internal edges, and six external (boundary) edges. The statement is that the partition function for the six-vertex model defined on the graph on the left is the same as the partition function for the six-vertex model defined on the graph on the right, for any fixed boundary conditions. In order to make sense of this statement, we must define weights on the "twisted" vertices, meaning those with no vertical edge. We do so via Figure 5.3, which gives the weights a^*, b^*, and c^* for each of the six "twisted"

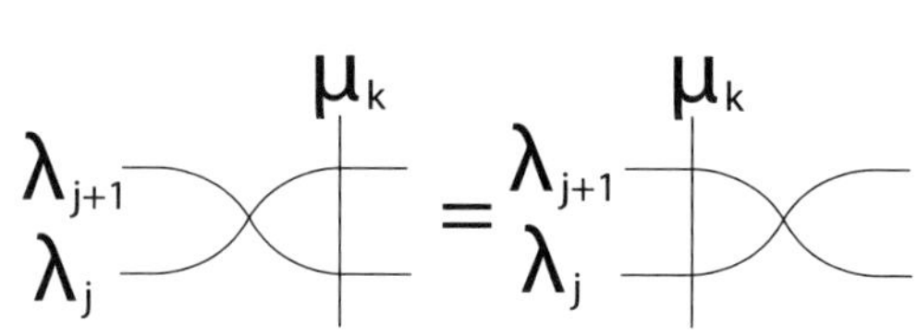

FIGURE 5.2. Graphical representation of the Yang-Baxter equation.

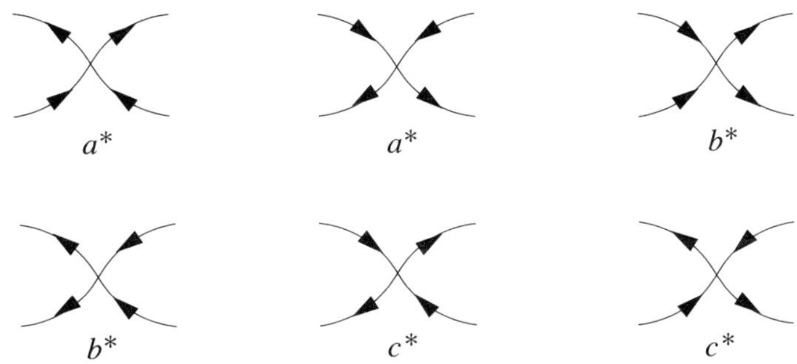

FIGURE 5.3. The weights given to the six "twisted" vertex types.

vertex types. Notice that this figure resembles Figure 4.1, but with each vertex rotated 45° counterclockwise. The weights a^*, b^*, and c^* are functions of the spectral parameters and are given as

$$
\begin{aligned}
a^*(\lambda_j, \lambda_{j+1}) = \sinh(\lambda_j - \lambda_{j+1}), \quad b^*(\lambda_j, \lambda_{j+1}) = \sinh(\lambda_j - \lambda_{j+1} + 2\gamma),\\
c^*(\lambda_j, \lambda_{j+1}) \equiv c = \sinh(2\gamma).
\end{aligned}
\tag{5.1.1}
$$

We are now in a position to state the Yang–Baxter equation as a theorem.

Theorem 5.1.2. *Fix some arrow configuration on the external edges of the graphs given on the left- and right-hand sides of Figure* 5.2. *Consider the six-vertex model on each of these graphs with the same fixed boundary conditions (the statistical mechanical model whose states are realized by drawing arrows on the three internal edges of the graph such that the ice rule is satisfied) with weights given by* (5.0.2) *and* (5.1.1). *Then the partition function of the model on the graph on the left-hand side of Figure* 5.2 *is equal to the partition function of the model on the graph on the right-hand side of Figure* 5.2.

PROOF. We will prove this theorem by simply inspecting the states of the model for each possible boundary condition. Before we dive into this task, we make a couple of observations.

Observation 1. The ice rule demands that the external edges have the same number of incoming as outgoing arrows. Any other boundary configuration will have no allowable states. Since we may choose the location of the three incoming arrows, there are $\binom{6}{3} = 20$ boundary conditions for which the model is nondegenerate.

Observation 2. The weight of any configuration is unchanged when all arrows are reversed, thus the partition function for the model with boundary conditions B is the same as the partition function for the model with boundary conditions obtained by reversing all arrows of B.

From these two observations, it follows that we need to inspect ten different boundary conditions.

Boundary conditions 1–4 *(trivial cases)*. For the first four boundary conditions, the Yang–Baxter equation is trivial. In the first three cases, shown in Figure 5.4, there is only one allowable configuration in each model, and one can easily see that they are comprised of the same vertex types.

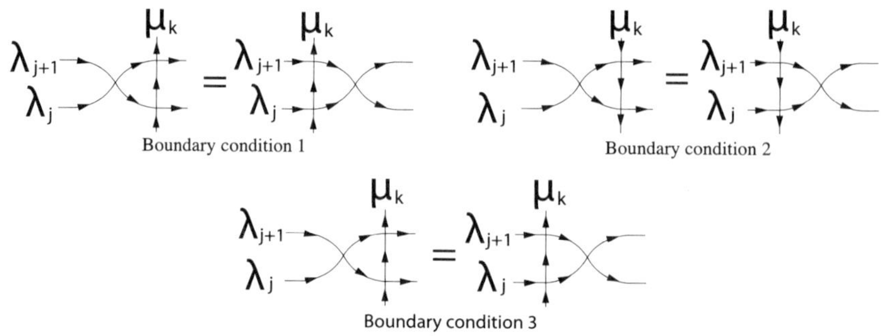

FIGURE 5.4. The first three boundary conditions, for which the Yang–Baxter equation is trivial.

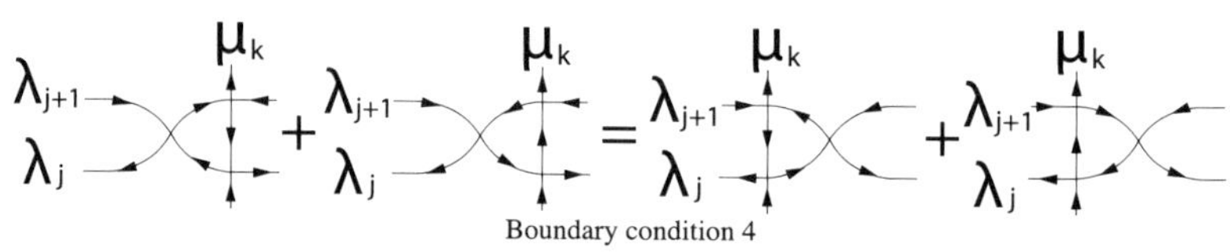

FIGURE 5.5. The fourth boundary condition, for which the Yang–Baxter equation is trivial.

In the fourth case, shown in Figure 5.5, there are two possible configurations for each model, and it is easy to see that, although not all vertex types are exactly the same on both sides of the equation, the vertex weights are the same. That is to say, the weight of the first configuration on the left-hand side of Figure 5.5 is c^3, which is the same as the weight of the first configuration on the right-hand side. Similarly, the weight of the second configuration on the left-hand side of Figure 5.5 is $a^*(\lambda_j, \lambda_{j+1})a(\lambda_{j+1}, \mu_k)b(\lambda_j, \mu_k)$, which is the same as the weight of the second configuration on the right-hand side.

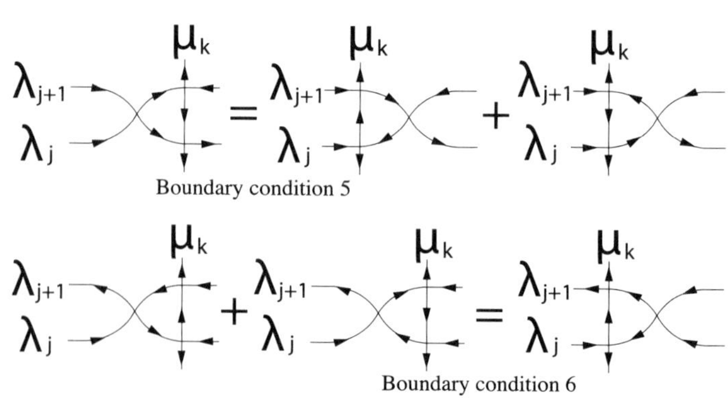

FIGURE 5.6. Boundary conditions 5 and 6, which are 180° rotations of each other.

Boundary conditions 5 *and* 6. This is the first non trivial case. The Yang–Baxter equation for these boundary conditions is shown in Figure 5.6. Notice that these equations have the property that if we rotate one of them by 180° and exchange λ_j with λ_{j+1}, we obtain the other equation. Since none of the weights change under rotation by 180°, and the spectral parameters are arbitrary, it follows that if the Yang–Baxter equation holds for boundary condition 5, then it also holds for boundary condition 6. Let us check that it holds for boundary condition 5.

Clearly, the weight of the left-hand side of the equation for boundary condition 5 in Figure 5.6 is $cb(\lambda_{j+1},\mu_k)b^*(\lambda_j,\lambda_{j+1})$. We can use the simple identity

$$\sinh(x+y)\sinh(x+z)=\tfrac{1}{2}\big(\cosh(2x+y+z)-\cosh(y-z)\big) \tag{5.1.2}$$

to show that the same is true of the sum of weights on the right-hand side. Indeed, we have

$$\begin{aligned}
&Z(\text{right-hand side}) \\
&\quad= a(\lambda_{j+1},\mu_k)a^*(\lambda_j,\lambda_{j+1})c+c^2b(\lambda_j,\mu_k)\\
&\quad= \sinh(2\gamma)\big(\sinh(\lambda_{j+1}-\mu_k-\gamma)\sinh(\lambda_{j+1}-\lambda_j)\\
&\qquad\qquad\qquad\qquad + \sinh(2\gamma)\sinh(\lambda_j-\mu_k-\gamma)\big)\\
&\quad= \frac{\sinh(2\gamma)}{2}\big(\cosh(\lambda_j-\mu_k-\gamma)-\cosh(2\lambda_{j+1}-\lambda_j-\mu_k-\gamma)\\
&\qquad\qquad\qquad\qquad + \cosh(\lambda_j-\mu_k+3\gamma)-\cosh(-\lambda_j+\mu_k+\gamma)\big)\\
&\quad= \frac{\sinh(2\gamma)}{2}\big(\cosh(\lambda_j-\mu_k+3\gamma)-\cosh(2\lambda_{j+1}-\lambda_j-\mu_k-\gamma)\big)\\
&\quad= \sinh(\gamma)\sinh(\lambda_{j+1}-\mu_k)+\gamma)\sinh(\lambda_j-\lambda_{j+1}+2\gamma)\\
&\quad= cb(\lambda_{j+1},\mu_k)b^*(\lambda_j,\lambda_{j+1}).
\end{aligned} \tag{5.1.3}$$

This proves that the Yang–Baxter equation holds for boundary condition 5, and thus for boundary condition 6 as well, because of the symmetry described above.

Boundary conditions 7 *and* 8. The Yang–Baxter equation for these boundary conditions is shown in Figure 5.7. As with the boundary conditions 5 and 6, these equations have the property that if we rotate one of them by 180° and exchange λ_j with λ_{j+1}, we obtain the other equation. Thus once again it follows that if

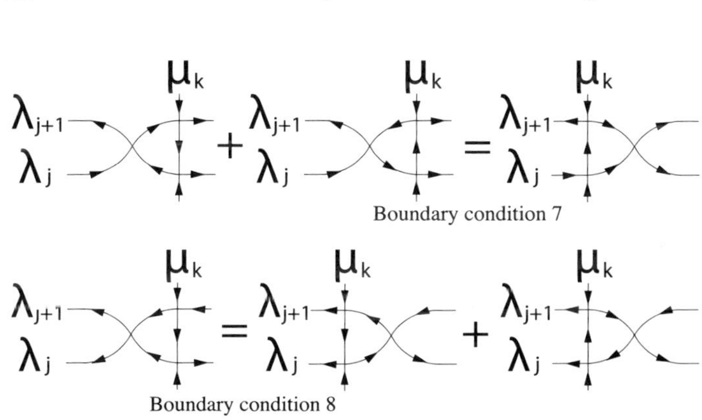

FIGURE 5.7. Boundary conditions 7 and 8, which are 180° rotations of each other.

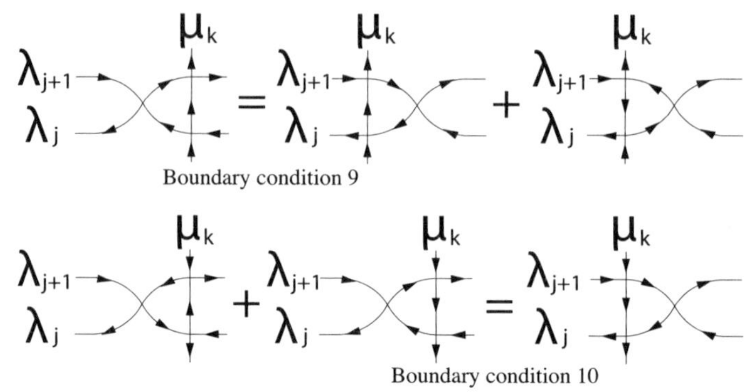

FIGURE 5.8. Boundary conditions 9 and 10, which are 180° rotations of each other.

the Yang–Baxter equation holds for boundary condition 7, then it also holds for boundary condition 8. We leave it to the reader to check that it holds for either boundary condition. The calculation is very similar to the calculation above for boundary condition 5.

Boundary conditions 9 *and* 10. The Yang–Baxter equation for these boundary conditions is shown in Figure 5.8. Once again, these equations are 180° rotations of each other. Thus we need only verify the Yang–Baxter equation for one of these cases. The calculation is very similar to the previous cases, and we once again leave it to the reader.

We have proved that the Yang–Baxter equation holds for ten of the twenty possible boundary conditions, but the remaining ten boundary conditions amount to a reversal of arrows in the above boundary conditions. Thus, according to Observation 2 above, we have proven that the Yang–Baxter equation holds for all possible boundary conditions. □

5.2. A proof of Proposition 5.1.1

We will prove that the partition function is a symmetric function of the row spectral parameters $\{\lambda_j\}$. An identical proof gives the same result for the column spectral parameters $\{\mu_k\}$. The group of permutations on n elements is generated by transposition of adjacent elements, so it is enough to prove that, for $j = 1, \dots, n-1$,

$$Z_n(\lambda_1, \dots, \lambda_n, \mu_1, \dots, \mu_n) = Z_n(\lambda_1, \dots, \lambda_{j-1}, \lambda_{j+1}, \lambda_j, \lambda_{j+2}, \dots, \lambda_n, \mu_1, \dots, \mu_n). \tag{5.2.1}$$

To prove this equality, we consider the six-vertex model on the $n \times n$ lattice with DWBC with one twisted vertex added between the jth and $(j+1)$st rows, see Figure 5.9. The effect of this twisted vertex is to interchange the rows. Because of the Yang–Baxter equation, we can move the twisted vertex along the row so that it sits between any adjacent columns, or on either boundary, without changing the partition function. This is what is meant by the string of equalities in Figure 5.9.

Notice that, because of DWBC, the vertex type of the twisted vertex is uniquely prescribed when the twisted vertex is on either boundary. Furthermore, if we remove the twisted vertex in either case, we have exactly the six-vertex model on an $n \times n$ lattice with DWBC. Therefore, the partition function for the model on

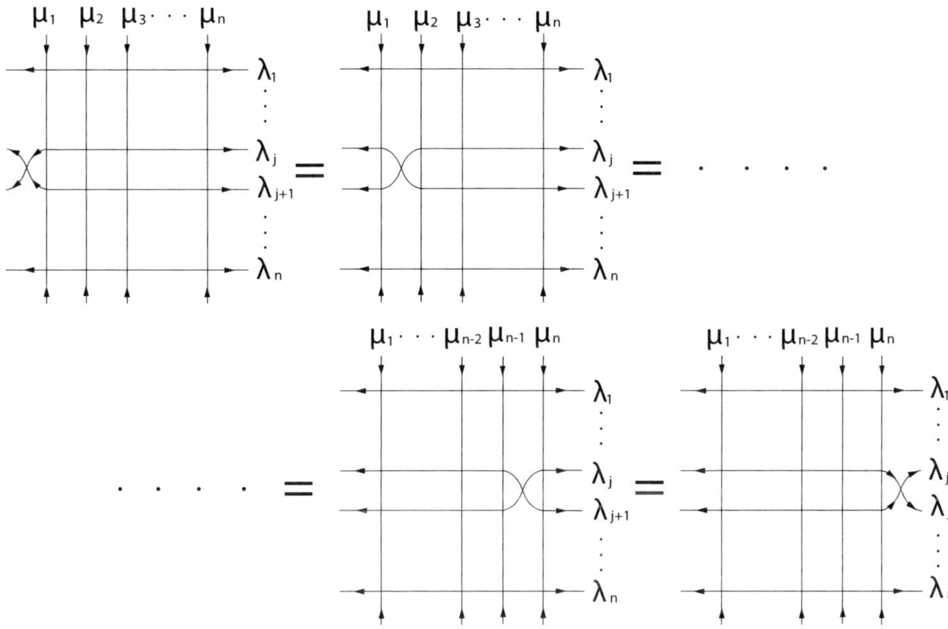

FIGURE 5.9. The Yang–Baxter equations allow us to interchange any adjacent rows of the lattice without changing the partition function.

the far left of Figure 5.9 is given by

$$Z(\text{far left model}) = b^*(\lambda_j, \lambda_{j+1}) Z_n(\lambda_1, \ldots, \lambda_n, \mu_1, \ldots, \mu_n), \tag{5.2.2}$$

whereas the partition function for the model on the far right of Figure 5.9 is given by

$$\begin{aligned} &Z(\text{far right model}) \\ &\quad = b^*(\lambda_j, \lambda_{j+1}) Z_n(\lambda_1, \ldots, \lambda_{j-1}, \lambda_{j+1}, \lambda_j, \lambda_{j+2}, \ldots, \lambda_n, \mu_1, \ldots, \mu_n). \end{aligned} \tag{5.2.3}$$

According to the Yang–Baxter equation these two quantities are equal, and it follows that

$$\begin{aligned} &Z_n(\lambda_1, \ldots, \lambda_n, \mu_1, \ldots, \mu_n) \\ &\quad = Z_n(\lambda_1, \ldots, \lambda_{j-1}, \lambda_{j+1}, \lambda_j, \lambda_{j+2}, \ldots, \lambda_n, \mu_1, \ldots, \mu_n). \end{aligned} \tag{5.2.4}$$

Thus Proposition 5.1.1 is proven. □

5.3. The recursion equation for Z_n

We can now prove that both sides of equation (5.0.5) satisfy recursion relation (5.0.11). From Proposition 5.1.1, and an inspection of the expression on the right-hand side of (5.0.5), we know that both sides are symmetric functions of the variables $\{\lambda_j\}$ and of $\{\mu_k\}$. Thus to prove the recursion (5.0.11), we may assume that $j = k = 1$.

First consider the partition function $Z_n(\lambda, \mu)$ on the left-hand side of (5.0.5). For $\lambda_1 = \mu_1 - \gamma$, we have that $b(\lambda_1, \mu_1) = 0$, thus the vertex in the upper left-hand

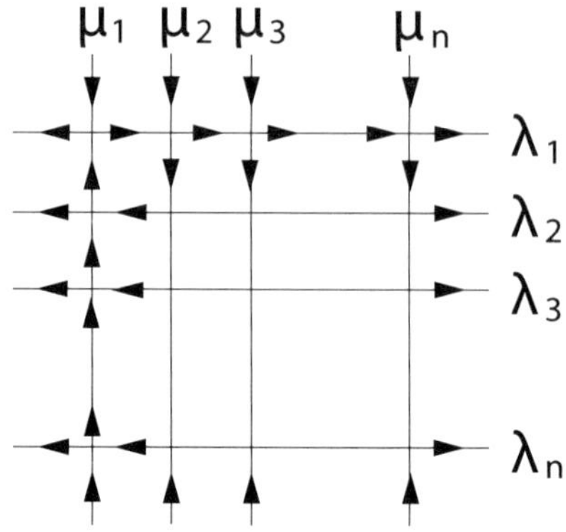

FIGURE 5.10. Graphical proof of recursion (5.0.11).

corner of the lattice cannot be of type 3 or 4. Because of the boundary conditions, the only possibility is that this vertex is of type 5. This uniquely prescribes the first row and column of the lattice, see Figure 5.10. Notice that what remains is an $(n-1)\times(n-1)$ lattice with DWBC. Multiplying by the uniquely prescribed weights given by vertices in the first row and first column proves that the left-hand side of (5.0.5) satisfies the recursion (5.0.11).

Now consider the expression on the right-hand side of equation (5.0.5). Denote this quantity $\widetilde{Z}_n(\lambda,\mu)$. Recall the notation $\hat{\lambda}^1=(\lambda_2,\dots,\lambda_n)$, and $\hat{\mu}^1=(\mu_2,\dots,\mu_n)$. We can multiply the factors $\sinh(\lambda_1-\mu_1+\gamma)$ and $\sinh(\lambda_1-\mu_1-\gamma)$ in the numerator of (5.0.5) into the first column of the determinant, obtaining

$$
\begin{aligned}
(5.3.1)\quad & \widetilde{Z}_n(\lambda,\mu) \\
&= \prod_{k=2}^{n}[\sinh(\lambda_1-\mu_k+\gamma)\sinh(\lambda_1-\mu_k-\gamma)]\prod_{j=2}^{n}[\sinh(\lambda_j-\mu_1-\gamma)\sinh(\lambda_j-\mu_1+\gamma)] \\
&\quad\times\frac{\prod_{j,k=2}^{n}[\sinh(\lambda_j-\mu_k+\gamma)\sinh(\lambda_j-\mu_k-\gamma)]}{\prod_{1\le j<k\le n}[\sinh(\lambda_k-\lambda_j)\sinh(\mu_j-\mu_k)]} \\
&\quad\times\begin{vmatrix}
\sinh(2\gamma) & \phi(\lambda_1-\mu_2) & \dots & \phi(\lambda_1-\mu_n) \\
\dfrac{\sinh(2\gamma)\sinh(\lambda_1-\mu_1-\gamma)\sinh(\lambda_1-\mu_1+\gamma)}{\sinh(\lambda_2-\mu_1-\gamma)\sinh(\lambda_2-\mu_1+\gamma)} & \phi(\lambda_2-\mu_2) & \dots & \phi(\lambda_2-\mu_n) \\
\vdots & \vdots & \ddots & \vdots \\
\dfrac{\sinh(2\gamma)\sinh(\lambda_1-\mu_1-\gamma)\sinh(\lambda_1-\mu_1+\gamma)}{\sinh(\lambda_n-\mu_1-\gamma)\sinh(\lambda_n-\mu_1+\gamma)} & \phi(\lambda_n-\mu_2) & \dots & \phi(\lambda_n-\mu_n)
\end{vmatrix}.
\end{aligned}
$$

It thus follows that

$$
\begin{aligned}
(5.3.2)\quad & \widetilde{Z}_n(\lambda,\mu)|_{\lambda_1=\mu_1-\gamma} \\
&= \prod_{k=2}^{n}[\sinh(\lambda_1-\mu_k+\gamma)\sinh(\lambda_1-\mu_k-\gamma)]\prod_{j=2}^{n}[\sinh(\lambda_j-\mu_1-\gamma)\sinh(\lambda_j-\mu_1-\gamma)] \\
&\qquad\times\frac{\prod_{j,k=2}^{n}[\sinh(\lambda_j-\mu_k+\gamma)\sinh(\lambda_j-\mu_k-\gamma)]}{\prod_{1\le j<k\le n}[\sinh(\lambda_k-\lambda_j)\sinh(\mu_j-\mu_k)]} \\
&\qquad\times\begin{vmatrix}\sinh(2\gamma) & \phi(\lambda_1-\mu_2) & \dots & \phi(\lambda_1-\mu_n)\\ 0 & \phi(\lambda_2-\mu_2) & \dots & \phi(\lambda_2-\mu_n)\\ \vdots & \vdots & \ddots & \vdots\\ 0 & \phi(\lambda_n-\mu_2) & \dots & \phi(\lambda_n-\mu_n)\end{vmatrix}.
\end{aligned}
$$

Expanding the determinant in the first column gives that

$$
\begin{aligned}
(5.3.3)\quad & \widetilde{Z}_n(\lambda,\mu)|_{\lambda_1=\mu_1-\gamma} \\
&= \prod_{k=2}^{n}[\sinh(\lambda_1-\mu_k+\gamma)\sinh(\lambda_1-\mu_k-\gamma)]\prod_{j=2}^{n}[\sinh(\lambda_j-\mu_1+\gamma)\sinh(\lambda_j-\mu_1-\gamma)] \\
&\quad\times\frac{\prod_{j,k=2}^{n}[\sinh(\lambda_j-\mu_k+\gamma)\sinh(\lambda_j-\mu_k-\gamma)]}{\prod_{k=2}^{n}[\sinh(\lambda_k-\lambda_1)\sinh(\mu_1-\mu_k)]\prod_{2\le j<k\le n}[\sinh(\lambda_k-\lambda_j)\sinh(\mu_j-\mu_k)]} \\
&\quad\times\sinh(2\gamma)\tau_{n-1}(\hat{\lambda}^1,\hat{\mu}^1) \\
&= \prod_{k=2}^{n}[\sinh(\lambda_1-\mu_k+\gamma)\sinh(\lambda_1-\mu_k-\gamma)]\prod_{j=2}^{n}[\sinh(\lambda_j-\mu_1+\gamma)\sinh(\lambda_j-\mu_1-\gamma)] \\
&\quad\times\frac{\prod_{j,k=2}^{n}[\sinh(\lambda_j-\mu_k+\gamma)\sinh(\lambda_j-\mu_k-\gamma)]}{\prod_{k=2}^{n}[\sinh(\lambda_k-\mu_1+\gamma)\sinh(\lambda_1-\mu_k+\gamma)]\prod_{2\le j<k\le n}[\sinh(\lambda_k-\lambda_j)\sinh(\mu_j-\mu_k)]} \\
&\quad\times\sinh(2\gamma)\tau_{n-1}(\hat{\lambda}^1,\hat{\mu}^1) \\
&= \prod_{k=2}^{n}\sinh(\lambda_1-\mu_k+\gamma)\prod_{j=2}^{n}\sinh(\lambda_j-\mu_1-\gamma) \\
&\qquad\times\frac{\prod_{j,k=2}^{n}[\sinh(\lambda_j-\mu_k+\gamma)\sinh(\lambda_j-\mu_k-\gamma)]}{\prod_{2\le j<k\le n}[\sinh(\lambda_k-\lambda_j)\sinh(\mu_k-\mu_j)]}\sinh(2\gamma)\tau_{n-1}(\hat{\lambda}^1,\hat{\mu}^1) \\
&= \sinh(2\gamma)\widetilde{Z}_{n-1}(\hat{\lambda}^1,\hat{\mu}^1)\prod_{k=2}^{n}\sinh(\lambda_1-\mu_k+\gamma)\prod_{j=2}^{n}\sinh(\lambda_j-\mu_1-\gamma),
\end{aligned}
$$

which is the recursion (5.0.11) for $j=k=1$. The recursion is thus proven, and thus we have proven Theorem 5.0.1, according to our discussion immediately following the statement of the theorem.

5.4. The inhomogeneous model on the free fermion line

Before we discuss the Izergin – Korepin formula in the homogeneous limit, let us look as the inhomogeneous model in the free fermion case

$$(5.4.1)\qquad a(\lambda,\mu)^2+b(\lambda,\mu)^2=c(\lambda,\mu)^2.$$

This is part of the disordered phase region, and we use the formula (5.0.10) with parametrization (5.0.9). Then equation (5.4.1) is satisfied precisely when $\gamma = \pi/4$. Indeed, in this case

$$\begin{aligned} a(\lambda,\mu)^2 + b(\lambda,\mu)^2 &= \sin(\pi/4 - \lambda + \mu)^2 + \sin(\pi/4 + \lambda - \mu)^2 \\ &= 1 \\ &= \sin(\pi/2)^2 = c(\lambda,\mu)^2. \end{aligned} \tag{5.4.2}$$

Using the identity

$$\sin(x + \pi/4)\sin(x - \pi/4) = -\tfrac{1}{2}\cos(2x), \tag{5.4.3}$$

we can rewrite equation (5.0.10) for $\gamma = \pi/4$ as

$$\begin{aligned} Z_n(\lambda,\mu) &= \frac{\prod_{j,k=1}^n [\sin(\pi/4 - \lambda_j + \mu_k)\sin(\pi/4 + \lambda_j - \mu_k)]}{\prod_{1\le j<k\le n}\sin(\lambda_j - \lambda_k)\prod_{1\le j<k\le n}\sin(\mu_j - \mu_k)} \\ &\quad \times \det\left(\frac{1}{\sin(\pi/4 - \lambda_j + \mu_k)\sin(\pi/4 + \lambda_j - \mu_k)}\right)_{j,k=1}^n \\ &= \frac{2^{-n^2+n}\prod_{j,k=1}^n \cos(2\lambda_j - 2\mu_k)}{\prod_{1\le j<k\le n}\sin(\lambda_j - \lambda_k)\prod_{1\le j<k\le n}\sin(\mu_j - \mu_k)} \\ &\quad \times \det\left(\frac{1}{\cos(2\lambda_j - 2\mu_k)}\right)_{j,k=1}^n . \end{aligned} \tag{5.4.4}$$

To evaluate the last determinant we use the addition formula

$$\cos(x + y) = \cos(x)\cos(y) - \sin(x)\sin(y), \tag{5.4.5}$$

to write

$$\begin{aligned} &\det\left(\frac{1}{\cos(2\lambda_j - 2\mu_k)}\right)_{j,k=1}^n \\ &\quad = \det\left(\frac{1}{\cos(2\lambda_j)\cos(2\mu_k) + \sin(2\lambda_j)\sin(2\mu_k)}\right)_{j,k=1}^n \\ &\quad = \prod_{j=1}^n \frac{1}{\cos(2\lambda_j)} \prod_{k=1}^N \frac{1}{\sin(2\mu_k)} \det\left(\frac{1}{\tan(2\lambda_j) + \cot(2\mu_k)}\right)_{j,k=1}^n . \end{aligned} \tag{5.4.6}$$

Now we use the Cauchy determinantal formula:

$$\det\left(\frac{1}{x_j + y_k}\right)_{j,k=1}^n = \frac{\prod_{j<k}^n (x_j - x_k)\prod_{j<k}^n (y_j - y_k)}{\prod_{j,k=1}^n (x_j + y_k)}. \tag{5.4.7}$$

By applying it to (5.4.6) we obtain that

$$\det\left(\frac{1}{\cos(2\lambda_j-2\mu_k)}\right)_{j,k=1}^n \tag{5.4.8}$$
$$=\prod_{j=1}^n\frac{1}{\cos(2\lambda_j)}\prod_{k=1}^n\frac{1}{\sin(2\mu_k)}$$
$$\times\frac{\prod_{j<k}^n[\tan(2\lambda_j)-\tan(2\lambda_k)]\prod_{j<k}^n[\cot(2\mu_j)-\cot(2\mu_k)]}{\prod_{j,k=1}^n[\tan(2\lambda_j)+\cot(2\mu_k)]}$$
$$=\frac{\prod_{j<k}^n\sin(2\lambda_j-2\lambda_k)\prod_{j<k}^n\sin(2\mu_k-2\mu_j)}{\prod_{j=1}^n\cos(2\lambda_j)\prod_{k=1}^n\sin(2\mu_k)\prod_{j<k}^n[\cos(2\lambda_j)\cos(2\lambda_k)]\prod_{j<k}^n[\sin(2\mu_j)\sin(2\mu_k)]}$$
$$\times\frac{\prod_{j,k=1}^n\cos(2\lambda_j)\sin(2\mu_k)}{\prod_{j,k=1}^n\cos(2\lambda_j-2\mu_k)}$$
$$=\frac{\prod_{k<j}^n\sin(2\lambda_j-2\lambda_k)\prod_{k<j}^n\sin(2\mu_k-2\mu_j)}{\prod_{j,k=1}^n\cos(2\lambda_j-2\mu_k)},$$

so that

$$\det\left(\frac{1}{\cos(2\lambda_j-2\mu_k)}\right)_{j,k=1}^n=\frac{\prod_{k<j}^n\sin(2\lambda_j-2\lambda_k)\prod_{k<j}^n\sin(2\mu_k-2\mu_j)}{\prod_{j,k=1}^n\cos(2\lambda_j-2\mu_k)}. \tag{5.4.9}$$

From (5.4.4) we obtain now that

$$Z_n(\lambda,\mu)=\frac{2^{-n^2+n}\prod_{j,k=1}^n\cos(2\lambda_j-2\mu_k)}{\prod_{1\le j<k\le n}\sin(\lambda_j-\lambda_k)\prod_{1\le j<k\le n}\sin(\mu_j-\mu_k)}\times\frac{\prod_{k<j}^n\sin(2\lambda_j-2\lambda_k)\prod_{k<j}^n\sin(2\mu_k-2\mu_j)}{\prod_{j,k=1}^n\cos(2\lambda_j-2\mu_k)}. \tag{5.4.10}$$

Applying the appropriate double angle formulas, after cancellations it reduces to

$$Z_n(\lambda,\mu)=\prod_{1\le j<k\le n}\cos(\lambda_j-\lambda_k)\prod_{1\le j<k\le n}\cos(\mu_j-\mu_k). \tag{5.4.11}$$

From this expression, it is trivial to take the homogeneous limit. If we take $\lambda_j\equiv\lambda_0$, $\mu_k\equiv\mu_0$, for all $1\le j,k\le n$, then we obtain

$$Z_n(\lambda_j\equiv\lambda_0,\mu_k\equiv\mu_0)=1. \tag{5.4.12}$$

5.5. The homogeneous limit

We would like to consider the homogeneous model, meaning we need to understand the Izergin–Korepin formula (5.0.5) when all of the spectral parameters coincide (the treatment of formulas (5.0.8) and (5.0.10) is similar). Some care must be taken in considering the homogeneous version of this formula, as both the determinant and the denominator of the formula for Z_n vanish when $\lambda_j=\lambda_k$ or $\mu_j=\mu_k$ for $j\neq k$. Because the weights depend only on the difference $\lambda_j-\mu_k$, there is no loss of generality in assuming that $\lambda_j\to t$ for $j=1,2,\dots,n$, and $\mu_k\to 0$ for $k=1,2,\dots,n$. We will take the limit first row by row, and then column by column.

Assume that $\lambda_1 = t$, λ_j is close to t for $j = 2, \dots n$, and $\mu_1 = 0$, μ_k is close to zero for $k = 2, \dots n$. Introduce the notations

$$t_k := t - \mu_k, \quad \tilde{\lambda}_j = \lambda_j - t. \tag{5.5.1}$$

We can expand the entries of the second row of the matrix τ_n as a Taylor series about t_j, so that

$$\begin{aligned} &\tau_n(t, \lambda_2, \dots, \lambda_n, 0, \mu_2, \dots, \mu_n) \\ &= \begin{vmatrix} \phi(t) & \phi(t_2) & \dots & \phi(t_n) \\ \phi(t)+\tilde{\lambda}_2\phi'(t)+\mathcal{O}(\tilde{\lambda}_2^2) & \phi(t_2)+\tilde{\lambda}_2\phi'(t_2)+\mathcal{O}(\tilde{\lambda}_2^2) & \dots & \phi(t_n)+\tilde{\lambda}_2\phi'(t_n)+\mathcal{O}(\tilde{\lambda}_2^2) \\ \phi(\lambda_3) & \phi(\lambda_3 - \mu_2) & \dots & \phi(\lambda_3 - \mu_n) \\ \vdots & \vdots & \ddots & \vdots \\ \phi(\lambda_n) & \phi(\lambda_n - \mu_2) & \dots & \phi(\lambda_n - \mu_n) \end{vmatrix}. \end{aligned} \tag{5.5.2}$$

Subtracting the first row from the second eliminates the constant term in the Taylor series, and we can factor out the common factor $\tilde{\lambda}_2 = \lambda_2 - t$ from the second row, obtaining

$$\begin{aligned} &\tau_n(t, \lambda_2, \dots, \lambda_n, 0, \mu_2, \dots, \mu_n) = (\lambda_2 - t) \\ &\times \begin{vmatrix} \phi(t) & \phi(t_2) & \dots & \phi(t_n) \\ \phi'(t) + \mathcal{O}(\tilde{\lambda}_2^2) & \phi'(t_2) + \mathcal{O}(\tilde{\lambda}_2^2) & \dots & \phi'(t_n) + \mathcal{O}(\tilde{\lambda}_2^2) \\ \phi(\lambda_3) & \phi(\lambda_3 - \mu_2) & \dots & \phi(\lambda_3 - \mu_n) \\ \vdots & \vdots & \ddots & \vdots \\ \phi(\lambda_n) & \phi(\lambda_n - \mu_2) & \dots & \phi(\lambda_n - \mu_n) \end{vmatrix}. \end{aligned} \tag{5.5.3}$$

Thus, using L'Hospital's rule we see that

$$\begin{aligned} &\lim_{\lambda_2 \to t} \frac{\tau_n(t, \lambda_2, \dots, \lambda_n, 0, \mu_2, \dots, \mu_n)}{\sinh(\lambda_2 - t)} \\ &= \begin{vmatrix} \phi(t) & \phi(t_2) & \dots & \phi(t_n) \\ \phi'(t) & \phi'(t_2) & \dots & \phi'(t_n) \\ \phi(\lambda_3) & \phi(\lambda_3 - \mu_2) & \dots & \phi(\lambda_3 - \mu_n) \\ \vdots & \vdots & \ddots & \vdots \\ \phi(\lambda_n) & \phi(\lambda_n - \mu_2) & \dots & \phi(\lambda_n - \mu_n) \end{vmatrix}. \end{aligned} \tag{5.5.4}$$

We proceed row by row. If we expand the entries of third row as a Taylor series about t_j, we can eliminate the first two terms of the series via row operations, and factor out a common factor of $\tilde{\lambda}_3^2/2!$ to obtain the limit

$$\begin{aligned} &\lim_{\lambda_3 \to t} \frac{\tau_n(t, t, \dots, \lambda_n, 0, \mu_2, \dots, \mu_n)}{\sinh^2(\lambda_3 - t)} \\ &= \frac{1}{2!} \begin{vmatrix} \phi(t) & \phi(t_2) & \dots & \phi(t_n) \\ \phi'(t) & \phi'(t_2) & \dots & \phi'(t_n) \\ \phi''(t) & \phi''(t_2) & \dots & \phi''(t_n) \\ \phi(\lambda_4) & \phi(\lambda_4 - \mu_2) & \dots & \phi(\lambda_4 - \mu_n) \\ \vdots & \vdots & \ddots & \vdots \\ \phi(\lambda_n) & \phi(\lambda_n - \mu_2) & \dots & \phi(\lambda_n - \mu_n) \end{vmatrix}. \end{aligned} \tag{5.5.5}$$

Continuing in this fashion for all the rows, we arrive at the formula

$$(5.5.6)\qquad \lim_{\lambda_1,\lambda_2,\dots,\lambda_n\to t} \frac{\tau_n(\lambda_1,\lambda_2,\dots,\lambda_n,0,\mu_2,\dots,\mu_n)}{\prod_{1\le j<k\le n}[\sinh(\lambda_k-\lambda_j)]}$$
$$=\frac{1}{\prod_{j=0}^{n-1} j!}\begin{vmatrix} \phi(t) & \phi(t_2) & \dots & \phi(t_n)\\ \phi'(t) & \phi'(t_2) & \dots & \phi'(t_n)\\ \phi''(t) & \phi''(t_2) & \dots & \phi''(t_n)\\ \vdots & \vdots & \ddots & \vdots\\ \phi^{(n-1)}(t) & \phi^{(n-1)}(t_2) & \dots & \phi^{(n-1)}(t_n)\end{vmatrix}.$$

Similarly, we can expand the entries of the columns as Taylor series and reduce the matrix accordingly. The result is

$$(5.5.7)\qquad \lim_{\substack{\lambda_1,\lambda_2,\dots,\lambda_n\to t\\ \mu_1,\mu_2,\dots,\mu_n\to 0}} \frac{\tau_n(\lambda_1,\lambda_2,\dots,\lambda_n,\mu_1,\mu_2,\dots,\mu_n)}{\prod_{1\le j<k\le n}[\sinh(\lambda_k-\lambda_j)\sinh(\mu_j-\mu_k)]}$$
$$=\frac{1}{\prod_{j=0}^{n-1} (j!)^2}\begin{vmatrix} \phi(t) & \phi'(t) & \dots & \phi^{(n-1)}(t)\\ \phi'(t) & \phi''(t) & \dots & \phi^{(n)}(t)\\ \phi''(t) & \phi'''(t) & \dots & \phi^{(n+1)}(t_n)\\ \vdots & \vdots & \ddots & \vdots\\ \phi^{(n-1)}(t) & \phi^{(n)}(t_2) & \dots & \phi^{(2n-2)}(t_n)\end{vmatrix}.$$

Combined with (5.0.5), this proves the Izergin – Korepin formula for the homogeneous six-vertex model,

$$(5.5.8)\qquad Z_n(t,\dots,t,0,\dots,0)=Z_n\big(a(t,\gamma),b(t,\gamma),c(t,\gamma)\big)$$
$$=\frac{[\sinh(t-\gamma)\sinh(t+\gamma)]^{n^2}}{\prod_{j=0}^{n-1} (j!)^2}\begin{vmatrix} \phi(t) & \phi'(t) & \dots & \phi^{(n-1)}(t)\\ \phi'(t) & \phi''(t) & \dots & \phi^{(n)}(t)\\ \phi''(t) & \phi'''(t) & \dots & \phi^{(n+1)}(t)\\ \vdots & \vdots & \ddots & \vdots\\ \phi^{(n-1)}(t) & \phi^{(n)}(t) & \dots & \phi^{(2n-2)}(t)\end{vmatrix}.$$

Similarly, we can consider the homogeneous limit of the formulas (5.0.8) and (5.0.10). In each case, we obtain

$$(5.5.9)\qquad Z_n(a,b,c)=\frac{(ab)^{n^2}}{\prod_{j=0}^{n-1}(j!)^2}\,\tau_n,$$

where

$$(5.5.10)\qquad \tau_n=\det\big(\phi^{(j+k-2)}(t)\big)_{j,k=1}^{n},\quad \phi(t)=\frac{c}{ab},$$

and the parametrizations of a, b, and c are slightly different in each of the phase regions, given by (4.4.2) – (4.4.4).

CHAPTER 6

Disordered Phase

6.1. Main results

In this chapter we consider the disordered phase region and we follow the paper [9] of Bleher and Fokin. The parameterization of the weights is as in (4.4.4):

$$a=\sin(\gamma-t),\quad b=\sin(\gamma+t),\quad c=\sin(2\gamma),\quad |t|<\gamma<\frac{\pi}{2}. \tag{6.1.1}$$

In order to apply the Riemann–Hilbert method, we first need to express the function

$$\phi(t)=\frac{c}{ab}, \tag{6.1.2}$$

as a Laplace type transform of some measure. We have the following proposition, which appears in the paper [78] of Zinn-Justin.

Proposition 6.1.1. *Let*

$$\phi(t)=\frac{\sin(2\gamma)}{\sin(\gamma+t)\sin(\gamma-t)}, \tag{6.1.3}$$

and

$$m(\lambda)=\frac{\sinh\big((\lambda/2)(\pi-2\gamma)\big)}{\sinh\big((\lambda/2)\pi\big)}, \tag{6.1.4}$$

where $|t|<\gamma<\pi/2$. *Then*

$$\phi(t)=\int_{-\infty}^{\infty}\mathrm{e}^{t\lambda}m(\lambda)\,\mathrm{d}\lambda. \tag{6.1.5}$$

PROOF. By the change of variables,

$$z=\frac{\pi\lambda}{2},\quad \widetilde{\gamma}=\frac{\pi-2\gamma}{\pi},\quad \tilde{t}=\frac{2t}{\pi}, \tag{6.1.6}$$

we obtain that

$$\int_{-\infty}^{\infty}\mathrm{e}^{t\lambda}m(\lambda)\,\mathrm{d}\lambda=\frac{2}{\pi}\int_{-\infty}^{\infty}\frac{\mathrm{e}^{\tilde{t}z}\sinh(\widetilde{\gamma}z)\,\mathrm{d}z}{\sinh z}. \tag{6.1.7}$$

By the Cauchy theorem,

$$\begin{aligned}\frac{2}{\pi}\int_{-\infty}^{\infty}\frac{\mathrm{e}^{\tilde{t}z}\sinh(\widetilde{\gamma}z)\,\mathrm{d}z}{\sinh z}&=\frac{2}{\pi}\int_{-\infty+\pi\mathrm{i}/2}^{\infty+\pi\mathrm{i}/2}\frac{\mathrm{e}^{\tilde{t}z}\sinh(\widetilde{\gamma}z)\,\mathrm{d}z}{\sinh z}\\&=\frac{1}{\pi}\int_{-\infty+\pi\mathrm{i}/2}^{\infty+\pi\mathrm{i}/2}\frac{\mathrm{e}^{\tilde{t}z}(\mathrm{e}^{\widetilde{\gamma}z}-\mathrm{e}^{-\widetilde{\gamma}z})\,\mathrm{d}z}{\sinh z},\end{aligned} \tag{6.1.8}$$

hence

$$\int_{-\infty}^{\infty} \mathrm{e}^{t\lambda} m(\lambda)\,\mathrm{d}\lambda = \phi_1(t) - \phi_2(t), \tag{6.1.9}$$

where

$$\phi_{1,2}(t) = \frac{1}{\pi} \int_{-\infty+\pi\mathrm{i}/2}^{\infty+\pi\mathrm{i}/2} \frac{\mathrm{e}^{(\tilde{t}\pm\widetilde{\gamma})z}\,\mathrm{d}z}{\sinh z}, \tag{6.1.10}$$

By the Cauchy theorem,

$$\begin{aligned}\phi_1(t) &= \frac{1}{\pi} \int_{-\infty+\pi\mathrm{i}/2}^{\infty+\pi\mathrm{i}/2} \frac{\mathrm{e}^{(\tilde{t}+\widetilde{\gamma})z}\,\mathrm{d}z}{\sinh z} \\ &= \frac{1}{\pi} \int_{-\infty+3\pi\mathrm{i}/2}^{\infty+3\pi\mathrm{i}/2} \frac{\mathrm{e}^{(\tilde{t}+\widetilde{\gamma})z}\,\mathrm{d}z}{\sinh z} + 2\mathrm{i} \operatorname*{Res}_{z=\pi\mathrm{i}} \frac{\mathrm{e}^{(\tilde{t}+\widetilde{\gamma})z}}{\sinh z} \\ &= \frac{1}{\pi} \int_{-\infty+3\pi\mathrm{i}/2}^{\infty+3\pi\mathrm{i}/2} \frac{\mathrm{e}^{(\tilde{t}+\widetilde{\gamma})z}\,\mathrm{d}z}{\sinh z} - 2\mathrm{i}\mathrm{e}^{(\tilde{t}+\widetilde{\gamma})\pi\mathrm{i}},\end{aligned} \tag{6.1.11}$$

and by the change of variable $z = z' + \pi\mathrm{i}$,

$$\phi_1(t) = -\frac{\mathrm{e}^{(\tilde{t}+\widetilde{\gamma})\pi\mathrm{i}}}{\pi} \int_{-\infty+\pi\mathrm{i}/2}^{\infty+\pi\mathrm{i}/2} \frac{\mathrm{e}^{(\tilde{t}+\widetilde{\gamma})z'}\,\mathrm{d}z'}{\sinh z'} - 2ie^{(\tilde{t}+\widetilde{\gamma})\pi\mathrm{i}}, \tag{6.1.12}$$

hence

$$\phi_1(t) = -\mathrm{e}^{(\tilde{t}+\widetilde{\gamma})\pi\mathrm{i}} \phi_1(t) - 2\mathrm{i}\mathrm{e}^{(\tilde{t}+\widetilde{\gamma})\pi\mathrm{i}}. \tag{6.1.13}$$

By solving this equation for $\phi_1(t)$, we obtain that

$$\phi_1(t) = -\frac{2\mathrm{i}\mathrm{e}^{(\tilde{t}+\widetilde{\gamma})\pi\mathrm{i}}}{\mathrm{e}^{(\tilde{t}+\widetilde{\gamma})\pi\mathrm{i}} + 1} = -\frac{2\mathrm{i}}{1 + \mathrm{e}^{-(\tilde{t}+\widetilde{\gamma})\pi\mathrm{i}}}. \tag{6.1.14}$$

Similarly,

$$\phi_2(t) = -\frac{2\mathrm{i}}{1 + \mathrm{e}^{-(\tilde{t}-\widetilde{\gamma})\pi\mathrm{i}}}, \tag{6.1.15}$$

hence

$$\begin{aligned}\int_{-\infty}^{\infty} \mathrm{e}^{t\lambda} m(\lambda)\,\mathrm{d}\lambda &= \frac{2\mathrm{i}}{1 + \mathrm{e}^{-(\tilde{t}-\widetilde{\gamma})\pi\mathrm{i}}} - \frac{2\mathrm{i}}{1 + \mathrm{e}^{-(\tilde{t}+\widetilde{\gamma})\pi\mathrm{i}}} \\ &= \frac{2\mathrm{i}(\mathrm{e}^{-(\tilde{t}+\widetilde{\gamma})\pi\mathrm{i}} - \mathrm{e}^{-(\tilde{t}-\widetilde{\gamma})\pi\mathrm{i}})}{1 + \mathrm{e}^{-(\tilde{t}+\widetilde{\gamma})\pi\mathrm{i}} + \mathrm{e}^{-(\tilde{t}-\widetilde{\gamma})\pi\mathrm{i}} + \mathrm{e}^{-2\tilde{t}\pi\mathrm{i}}} \\ &= \frac{4\sin(\pi\widetilde{\gamma})}{\mathrm{e}^{\tilde{t}\pi\mathrm{i}} + \mathrm{e}^{-\widetilde{\gamma}\pi\mathrm{i}} + \mathrm{e}^{\widetilde{\gamma}\pi\mathrm{i}} + \mathrm{e}^{-\tilde{t}\pi\mathrm{i}}} = \frac{2\sin(\pi\widetilde{\gamma})}{\cos(\tilde{t}\pi) + \cos(\widetilde{\gamma}\pi)}.\end{aligned} \tag{6.1.16}$$

Thus, by (6.1.6),

$$\begin{aligned}\int_{-\infty}^{\infty} \mathrm{e}^{t\lambda} m(\lambda)\,\mathrm{d}\lambda &= \frac{2\sin(\pi - 2\gamma)}{\cos(2t) + \cos(\pi - 2\gamma)} = \frac{2\sin(2\gamma)}{\cos(2t) - \cos(2\gamma)} \\ &= \frac{\sin(2\gamma)}{\sin(\gamma + t)\sin(\gamma - t)}.\end{aligned} \tag{6.1.17}$$

□

According to the Proposition 6.1.1 along with equations (4.4.5) and (4.4.14), we find that the partition function is equal to

$$Z_n = (ab)^{n^2} \prod_{k=0}^{n-1} \frac{h_k}{(k!)^2}, \tag{6.1.18}$$

where h_k are normalizing constants of monic orthogonal polynomials $P_k(x)$ on the line with the weight

$$w(x) = \frac{\mathrm{e}^{tx} \sinh\big((\pi - 2\gamma)x/2\big)}{\sinh(\pi x/2)}, \tag{6.1.19}$$

so that

$$\int_{-\infty}^{\infty} P_j(x) P_k(x) w(x) \, \mathrm{d}x = h_k \delta_{jk}. \tag{6.1.20}$$

The main technical result to be proved in this chapter is an asymptotic expansion of the normalizing constants h_n as $n \to \infty$, and is given in the following proposition.

Proposition 6.1.2. *As $n \to \infty$, the normalizing constants h_n satisfy*

$$\frac{h_n}{(n!)^2} = G^{2n+1}\left(1 + \frac{\kappa}{n} + \varepsilon_n + \mathcal{O}(n^{-2})\right), \tag{6.1.21}$$

where

$$\begin{aligned} G &= \frac{\pi}{2\gamma \cos(\pi\zeta/2)}, \qquad \zeta = \frac{t}{\gamma}; \\ \kappa &= \frac{1}{12} - \frac{2\gamma^2}{3\pi(\pi - 2\gamma)}, \end{aligned} \tag{6.1.22}$$

and

$$\varepsilon_n = \left[\cos(n\omega) + \tan\left(\frac{\pi\zeta}{2}\right)\sin(n\omega)\right] \sum_{j:\kappa_j<2} c_j n^{-\kappa_j}. \tag{6.1.23}$$

Here

$$\omega = -\pi(1+\zeta), \tag{6.1.24}$$

and

$$\kappa_j = 1 + \frac{2j}{\pi/(2\gamma) - 1} > 1, \quad j = 1, 2, \dots. \tag{6.1.25}$$

The numbers c_j are equal to

$$c_j = d_j \sin\left(\frac{\pi j}{1 - 2\gamma/\pi}\right), \quad j = 1, 2, \dots, \tag{6.1.26}$$

where $d_j \neq 0$.

From this proposition and (6.1.18), we get the asymptotics of Z_n. Note that the theorem below is a slight improvement on the one given in [9]. The error term in [9] is of the order $\mathcal{O}(n^{-\varepsilon})$ for some $\varepsilon > 0$, whereas we can now improve it to $\mathcal{O}(n^{-1})$. Also, in [9] the constant term C was unknown. In the theorem below, although we are unable to determine C precisely, we are able to determine its dependence on t.

Theorem 6.1.3. *As $n \to \infty$,*

$$Z_n = Cn^{\kappa} F^{n^2}\big(1+\mathcal{O}(n^{-1})\big), \tag{6.1.27}$$

where $C>0$ is a constant,

$$\kappa = \frac{1}{12} - \frac{2\gamma^2}{3\pi(\pi-2\gamma)}, \tag{6.1.28}$$

and

$$F = \frac{\pi\sin(\gamma-t)\sin(\gamma+t)}{2\gamma\cos(\pi\zeta/2)}, \qquad \zeta = \frac{t}{\gamma}. \tag{6.1.29}$$

The constant C is of the form

$$C = \left[\cos\left(\frac{\pi t}{2\gamma}\right)\right]^{\kappa} D(\gamma), \tag{6.1.30}$$

where $D(\gamma)$ may depend on γ, but not on t.

PROOF. From (6.1.18) and (6.1.21), we obtain that

$$Z_n = (ab)^{n^2}\prod_{k=0}^{n-1}\frac{h_k}{(k!)^2} = h_0(ab)^{n^2}\prod_{k=0}^{n-1}G^{2k+1}\prod_{k=1}^{n-1}\left(1+\frac{\kappa}{k}+\varepsilon_k+\mathcal{O}(k^{-2})\right). \tag{6.1.31}$$

Since

$$abG = F,$$

we obtain that

$$Z_n = h_0F^{n^2}\exp\left[\sum_{k=1}^{n-1}\left(\frac{\kappa}{k}+\varepsilon_k+\mathcal{O}(k^{-2})\right)\right]. \tag{6.1.32}$$

By the Euler summation formula,

$$\sum_{k=1}^{n-1}\frac{\kappa}{k} = \kappa\big(\log n + \gamma_{\mathrm{e}} + \mathcal{O}(n^{-1})\big), \tag{6.1.33}$$

where γ_{e} is the Euler constant. Hence we obtain that

$$Z_n = C_0 n^{\kappa}F^{n^2}\exp\left(\sum_{k=1}^{n-1}\varepsilon_k\right)\big(1+\mathcal{O}(n^{-1})\big), \tag{6.1.34}$$

where $C_0>0$ is some constant. We estimate $\sum_{k=1}^{n-1}\varepsilon_k$ by the summation by parts formula,

$$\sum_{k=1}^{n-1}a_kb_k = a_{n-1}b_{n-1} - \sum_{k=1}^{n-2}B_k(a_{k+1}-a_k), \qquad B_k = b_1+\cdots+b_k. \tag{6.1.35}$$

As an example, let us evaluate one term in $\sum_{k=1}^{n-1}\varepsilon_k$,

$$\sum_{k=1}^{n-1}\cos(k\omega)k^{-\kappa_j}. \tag{6.1.36}$$

We take $b_k=\cos(k\omega)$ and $a_k=k^{-\kappa_j}$. Then

$$B_k = \cos\omega+\cdots+\cos(k\omega) = -\frac{1}{2}+\frac{\sin(k+\frac{1}{2})\omega}{2\sin(\omega/2)} \tag{6.1.37}$$

is bounded and

$$a_{k+1} - a_k = \mathcal{O}(k^{-\kappa_j - 1}),$$

hence as $n \to \infty$,

$$\sum_{k=1}^{n-1} \cos(k\omega) k^{-\kappa_j} = C_1 + \mathcal{O}(n^{-\kappa_j}). \tag{6.1.38}$$

A similar formula holds for $\sum_{k=1}^{n-1} \varepsilon_k$,

$$\sum_{k=1}^{n-1} \varepsilon_k = C_2 + \mathcal{O}(n^{-\kappa_j}). \tag{6.1.39}$$

Since $\kappa_j > 1$, we obtain from (6.1.34) that

$$Z_n = C n^{\kappa} F^{n^2} \big(1 + \mathcal{O}(n^{-1})\big), \tag{6.1.40}$$

where $C > 0$. This proves the asymptotics (6.1.27) up to the constant term. A proof of the formula (6.1.30) is given in Section 6.8. □

6.2. Rescaling of the weight

We can evaluate the asymptotics of the normalizing constants h_k as $k \to \infty$ using the Riemann–Hilbert method described in Chapter 2. We start by rescaling the weight (6.1.19) so that it is in a form similar to (2.0.1). Consider the rescaled weight

$$w_n(x) = w\left(\frac{nx}{\gamma}\right) = \frac{\mathrm{e}^{n\zeta x} \sinh(n(\pi/(2\gamma) - 1)x)}{\sinh\big(n\pi x/(2\gamma)\big)}, \tag{6.2.1}$$

and the rescaled monic orthogonal polynomials

$$P_{nk}(x) = \left(\frac{\gamma}{n}\right)^k P_k\left(\frac{nx}{\gamma}\right). \tag{6.2.2}$$

Then

$$\int_{-\infty}^{\infty} P_{nj}(x) P_{nk}(x) w_n(x)\,\mathrm{d}x = h_{nk} \delta_{jk}, \tag{6.2.3}$$

where

$$h_{nk} = \left(\frac{\gamma}{n}\right)^{2k+1} h_k. \tag{6.2.4}$$

Define

$$V_n(x) = -\zeta x - \frac{1}{n} \log \frac{\sinh(n(\pi/(2\gamma) - 1)x)}{\sinh\big(n\pi x/(2\gamma)\big)}, \tag{6.2.5}$$

so that

$$w_n(x) = \mathrm{e}^{-nV_n(x)}. \tag{6.2.6}$$

Observe that as $n \to \infty$,

$$V_n(x) \to V(x) \equiv -\zeta x + |x|. \tag{6.2.7}$$

We will evaluate the equilibrium measure first for V, and then for V_n.

6.3. Equilibrium measure

6.3.1. Equilibrium measure for V. $V(x)$ is obviously a convex function, and

$$V'(x) = -\zeta + \operatorname{sgn}(x). \tag{6.3.1}$$

Since $V(x)$ is convex, the equilibrium measure $\rho(x)\,dx$ is supported by one interval $[\alpha, \beta]$. The resolvent

$$\omega(z) = \int_\alpha^\beta \frac{\rho(x)\,dx}{z-x}, \tag{6.3.2}$$

is analytic on $\mathbb{C}\setminus[\alpha,\beta]$ and it solves the Euler–Lagrange equation,

$$\omega_+(x) + \omega_-(x) = -\zeta + \operatorname{sgn}(x), \quad x \in [\alpha, \beta]. \tag{6.3.3}$$

A direct calculation shows that the function

$$\omega(z) = \frac{1-\zeta}{2} + \frac{2}{\pi i}\log\left[\frac{\sqrt{\beta(z-\alpha)} - i\sqrt{-\alpha(z-\beta)}}{\sqrt{z(\beta-\alpha)}}\right], \tag{6.3.4}$$

where the square roots and the logarithm are taken on the principal sheet, is analytic on $\mathbb{C}\setminus[\alpha,\beta]$, and it solves equation (6.3.3). From the asymptotics,

$$\omega(z) = z^{-1} + \mathcal{O}(z^{-2}), \quad z \to \infty, \tag{6.3.5}$$

one finds that

$$\alpha = -\pi\tan\frac{\pi}{4}(1-\zeta), \quad \beta = \pi\tan\frac{\pi}{4}(1+\zeta), \tag{6.3.6}$$

cf. [78]. Observe that $\alpha < 0 < \beta$, and

$$\beta + \alpha = 2\pi\tan\frac{\pi\zeta}{2}, \quad \beta - \alpha = \frac{2\pi}{\cos(\pi\zeta/2)}, \quad (-\alpha)\beta = \pi^2. \tag{6.3.7}$$

From (6.3.4) we obtain that

$$\omega(x \pm i0) = \frac{-\zeta + \operatorname{sgn}(x)}{2} \pm \frac{2}{\pi i}\log\left[\frac{\sqrt{\beta(x-\alpha)} + \sqrt{-\alpha(\beta-x)}}{\sqrt{|x|(\beta-\alpha)}}\right], \quad \alpha < x < \beta, \tag{6.3.8}$$

so that the density function $\rho(x)$ is equal to

$$\begin{aligned}\rho(x) &= \frac{1}{2\pi i}[\omega(x-i0) - \omega(x+i0)] \\ &= \frac{2}{\pi^2}\log\left[\frac{\sqrt{\beta(x-\alpha)} + \sqrt{-\alpha(\beta-x)}}{\sqrt{|x|(\beta-\alpha)}}\right], \quad \alpha < x < \beta.\end{aligned} \tag{6.3.9}$$

Observe that $\rho(x)$ has a logarithmic singularity at the origin. From (6.3.8),

$$\omega(\alpha) = \frac{-\zeta - 1}{2}, \quad \omega(\beta) = \frac{-\zeta + 1}{2}. \tag{6.3.10}$$

Consider now the g-function,

$$g(z) = \int_\alpha^\beta \rho(x)\log(z-x)\,dx, \quad z \in \mathbb{C}\setminus(-\infty, \beta], \tag{6.3.11}$$

where for $\log z$ the principal branch is taken. It can be evaluated as follows.

Proposition 6.3.1. *The g-function* (6.3.11) *is given by*

$$g(z) = z\omega(z) + 2\log\big(\sqrt{z-\alpha} + \sqrt{z-\beta}\big) - (1 + 2\log 2). \tag{6.3.12}$$

PROOF. We have:

$$\int \omega(z)\,\mathrm{d}z = z\omega(z) - \int z\omega'(z)\,\mathrm{d}z. \tag{6.3.13}$$

From (6.3.4),

(6.3.14)

$$\begin{aligned}\omega'(z) &= \frac{2}{\mathrm{i}\pi}\left[\frac{\sqrt{\beta}/(2\sqrt{z-\alpha}) - \mathrm{i}\sqrt{-\alpha}/(2\sqrt{z-\beta})}{\sqrt{\beta(z-\alpha)} - \mathrm{i}\sqrt{-\alpha(z-\beta)}} - \frac{1}{2z}\right] \\ &= \frac{1}{\mathrm{i}\pi}\left[\frac{\big(\sqrt{\beta}/\sqrt{z-\alpha} - \mathrm{i}\sqrt{-\alpha}/\sqrt{z-\beta}\big)\big(\sqrt{\beta(z-\alpha)} + \mathrm{i}\sqrt{-\alpha(z-\beta)}\big)}{(\beta-\alpha)z} - \frac{1}{z}\right] \\ &= \frac{1}{\pi}\frac{\sqrt{\beta(-\alpha)}}{(\beta-\alpha)z}\left(\sqrt{\frac{z-\beta}{z-\alpha}} - \sqrt{\frac{z-\alpha}{z-\beta}}\right) \\ &= -\frac{\sqrt{\beta(-\alpha)}}{\pi z\sqrt{(z-\alpha)(z-\beta)}},\end{aligned}$$

hence

$$\begin{aligned}\int \omega(z)\,\mathrm{d}z &= z\omega(z) + \frac{\sqrt{\beta(-\alpha)}}{\pi}\int \frac{\mathrm{d}z}{\sqrt{(z-\alpha)(z-\beta)}} \\ &= z\omega(z) + \frac{2\sqrt{\beta(-\alpha)}}{\pi}\log\big(\sqrt{z-\alpha} + \sqrt{z-\beta}\big).\end{aligned} \tag{6.3.15}$$

From (6.3.7), $\sqrt{\beta(-\alpha)} = \pi$, hence

$$g(z) = z\omega(z) + 2\log\big(\sqrt{z-\alpha} + \sqrt{z-\beta}\big) + C. \tag{6.3.16}$$

As $z \to \infty$,

$$\begin{aligned}g(z) &= \log z + \mathcal{O}(z^{-1}) \\ &= z\big[z^{-1} + \mathcal{O}(z^{-2})\big] + 2\big[\log(2\sqrt{z}) + \mathcal{O}(z^{-1})\big] + C \\ &= \log z + 1 + 2\log 2 + C + \mathcal{O}(z^{-1}),\end{aligned} \tag{6.3.17}$$

hence $C = -1 - 2\log 2$, and (6.3.12) follows. □

Let us recall the properties of the g-function:

(1) $g(z)$ is analytic on $\mathbb{C} \setminus (-\infty, \beta]$, and

$$g'(z) = \omega(z). \tag{6.3.18}$$

(2) For any $x \in [\alpha, \beta]$,

$$g_+(x) + g_-(x) - V(x) - l = 0, \quad x \in [\alpha, \beta]. \tag{6.3.19}$$

(3) For any $x \in \mathbb{R} \setminus [\alpha, \beta]$,

$$g_+(x) + g_-(x) - V(x) - l < 0, \quad x \in \mathbb{R} \setminus [\alpha, \beta]. \tag{6.3.20}$$

(4) By (6.3.11),

$$g_+(x)-g_-(x)=\begin{cases}2\pi\mathrm{i}, & x\le\alpha,\\ 2\pi\mathrm{i}\displaystyle\int_x^\beta \rho(s)\,\mathrm{d}s, & \alpha\le x\le\beta,\\ 0, & x\ge\beta.\end{cases}\tag{6.3.21}$$

From (6.3.12) and (6.3.21) we obtain that

$$\begin{aligned}\int_x^\beta \rho(s)\,\mathrm{d}s&=\frac{1}{2\pi\mathrm{i}}[g_+(x)-g_-(x)]\\&=-x\rho(x)+\frac{2}{\pi}\arctan\sqrt{\frac{\beta-x}{x-\alpha}},\quad \alpha<x<\beta.\end{aligned}\tag{6.3.22}$$

In particular,

$$\int_0^\beta \rho(s)\,\mathrm{d}s=\frac{1+\zeta}{2}.\tag{6.3.23}$$

6.3.2. Equilibrium measure for V_n. From (6.2.5) we obtain that

$$V_n'(x)=\frac{\pi}{2\gamma}\coth\frac{\pi nx}{2\gamma}-\frac{\pi-2\gamma}{2\gamma}\coth\frac{(\pi-2\gamma)nx}{2\gamma}-\zeta.\tag{6.3.24}$$

The function $V_n'(x)$ is increasing, hence $V_n(x)$ is convex. Its equilibrium measure, $\rho_n(x)\,\mathrm{d}x$, is supported by one interval $[\alpha_n,\beta_n]$. As $n\to\infty$, the equilibrium measure for V_n converges to the one for V. Our goal is to obtain some asymptotic formulas for ρ_n, as $n\to\infty$

It is convenient to write (6.3.24) in the form

$$V_n'(x)=\operatorname{sgn}x-\zeta+f(nx)=V'(x)+f(nx),\tag{6.3.25}$$

where

$$f(x)=\frac{\pi}{2\gamma}\coth\frac{\pi x}{2\gamma}-\frac{\pi-2\gamma}{2\gamma}\coth\frac{(\pi-2\gamma)x}{2\gamma}-\operatorname{sgn}x.\tag{6.3.26}$$

The function $f(x)$ possesses the following properties:

(1) $f(x)$ satisfies the estimate,

$$|f(x)|\le C_0e^{-C|x|},\tag{6.3.27}$$

with some $C_0, C>0$.

(2) $f(x)$ is an odd function.

(3) The function $f(x)+\operatorname{sgn}x$ is real analytic.

We will also be interested in the function

$$k(x):=\int_{-\infty}^{\infty}\frac{f(u)\,\mathrm{d}u}{x-u},\tag{6.3.28}$$

which subsequently has the following properties.

(1) The function $k(u)$ is even and is of the form

$$k(x)=-2\log|x|+k_0(x),\tag{6.3.29}$$

where $k_0(x)$ is real analytic.

(2) As $x \to \infty$,

$$k(x) = \frac{C}{x^2} + \mathcal{O}(x^{-4}), \tag{6.3.30}$$

where

$$C = \int_{-\infty}^{\infty} x f(x)\, dx = -\frac{2\pi\gamma^2}{3(\pi - 2\gamma)}. \tag{6.3.31}$$

Consider the resolvent,

$$\omega_n(z) = \int_{\alpha_n}^{\beta_n} \frac{dx\, \rho_n(x)}{z - x}, \quad z \in \mathbb{C} \setminus [\alpha_n, \beta_n]. \tag{6.3.32}$$

It solves the equation,

$$\omega_{n+}(x) + \omega_{n-}(x) = V_n'(x), \quad \alpha_n < x < \beta_n. \tag{6.3.33}$$

The solution to this equation is

$$\omega_n(z) = \frac{\sqrt{(z-\alpha_n)(z-\beta_n)}}{2\pi} \int_{\alpha_n}^{\beta_n} \frac{V_n'(x)}{(z-x)\sqrt{(x-\alpha_n)(\beta_n - x)}}\, dx, \quad z \in \mathbb{C} \setminus [\alpha_n, \beta_n], \tag{6.3.34}$$

6.3.3. Evaluation of the endpoints. From (6.3.32) we have that

$$\omega_n(z) = \frac{1}{z} + \mathcal{O}(z^{-2}), \quad z \to \infty. \tag{6.3.35}$$

By evaluating the large z asymptotics of the integral on the right in (6.3.34), we obtain the equations,

$$\frac{1}{2\pi} \int_{\alpha_n}^{\beta_n} \frac{V_n'(x)}{\sqrt{(x-\alpha_n)(\beta_n - x)}}\, dx = 0, \tag{6.3.36}$$

and

$$\frac{1}{2\pi} \int_{\alpha_n}^{\beta_n} \frac{xV_n'(x)}{\sqrt{(x-\alpha_n)(\beta_n - x)}}\, dx = 1. \tag{6.3.37}$$

From these two equations we will obtain the following asymptotics of α_n, β_n as $n \to \infty$.

Proposition 6.3.2. *As $n \to \infty$,*

$$\begin{aligned} \alpha_n &= \alpha - n^{-2} \frac{\gamma^2}{3(\pi - 2\gamma)\cos(\pi\zeta/2)} + \mathcal{O}(n^{-3}), \\ \beta_n &= \beta + n^{-2} \frac{\gamma^2}{3(\pi - 2\gamma)\cos(\pi\zeta/2)} + \mathcal{O}(n^{-3}), \end{aligned} \tag{6.3.38}$$

where α, β are given in (6.3.6).

PROOF. Consider the expressions on the left in (6.3.36) and (6.3.37) as functions of α_n, β_n:

$$F_n(\alpha_n, \beta_n) = \frac{1}{2\pi} \int_{\alpha_n}^{\beta_n} \frac{V_n'(x)}{\sqrt{(x-\alpha_n)(\beta_n - x)}}\, dx, \tag{6.3.39}$$

and

$$G_n(\alpha_n,\beta_n)=\frac{1}{2\pi}\int_{\alpha_n}^{\beta_n}\frac{xV_n'(x)}{\sqrt{(x-\alpha_n)(\beta_n-x)}}\,dx. \tag{6.3.40}$$

Then equations (6.3.36), (6.3.37) read

$$F_n(\alpha_n,\beta_n)=0,\qquad G_n(\alpha_n,\beta_n)=1. \tag{6.3.41}$$

We will apply the implicit function theorem to prove the existence of a solution (α_n,β_n) to this system for large n in the $\mathcal{O}(1/n)$-neighborhood of the point (α,β).

Introduce the auxiliary functions,

$$\begin{aligned}F(\alpha_n,\beta_n)&=\frac{1}{2\pi}\int_{\alpha_n}^{\beta_n}\frac{V'(x)}{\sqrt{(x-\alpha_n)(\beta_n-x)}}\,dx\\&=\frac{1}{2\pi}\int_{\alpha_n}^{\beta_n}\frac{(-\zeta+\operatorname{sgn}(x))}{\sqrt{(x-\alpha_n)(\beta_n-x)}}\,dx,\\G(\alpha_n,\beta_n)&=\frac{1}{2\pi}\int_{\alpha_n}^{\beta_n}\frac{xV'(x)}{\sqrt{(x-\alpha_n)(\beta_n-x)}}\,dx\\&=\frac{1}{2\pi}\int_{\alpha_n}^{\beta_n}\frac{x(-\zeta+\operatorname{sgn}(x))}{\sqrt{(x-\alpha_n)(\beta_n-x)}}\,dx,\end{aligned} \tag{6.3.42}$$

and

$$\begin{aligned}F_n^0(\alpha_n,\beta_n)&=\frac{1}{2\pi}\int_{\alpha_n}^{\beta_n}\frac{f(nx)}{\sqrt{(x-\alpha_n)(\beta_n-x)}}\,dx,\\G_n^0(\alpha_n,\beta_n)&=\frac{1}{2\pi}\int_{\alpha_n}^{\beta_n}\frac{xf(nx)}{\sqrt{(x-\alpha_n)(\beta_n-x)}}\,dx.\end{aligned} \tag{6.3.43}$$

Then, by (6.3.25),

$$\begin{aligned}F_n(\alpha_n,\beta_n)&=F(\alpha_n,\beta_n)+F_n^0(\alpha_n,\beta_n),\\G_n(\alpha_n,\beta_n)&=G(\alpha_n,\beta_n)+G_n^0(\alpha_n,\beta_n).\end{aligned} \tag{6.3.44}$$

Since α and β are the end-points of the support of the equilibrium measure for V, they solve the equations,

$$F(\alpha,\beta)=0,\quad G(\alpha,\beta)=1. \tag{6.3.45}$$

Furthermore, the functions $F(\alpha_n,\beta_n)$, $G(\alpha_n,\beta_n)$ can be calculated explicitly as

$$\begin{aligned}F(\alpha_n,\beta_n)&=-\frac{\zeta}{2}+\frac{1}{\pi}\arcsin\frac{\beta_n+\alpha_n}{\beta_n-\alpha_n},\\G(\alpha_n,\beta_n)&=-\frac{\zeta(\beta_n+\alpha_n)}{4}+\frac{\sqrt{\beta_n(-\alpha_n)}}{\pi}+\frac{\beta_n+\alpha_n}{2\pi}\arcsin\frac{\beta_n+\alpha_n}{\beta_n-\alpha_n}.\end{aligned} \tag{6.3.46}$$

By differentiation, we obtain the partial derivatives of F and G at $\alpha_n=\alpha$, $\beta_n=\beta$ as

$$\begin{aligned}F_{\alpha_n}(\alpha,\beta)&=\frac{1}{2\pi^2}\left(1+\sin\frac{\pi\zeta}{2}\right),&\quad F_{\beta_n}(\alpha,\beta)&=\frac{1}{2\pi^2}\left(1-\sin\frac{\pi\zeta}{2}\right),\\G_{\alpha_n}(\alpha,\beta)&=-\frac{1}{2\pi}\cos\frac{\pi\zeta}{2},&\quad G_{\beta_n}(\alpha,\beta)&=\frac{1}{2\pi}\cos\frac{\pi\zeta}{2},\end{aligned} \tag{6.3.47}$$

and the Jacobian

$$F_{\alpha_n}(\alpha,\beta)G_{\beta_n}(\alpha,\beta)-F_{\beta_n}(\alpha,\beta)G_{\alpha_n}(\alpha,\beta)=\frac{1}{2\pi^3}\cos\frac{\pi\zeta}{2}\neq 0. \tag{6.3.48}$$

Let us evaluate now the functions $F_n^0(\alpha_n,\beta_n)$, $G_n^0(\alpha_n,\beta_n)$. As $x\to 0$,

$$\frac{1}{\sqrt{(x-\alpha_n)(\beta_n-x)}}=\frac{1}{\sqrt{(-\alpha_n)\beta_n}}\left[1-\frac{(\beta_n+\alpha_n)x}{2(-\alpha_n)\beta_n}+\mathcal{O}(x^2)\right]. \tag{6.3.49}$$

Since the function $f(x)$ is exponentially decaying as $x\to\infty$, we obtain that as $n\to\infty$,

$$\begin{aligned} F_n^0(\alpha_n,\beta_n)&=\frac{1}{2\pi}\int_{\alpha_n}^{\beta_n}\frac{f(nx)}{\sqrt{(x-\alpha_n)(\beta_n-x)}}\,\mathrm{d}x\\ &=\frac{1}{2\pi n}\int_{n\alpha_n}^{n\beta_n}\frac{f(x)}{\sqrt{(x/n-\alpha_n)(\beta_n-x/n)}}\,dx\\ &=\frac{1}{2\pi n\sqrt{(-\alpha_n)\beta_n}}\int_{-\infty}^{\infty}f(x)\left[1-\frac{(\beta_n+\alpha_n)x}{2n(-\alpha_n)\beta_n}\right]\mathrm{d}x+\mathcal{O}(n^{-3}). \end{aligned} \tag{6.3.50}$$

Since f is odd,

$$\int_{-\infty}^{\infty}f(x)\,\mathrm{d}x=0, \tag{6.3.51}$$

hence

$$F_n^0(\alpha_n,\beta_n)=-\frac{\beta_n+\alpha_n}{4\pi n^2[(-\alpha_n)\beta_n]^{3/2}}\int_{-\infty}^{\infty}xf(x)\,\mathrm{d}x+\mathcal{O}(n^{-3}). \tag{6.3.52}$$

Similarly,

$$\begin{aligned} G_n^0(\alpha_n,\beta_n)&=\frac{1}{2\pi}\int_{\alpha_n}^{\beta_n}\frac{xf(nx)}{\sqrt{(x-\alpha_n)(\beta_n-x)}}\,\mathrm{d}x\\ &=\frac{1}{2\pi n^2[(-\alpha_n)\beta_n]^{1/2}}\int_{-\infty}^{\infty}xf(x)\,\mathrm{d}x+\mathcal{O}(n^{-3}). \end{aligned} \tag{6.3.53}$$

By (6.3.26),

$$\begin{aligned} &\int_{-\infty}^{\infty}xf(x)\,\mathrm{d}x\\ &=\int_0^{\infty}x\left[\frac{\pi}{\gamma}\coth\frac{\pi x}{2\gamma}-\frac{\pi-2\gamma}{\gamma}\coth\frac{(\pi-2\gamma)x}{\gamma}-2\,\mathrm{sgn}\,x\right]\mathrm{d}x\\ &=2\int_{-\infty}^{\infty}x\left[\frac{\pi}{\gamma}\frac{1}{\mathrm{e}^{x\pi/\gamma}-1}-\left(\frac{\pi}{\gamma}-2\right)\frac{1}{\mathrm{e}^{x(\pi/\gamma-2)}-1}\right]\mathrm{d}x\\ &=2\left(\frac{\gamma}{\pi}-\frac{\gamma}{\pi-2\gamma}\right)\int_{-\infty}^{\infty}\frac{u}{\mathrm{e}^u-1}\,\mathrm{d}u\\ &=2\left(\frac{\gamma}{\pi}-\frac{\gamma}{\pi-2\gamma}\right)\frac{\pi^2}{6}=\frac{2\pi\gamma^2}{3(\pi-2\gamma)}, \end{aligned} \tag{6.3.54}$$

hence

$$\begin{aligned} F_n^0(\alpha_n,\beta_n)&=\frac{n^{-2}(\beta_n+\alpha_n)\gamma^2}{6(\pi-2\gamma)[(-\alpha_n)\beta_n]^{3/2}}+\mathcal{O}(n^{-3}),\\ G_n^0(\alpha_n,\beta_n)&=-\frac{n^{-2}\gamma^2}{3(\pi-2\gamma)[(-\alpha_n)\beta_n]^{1/2}}+\mathcal{O}(n^{-3}). \end{aligned} \tag{6.3.55}$$

In particular, by using formulas (6.3.7), we obtain that

$$(6.3.56)\qquad \begin{aligned} F_n^0(\alpha,\beta) &= \frac{n^{-2}\gamma^2 \tan\frac{\pi\zeta}{2}}{3\pi^2(\pi-2\gamma)} + \mathcal{O}(n^{-3}),\\ G_n^0(\alpha,\beta) &= -\frac{n^{-2}\gamma^2}{3\pi(\pi-2\gamma)} + \mathcal{O}(n^{-3}). \end{aligned}$$

Let us return now to the functions F_n, G_n. From (6.3.46) and (6.3.55) we have that

$$(6.3.57)\qquad \begin{aligned} F_n(\alpha_n,\beta_n) &= -\frac{\zeta}{2} + \frac{1}{\pi}\arcsin\frac{\beta_n+\alpha_n}{\beta_n-\alpha_n}\\ &\quad + \frac{n^{-2}(\beta_n+\alpha_n)\gamma^2}{6(\pi-2\gamma)[(-\alpha_n)\beta_n]^{3/2}} + \mathcal{O}(n^{-3}),\\ G_n(\alpha_n,\beta_n) &= -\frac{\zeta(\beta_n+\alpha_n)}{4} + \frac{\sqrt{\beta_n(-\alpha_n)}}{\pi} + \frac{\beta_n+\alpha_n}{2\pi}\arcsin\frac{\beta_n+\alpha_n}{\beta_n-\alpha_n}\\ &\quad - \frac{n^{-2}\gamma^2}{3(\pi-2\gamma)[(-\alpha_n)\beta_n]^{1/2}} + \mathcal{O}(n^{-3}). \end{aligned}$$

These estimates are valid in a fixed complex neighborhood of the point (α,β). Also, by (6.3.45) and (6.3.56),

$$(6.3.58)\qquad \begin{aligned} F_n(\alpha,\beta) &= \frac{n^{-2}\gamma^2\tan(\pi\zeta/2)}{3\pi^2(\pi-2\gamma)} + \mathcal{O}(n^{-3}),\\ G_n(\alpha,\beta) &= 1 - \frac{n^{-2}\gamma^2}{3\pi(\pi-2\gamma)} + \mathcal{O}(n^{-3}). \end{aligned}$$

Since equations (6.3.57) are valid in a fixed neighborhood of (α,β), we obtain from (6.3.48) that in a fixed smaller neighborhood of (α,β), the Jacobian

$$(6.3.59)\qquad (F_n)_{\alpha_n}(\alpha_n,\beta_n)(G_n)_{\beta_n}(\alpha_n,\beta_n) - (F_n)_{\beta_n}(\alpha_n,\beta_n)(G_n)_{\alpha_n}(\alpha_n,\beta_n)$$

is separated from 0 uniformly in n. By the implicit function theorem, these properties ensure the existence of a unique solution (α_n,β_n) to system of equations (6.3.40) in the $\mathcal{O}(n^{-1})$ neighborhood of (α,β).

In the linear approximation, system (6.3.40) reduces to

$$(6.3.60)\qquad \begin{cases} A(\alpha_n-\alpha) + B(\beta_n-\beta) = -\dfrac{n^{-2}\gamma^2\tan(\pi\zeta/2)}{3\pi^2(\pi-2\gamma)} + \mathcal{O}(n^{-3}),\\ C(\alpha_n-\alpha) + D(\beta_n-\beta) = \dfrac{n^{-2}\gamma^2}{3\pi(\pi-2\gamma)} + \mathcal{O}(n^{-3}). \end{cases}$$

where, according to (6.3.47),

$$(6.3.61)\qquad \begin{aligned} A &= \frac{1}{2\pi^2}\left(1+\sin\frac{\pi\zeta}{2}\right) + \mathcal{O}(n^{-2}), & B &= \frac{1}{2\pi^2}\left(1-\sin\frac{\pi\zeta}{2}\right) + \mathcal{O}(n^{-2}),\\ C &= -\frac{1}{2\pi}\cos\frac{\pi\zeta}{2} + \mathcal{O}(n^{-2}), & D &= \frac{1}{2\pi}\cos\frac{\pi\zeta}{2} + \mathcal{O}(n^{-2}). \end{aligned}$$

By solving this system, we obtain formulas (6.3.38). □

6.3.4. Evaluation of the density. Consider now the density function $\rho_n(x)$, which approaches the density $\rho(x)$ as $n \to \infty$. Notice that the density $\rho(x)$ has a logarithmic singularity at the origin due to the fact that $V(x)$ vanishes there. For finite n, however, $\rho_n(x)$ is smooth, and we are interested in the large n asymptotics of $\rho_n(x)$ near the origin. The Plemelj – Sokhotsky formula gives

$$\rho_n(x) = -\frac{1}{2\pi \mathrm{i}}[\omega_{n+}(x) - \omega_{n-}(x)], \quad x \in [\alpha_n, \beta_n]. \tag{6.3.62}$$

From this and (6.3.34) we obtain, for $x \in [\alpha_n, \beta_n]$,

$$\rho_n(x) = -\frac{\sqrt{(x-\alpha_n)(\beta_n - x)}}{2\pi^2} \,\mathrm{P.V.} \int_{\alpha_n}^{\beta_n} \frac{V_n'(u)\,\mathrm{d}u}{(x-u)\sqrt{(u-\alpha_n)(\beta_n-u)}}, \tag{6.3.63}$$

where P. V. means that we take the Cauchy principal value for the integral. Let us write

$$\rho_n(x) = \rho_n^0(x) + \rho_n^1(x) \tag{6.3.64}$$

where

$$\rho_n^0(x) = -\frac{\sqrt{(x-\alpha_n)(\beta_n - x)}}{2\pi^2} \,\mathrm{P.V.} \int_{\alpha_n}^{\beta_n} \frac{(\operatorname{sgn} u - \zeta)\,\mathrm{d}u}{(x-u)\sqrt{(u-\alpha_n)(\beta_n-u)}}, \tag{6.3.65}$$

$$\rho_n^1(x) = -\frac{\sqrt{(x-\alpha_n)(\beta_n - x)}}{2\pi^2} \,\mathrm{P.V.} \int_{\alpha_n}^{\beta_n} \frac{f(nu)\,\mathrm{d}u}{(x-u)\sqrt{(u-\alpha_n)(\beta_n-u)}}, \tag{6.3.66}$$

where the function f is defined in (6.3.26). Similar to (6.3.9), the function $\rho_0(x)$ can be evaluated explicitly, giving

$$\rho_n^0(x) = \frac{2}{\pi^2}\log\left[\frac{\sqrt{\beta_n(x-\alpha_n)} + \sqrt{-\alpha_n(\beta_n - \alpha_n)}}{\sqrt{|x|(\beta_n-\alpha_n)}}\right]. \tag{6.3.67}$$

The large n behavior of $\rho_n^1(x)$ is described in the following proposition, in terms of the function $k(x)$ defined in (6.3.28).

Proposition 6.3.3. *As $n \to \infty$, ρ_n^1 satisfies*

$$\rho_n^1(x) = -\frac{1}{2\pi^2}k(nx) + \mathcal{O}(n^{-2}), \tag{6.3.68}$$

uniformly for $x \in [\alpha_n, \beta_n]$. Furthermore, the error is uniform in a complex neighborhood of the interval $[\alpha_n, \beta_n]$ in the following sense. If we denote

$$\varepsilon_n(x) := \rho_n^1(x) + \frac{1}{2\pi^2}k(nx), \tag{6.3.69}$$

then there exist complex domains $\{\Omega_r, r>0\}$ such that $[\alpha_n + r, \beta_n - r] \subset \Omega_r$ such that ε_n can be analytically continued on Ω_r, and

$$\sup_{z\in\Omega_r} |\varepsilon_n(z)| = \mathcal{O}(n^{-2}). \tag{6.3.70}$$

PROOF. Fix $0 < r < \min\{-\alpha, \beta\}$. We first consider the case $x \in [\alpha + r, \beta - r]$. From (6.3.28) and (6.3.66) we have

$$\begin{aligned}\rho_n^1(x) + \frac{k(nx)}{2\pi^2} = &-\frac{1}{2\pi^2}\int_{\alpha_n}^{\beta_n}\left[\frac{\sqrt{(x-\alpha_n)(\beta_n-x)}}{\sqrt{(u-\alpha_n)(\beta_n-u)}} - 1\right]\frac{f(nu)\,\mathrm{d}u}{x-u} \\ &+ \frac{1}{2\pi^2}\int_{\mathbb{R}\setminus[\alpha_n,\beta_n]} \frac{f(nu)\,\mathrm{d}u}{x-u}.\end{aligned} \tag{6.3.71}$$

By (6.3.27), the second integral is exponentially small as $n \to \infty$. In the first integral, we can take the integral over the interval $[\alpha + r/2, \beta - r/2]$, changing the integral by an exponentially small term as $n \to \infty$, giving

$$\rho_n^1(x) + \frac{k(nx)}{2\pi^2} = -\frac{1}{2\pi^2}\int_{\alpha+r/2}^{\beta-r/2}\left[\frac{\sqrt{(x-\alpha_n)(\beta_n-x)}}{\sqrt{(u-\alpha_n)(\beta_n-u)}} - 1\right]\frac{f(nu)\,\mathrm{d}u}{x-u} + \mathcal{O}(\mathrm{e}^{-cn}), \tag{6.3.72}$$

for some $c > 0$. Notice that the integrand in (6.3.72) is analytic and uniformly bounded in a fixed complex neighborhood of $(u, x) \in [\alpha + r/2, \beta - r/2] \times [\alpha + r, \beta - r]$. Taking the Taylor series at $u = x$, we obtain that

$$\frac{\sqrt{(x-\alpha_n)(\beta_n-x)}}{\sqrt{(u-\alpha_n)(\beta_n-u)}} - 1 = -\frac{\beta_n + \alpha_n - 2x}{2(x-\alpha_n)(\beta_n-x)}(x-u) + \mathcal{O}((x-u)^2). \tag{6.3.73}$$

Since f is an odd function,

$$\int_{-\infty}^{\infty} f(u)\,\mathrm{d}u = 0. \tag{6.3.74}$$

This implies

$$\int_{\alpha+\frac{r}{2}}^{\beta-\frac{r}{2}}\left[\frac{\sqrt{(x-\alpha_n)(\beta_n-x)}}{\sqrt{(u-\alpha_n)(\beta_n-u)}} - 1\right]\frac{f(nu)\,\mathrm{d}u}{x-u} = \mathcal{O}(n^{-2}). \tag{6.3.75}$$

This proves (6.3.70) in the case $x \in [\alpha + r, \beta - r]$.

Consider now the case $x \in [\alpha_n, \beta_n] \setminus [\alpha + r, \beta - r]$. Suppose $x \in [\beta - r, \beta_n]$. From (6.3.66),

$$\begin{aligned}\rho_n^1(x) = &-\frac{\sqrt{(x-\alpha_n)(\beta_n-x)}}{2\pi^2}\int_{\alpha_n}^{\alpha+2r}\frac{f(nu)\,\mathrm{d}u}{(x-u)\sqrt{(u-\alpha_n)(\beta_n-u)}}\\ &-\frac{\sqrt{(x-\alpha_n)(\beta_n-x)}}{2\pi^2}\int_{\alpha+2r}^{\beta-2r}\frac{f(nu)\,\mathrm{d}u}{(x-u)\sqrt{(u-\alpha_n)(\beta_n-u)}}\\ &-\frac{\sqrt{r_n(x)}}{2\pi^2}\,\mathrm{P.V.}\int_{\beta-2r}^{\beta_n}\frac{f(nu)\,\mathrm{d}u}{(x-u)\sqrt{(u-\alpha_n)(\beta_n-u)}}.\end{aligned} \tag{6.3.76}$$

The first term is exponentially small as $n \to \infty$ (because f is exponentially decaying), and the second one is $\mathcal{O}(n^{-2})$ (because f is odd and the integration is with respect to a smooth kernel). Let us consider the third term. We can rewrite it as

$$\begin{aligned}&-\frac{\sqrt{(x-\alpha_n)(\beta_n-x)}}{2\pi^2}\,\mathrm{P.V.}\int_{\beta-2r}^{\beta_n}\frac{[f(nu)-f(n\beta_n)]\,\mathrm{d}u}{(x-u)\sqrt{(u-\alpha_n)(\beta_n-u)}}\\ &\frac{\sqrt{(x-\alpha_n)(\beta_n-x)}}{2\pi^2}\,\mathrm{P.V.}\int_{\beta-2r}^{\beta_n}\frac{f(n\beta_n)\,\mathrm{d}u}{(x-u)\sqrt{(u-\alpha_n)(\beta_n-u)}}.\end{aligned} \tag{6.3.77}$$

The second term is evaluated explicitly as const. $f(n\beta_n)$, and it is exponentially small as $n \to \infty$. We can represent the first term as a half-sum of contour integrals over two contours, $\Gamma_\pm$, where Γ_+ (Γ_-) goes from $\beta - 2r$ to $x - \delta$, where $\delta = \frac{1}{3}(\beta_n - x)$, then along the upper (respectively, lower) semicircle of radius δ centered at x, and then from $x + \delta$ to β_n. The both integrals are exponentially small as $n \to \infty$, hence the third term in (6.3.76) is exponentially small, and $\rho_n^1(x) = \mathcal{O}(n^{-2})$ when

$x \in [\beta - r, \beta_n]$. From (6.3.30) we obtain that $k(nx) = \mathcal{O}(n^{-2})$ when $x \in [\beta - r, \beta_n]$. This proves (6.3.70) for $x \in [\beta - r, \beta_n]$. Similarly, it holds for $\mu \in [\alpha_n, \alpha + r]$. Proposition 6.3.3 is proved. □

This proposition implies that

$$\rho_n(x) = \rho_n^0(x) - \frac{1}{2\pi^2}k(nx) + \mathcal{O}(n^{-2}), \quad x \in [\alpha_n, \beta_n], \tag{6.3.78}$$

and that this equation may be extended to the complex plane. Indeed for any $0 < r < \frac{1}{2}\min\{-\alpha, \beta\}$, we have

$$\rho_n(x) = \rho(x) - \frac{1}{2\pi^2}k(nu) + \mathcal{O}(n^{-2}), \quad x \in [\alpha_n + r, \beta_n - r], \tag{6.3.79}$$

and

$$\rho_n(x) = \rho_n^0(x) + \mathcal{O}(n^{-2}), \quad x \in [\alpha_n, \alpha_n + r] \cup [\beta_n - r, \beta_n]. \tag{6.3.80}$$

From (6.3.9), (6.3.29), and (6.3.79), we obtain for $x \in [\alpha_n + r, \beta_n - r]$

$$\begin{aligned} \rho_n(x) &= a(x) - \frac{1}{\pi^2}\log(|x|) + \frac{1}{\pi^2}\log(|nx|) - k_0(nx) + \mathcal{O}(n^{-2}) \\ &= \frac{1}{\pi^2}\log n + a(x) - k_0(nx) + \mathcal{O}(n^{-2}), \end{aligned} \tag{6.3.81}$$

where

$$a(x) = \frac{2}{\pi^2}\log\left(\frac{\sqrt{\beta(x-\alpha)} + \sqrt{-\alpha(\beta - x)}}{\sqrt{\beta - \alpha}}\right). \tag{6.3.82}$$

6.3.5. Evaluation of the resolvent. We have

$$\omega_n(z) = \mp\frac{\mathrm{i}\log n}{\pi} + b(z) - \frac{1}{2\pi^2}m_0(nz) + \mathcal{O}(n^{-2}), \quad \pm\operatorname{Im} z > 0, \tag{6.3.83}$$

where

$$b(z) = \frac{1-\zeta}{2} + \frac{2}{\mathrm{i}\pi}\log\left(\frac{\sqrt{\beta(z-\alpha)} - \mathrm{i}\sqrt{-\alpha(z-\beta)}}{\sqrt{\beta-\alpha}}\right), \tag{6.3.84}$$

with a cut on $[\alpha, \beta]$, and

$$m_0(z) = \pm\pi\mathrm{i}\left[\int_{-\infty}^{\infty}\frac{f(x)\,\mathrm{d}x}{z-x} - 2\log z\right], \quad \pm\operatorname{Im} z > 0, \tag{6.3.85}$$

where the principal branch of the logarithm is taken. Notice that the functions $b(z)$ and $m_0(nz)$ have jumps across the interval $[\alpha, \beta]$, and satisfy the mirror symmetry

$$b(\bar{z}) = \overline{b(z)}, \quad m_0(\bar{z}) = \overline{m_0(z)}. \tag{6.3.86}$$

Using the formulas (6.3.7), it is easy to see

$$b_{\pm}(0) = \frac{1-\zeta}{2} \pm \frac{1}{\mathrm{i}\pi}\log\left(2\pi\cos\left(\frac{\pi\zeta}{2}\right)\right). \tag{6.3.87}$$

6.3.6. Evaluation of the Lagrange multiplier. The Lagrange multiplier l_n is equal to

$$l_n = 2g_n(\beta_n) - V_n(\beta_n). \tag{6.3.88}$$

From (6.2.5) we obtain that as $n \to \infty$,

$$V_n(\beta_n) = (1-\zeta)\beta_n + \mathcal{O}(\mathrm{e}^{-\varepsilon n}), \quad \varepsilon > 0. \tag{6.3.89}$$

By (6.3.38),

$$\alpha_n = \alpha - n^{-2}\Delta + \mathcal{O}(n^{-3}), \quad \beta_n = \alpha + n^{-2}\Delta + \mathcal{O}(n^{-3}), \tag{6.3.90}$$

where

$$\Delta = \frac{\gamma^2}{3(\pi - 2\gamma)\cos(\pi\zeta/2)}, \tag{6.3.91}$$

hence (6.3.89) gives that

$$V_n(\beta_n) = (1-\zeta)\beta + n^{-2}(1-\zeta)\Delta + \mathcal{O}(n^{-3}). \tag{6.3.92}$$

The difficult part is to calculate the asymptotic behavior of $g_n(\beta_n)$.

The g-function satisfies the relations,

$$g_n'(z) = \omega_n(z), \qquad g_n(z) = \log z + \mathcal{O}(z^{-1}), \quad z \to \infty. \tag{6.3.93}$$

hence the value of $g_n(\beta_n)$ can be calculated as

$$g_n(\beta_n) = \log \beta_n + \int_\infty^{\beta_n} \left[\omega_n(z) - \frac{1}{z}\right] \mathrm{d}z. \tag{6.3.94}$$

Indeed,

$$\begin{aligned} \int_\infty^{\beta_n} \left[\omega_n(z) - \frac{1}{z}\right] \mathrm{d}z &= \lim_{A\to\infty} \int_A^{\beta_n} \left[\omega_n(z) - \frac{1}{z}\right] \mathrm{d}z \\ &= g_n(\beta_n) - \log\beta_n - \lim_{A\to\infty} [g_n(A) - \log A] \\ &= g_n(\beta_n) - \log\beta_n, \end{aligned} \tag{6.3.95}$$

which implies (6.3.94). Let us consider the resolvent, $\omega_n(z)$.

By (6.3.34),

$$\omega_n(z) = \omega_n^0(z) + \omega_n^1(z), \tag{6.3.96}$$

where

$$\omega_n^0(z) = \frac{\sqrt{(z-\alpha_n)(z-\beta_n)}}{2\pi} \int_{\alpha_n}^{\beta_n} \frac{-\zeta + \operatorname{sgn}(x)}{(z-x)\sqrt{(x-\alpha_n)(\beta_n - x)}} \mathrm{d}x, \tag{6.3.97}$$

and

$$\omega_n^1(z) = \frac{\sqrt{(z-\alpha_n)(z-\beta_n)}}{2\pi} \int_{\alpha_n}^{\beta_n} \frac{f(nx)}{(z-x)\sqrt{(x-\alpha_n)(\beta_n - x)}} \mathrm{d}x. \tag{6.3.98}$$

The function $\omega_n^0(z)$ is explicitly calculated as

$$\omega_n^0(z) = \frac{1-\zeta}{2} + \frac{2}{\pi \mathrm{i}} \log\left[\frac{\sqrt{\beta_n(z-\alpha_n)} - \mathrm{i}\sqrt{(-\alpha_n)(z-\beta_n)}}{\sqrt{z(\beta_n-\alpha_n)}}\right]. \tag{6.3.99}$$

Observe that for $z \geq \beta_n$,

$$\omega_n^0(z) = \frac{1-\zeta}{2} - \frac{2}{\pi} \arctan\sqrt{\frac{(-\alpha_n)(z-\beta_n)}{\beta_n(z-\alpha_n)}}, \tag{6.3.100}$$

and as $z \to \infty$,

$$\omega_n^0(z) = A_0 + \frac{A_1}{z} + \mathcal{O}(z^{-2}), \tag{6.3.101}$$

where

$$\begin{aligned} A_0 &= \frac{1-\zeta}{2} - \frac{2}{\pi} \arctan \sqrt{\frac{(-\alpha_n)}{\beta_n}} = \mathcal{O}(n^{-2}), \\ A_1 &= \frac{\sqrt{(-\alpha_n)\beta_n}}{\pi} = 1 + \mathcal{O}(n^{-2}). \end{aligned} \tag{6.3.102}$$

We have the explicit integration formula

$$\begin{aligned} &\int_\infty^{\beta_n} \left[\omega_n^0(z) - A_0 - \frac{A_1}{z} \right] dz \\ &\quad = \frac{2\beta_n}{\pi} \arctan \sqrt{\frac{(-\alpha_n)}{\beta_n}} + \frac{\sqrt{(-\alpha_n)\beta_n}}{\pi} \log \frac{\beta_n - \alpha_n}{4\beta_n} - \frac{\sqrt{(-\alpha_n)\beta_n}}{\pi}. \end{aligned} \tag{6.3.103}$$

By substituting formulas (6.3.90) for α_n and β_n, we obtain that

$$\int_\infty^{\beta_n} \left[\omega_n^0(z) - A_0 - \frac{A_1}{z} \right] dz = d_0 + \frac{d_1}{n^2} + \mathcal{O}(n^{-3}), \tag{6.3.104}$$

where

$$d_0 = \frac{2\beta}{\pi} \arctan \sqrt{\frac{(-\alpha)}{\beta}} + \frac{\sqrt{(-\alpha)\beta}}{\pi} \log \frac{\beta - \alpha}{4\beta} - \frac{\sqrt{(-\alpha)\beta}}{\pi}. \tag{6.3.105}$$

Applying formulas (6.3.6) for α, β, we obtain that

$$d_0 = \frac{\beta(1-\zeta)}{2} + \log \frac{\beta - \alpha}{4\beta} - 1. \tag{6.3.106}$$

Similarly,

$$d_1 = \left(\frac{1-\zeta}{2} + \frac{\beta - \alpha}{2\pi^2} \log \frac{\beta - \alpha}{4\beta} + \frac{3\alpha + \beta}{2\pi^2} \right) \Delta. \tag{6.3.107}$$

Consider now the function $\omega_n^1(z)$ in (6.3.98).

The function $\omega_n^1(z)$ for large z can be written as

$$\omega_n^1(z) = \sqrt{(z-\alpha_n)(z-\beta_n)} \sum_{k=0}^\infty \frac{c_k(n)}{z^{k+1}}, \tag{6.3.108}$$

where

$$c_k(n) = \frac{1}{2\pi} \int_{\alpha_n}^{\beta_n} \frac{x^k f(nx)\, dx}{\sqrt{(x-\alpha_n)(\beta_n - x)}}. \tag{6.3.109}$$

Since the function $f(x)$ is exponentially decaying, we have the estimate,

$$c_k(n) = n^{-(k+1)} \int_{n\alpha_n}^{n\beta_n} \frac{x^k f(x)\, dx}{\sqrt{(n^{-1}x - \alpha_n)(\beta_n - n^{-1}x)}} = \mathcal{O}(n^{-(k+1)}). \tag{6.3.110}$$

We split the function $\omega_n^1(z)$ as

$$\omega_n^1(z) = \omega_n^2(z) + \omega_n^3(z), \tag{6.3.111}$$

where

$$\omega_n^2(z) = \sqrt{(z-\alpha_n)(z-\beta_n)} \sum_{k=0}^{1} \frac{c_k(n)}{z^{k+1}}, \tag{6.3.112}$$

and

$$\omega_n^3(z) = \sqrt{(z-\alpha_n)(z-\beta_n)} \sum_{k=2}^{\infty} \frac{c_k(n)}{z^{k+1}}. \tag{6.3.113}$$

Then

$$\omega_n^3(z) = \frac{\sqrt{(z-\alpha_n)(z-\beta_n)}}{2\pi z^2} \int_{\alpha_n}^{\beta_n} \frac{x^2 f(nx)\,\mathrm{d}x}{(z-x)\sqrt{(x-\alpha_n)(\beta_n-x)}}, \tag{6.3.114}$$

and we have the estimate,

$$|\omega_n^3(z)| \le \frac{C}{z^2 n^3}, \quad z \ge \beta_n, \tag{6.3.115}$$

hence

$$\int_{\infty}^{\beta_n} \omega_n^3(z)\,\mathrm{d}z = \mathcal{O}(n^{-3}). \tag{6.3.116}$$

In addition,

$$\omega_n^0(z) + \omega_n^2(z) = \omega_n(z) - \omega_n^3(z) = \frac{1}{z} + \mathcal{O}(z^{-2}), \quad z \to \infty. \tag{6.3.117}$$

Let us calculate $\omega_n^2(z)$.

Comparing (6.3.109) with (6.3.43), we obtain that

$$c_0(n) = F_n^0(\alpha_n, \beta_n), \quad c_1(n) = G_n^0(\alpha_n, \beta_n), \tag{6.3.118}$$

hence by (6.3.56),

$$\begin{aligned} c_0(n) &= \frac{n^{-2}\gamma^2 \tan(\pi\zeta/2)}{3\pi^2(\pi - 2\gamma)} + \mathcal{O}(n^{-3}), \\ c_1(n) &= -\frac{n^{-2}\gamma^2}{3\pi(\pi - 2\gamma)} + \mathcal{O}(n^{-3}). \end{aligned} \tag{6.3.119}$$

As $z \to \infty$,

$$\omega_n^2(z) = \sqrt{(z-\alpha_n)(z-\beta_n)} \left(\frac{c_0(n)}{z} + \frac{c_1(n)}{z^2} \right) = B_0 + \frac{B_1}{z} + \mathcal{O}(z^{-2}), \tag{6.3.120}$$

where

$$B_0 = c_0(n), \quad B_1 = c_1(n) - \frac{c_0(n)(\alpha_n + \beta_n)}{2}. \tag{6.3.121}$$

By (6.3.101), (6.3.117), and (6.3.120),

$$A_0 + B_0 = 0, \quad A_1 + B_1 = 1. \tag{6.3.122}$$

We have the explicit integration formulas

$$(6.3.123)\quad I_1 \equiv \int_\infty^{\beta_n} \left[\frac{\sqrt{(z-\alpha_n)(z-\beta_n)}}{z} - 1 + \frac{\alpha_n+\beta_n}{2z} \right] \mathrm{d}z$$
$$= -\frac{(\alpha_n+\beta_n)}{2} \log \frac{\beta_n-\alpha_n}{4\beta_n} - \frac{\beta_n-\alpha_n}{2}$$
$$+ \sqrt{(-\alpha_n)\beta_n} \left[\frac{\pi}{2} - \arctan \frac{\alpha_n+\beta_n}{2\sqrt{(-\alpha_n)\beta_n}} \right],$$

and

$$(6.3.124)\quad I_2 \equiv \int_\infty^{\beta_n} \left[\frac{\sqrt{(z-\alpha_n)(z-\beta_n)}}{z^2} - \frac{1}{z} \right] \mathrm{d}z$$
$$= 1 + \log \frac{\beta_n-\alpha_n}{4\beta_n} + \frac{\alpha_n+\beta_n}{2\sqrt{(-\alpha_n)\beta_n}} \left[\frac{\pi}{2} - \arctan \frac{\alpha_n+\beta_n}{2\sqrt{(-\alpha_n)\beta_n}} \right].$$

When we substitute $\alpha_n = \alpha + \mathcal{O}(n^{-2})$ and $\beta_n = \beta + \mathcal{O}(n^{-2})$, where α and β are defined as in (6.3.6), we obtain that

$$(6.3.125)\quad I_1 = -\frac{(\alpha+\beta)}{2} \log \frac{\beta-\alpha}{4\beta} - \frac{\beta-\alpha}{2} + \frac{\pi^2(1-\zeta)}{2} + \mathcal{O}(n^{-2}),$$

and

$$(6.3.126)\quad I_2 = 1 + \log \frac{\beta-\alpha}{4\beta} + \frac{(\alpha+\beta)(1-\zeta)}{4} + \mathcal{O}(n^{-2}).$$

Combining this with (6.3.119), we obtain that

$$(6.3.127)\quad \int_\infty^{\beta_n} \left[\omega_n^2(z) - B_0 - \frac{B_1}{z} \right] \mathrm{d}z = c_0(n) I_1 + c_1(n) I_2 = \frac{d_2}{n^2} + \mathcal{O}(n^{-4}),$$

where

$$(6.3.128)\quad d_2 = -\frac{\gamma^2}{3\pi(\pi-2\gamma)} - \left[(\beta-\alpha) \log \frac{\beta-\alpha}{4\beta} + (\beta+\alpha) \right] \frac{\Delta}{2\pi^2}.$$

From (6.3.94) and (6.3.122), we obtain that

$$(6.3.129)\quad g_n(\beta_n) = \log \beta_n + \int_\infty^{\beta_n} \left[\omega_n(z) - \frac{1}{z} \right] \mathrm{d}z$$
$$= \log \beta_n + \int_\infty^{\beta_n} \left[\omega_n^0(z) - A_0 - \frac{A_1}{z} \right] \mathrm{d}z$$
$$+ \int_\infty^{\beta_n} \left[\omega_n^2(z) - B_0 - \frac{B_1}{z} \right] \mathrm{d}z + \int_\infty^{\beta_n} \omega_n^3(z)\, \mathrm{d}z.$$

We have that

$$(6.3.130)\quad \log \beta_n = \log \beta + \frac{n^{-2}\Delta}{\beta} + \mathcal{O}(n^{-3}),$$

hence substituting (6.3.104), (6.3.115), and (6.3.127) into (6.3.129), we obtain that

$$(6.3.131)\quad g_n(\beta_n) = \log \beta + d_0 + n^{-2}\left(\frac{\Delta}{\beta} + d_1 + d_2 \right) + \mathcal{O}(n^{-3}).$$

Applying formulas (6.3.106), (6.3.107), (6.3.128) for d_0, d_1, d_2, we obtain that

$$\log\beta + d_0 = \frac{\beta(1-\zeta)}{2} + \log\frac{\beta-\alpha}{4} - 1, \tag{6.3.132}$$

and

$$d_1 + d_2 = -\frac{\gamma^2}{3\pi(\pi-2\gamma)} + \left(\frac{1-\zeta}{2} + \frac{\alpha}{\pi^2}\right)\Delta. \tag{6.3.133}$$

Therefore,

$$\begin{aligned} g_n(\beta_n) &= \frac{\beta(1-\zeta)}{2} + \log(\beta-\alpha) - 2\log 2 - 1 \\ &\quad + n^{-2}\left[-\frac{\gamma^2}{3\pi(\pi-2\gamma)} + \left(\frac{1-\zeta}{2} + \frac{\alpha}{\pi^2} + \frac{1}{\beta}\right)\Delta\right] + \mathcal{O}(n^{-3}). \end{aligned} \tag{6.3.134}$$

Since $(-\alpha)\beta = \pi^2$, this reduces to

$$\begin{aligned} g_n(\beta_n) &= \frac{\beta(1-\zeta)}{2} + \log(\beta-\alpha) - 2\log 2 - 1 \\ &\quad + n^{-2}\left[-\frac{\gamma^2}{3\pi(\pi-2\gamma)} + \left(\frac{1-\zeta}{2}\right)\Delta\right] + \mathcal{O}(n^{-3}). \end{aligned} \tag{6.3.135}$$

Thus,

$$\begin{aligned} l_n = 2g_n(\beta_n) - V(\beta_n) &= \Big[\beta(1-\zeta) + 2\log(\beta-\alpha) - 4\log 2 - 2\Big] \\ &\quad + n^{-2}\left[-\frac{2\gamma^2}{3\pi(\pi-2\gamma)} + (1-\zeta)\Delta\right] \\ &\quad - [\beta(1-\zeta) + n^{-2}(1-\zeta)\Delta] + \mathcal{O}(n^{-3}). \end{aligned} \tag{6.3.136}$$

By simplifying, we conclude that

$$l_n = 2\log(\beta-\alpha) - 4\log 2 - 2 - n^{-2}\frac{2\gamma^2}{3\pi(\pi-2\gamma)} + \mathcal{O}(n^{-3}). \tag{6.3.137}$$

Notice that

$$\mathrm{e}^{nl_n} = \frac{(\beta-\alpha)^{2n}}{2^{4n}\mathrm{e}^{2n}}\left(1 - \frac{2\gamma^2}{3n\pi(\pi-2\gamma)} + \mathcal{O}(n^{-2})\right). \tag{6.3.138}$$

6.4. Riemann – Hilbert analysis

As described in Section 2.2.1, the orthogonal polynomials (6.2.3) are described by the 2×2 matrix-valued solution $\mathbf{Y}_n(z)$ to the following Riemann – Hilbert problem.

(1) $\mathbf{Y}_n(z)$ is analytic on $\mathbb{C}\setminus\mathbb{R}$.
(2) For $x \in \mathbb{R}$, $\mathbf{Y}_n$ takes limiting values from above ($\mathbf{Y}_{n+}$) and below ($\mathbf{Y}_{n-}$) the real line, and these are related by the equation

$$\mathbf{Y}_{n+}(x) = \mathbf{Y}_{n-}(x)j_Y(x), \quad j_Y(x) = \begin{pmatrix} 1 & \mathrm{e}^{-nV_n(x)} \\ 0 & 1 \end{pmatrix}, \tag{6.4.1}$$

where $V_n(x)$ is defined in (6.2.5).

(3) As $z \to \infty$,

$$\mathbf{Y}_n(z) \sim \left(I + \sum_{k=1}^{\infty} \frac{\mathbf{Y}_k}{z^k}\right) \begin{pmatrix} z^n & 0 \\ 0 & z^{-n} \end{pmatrix}, \tag{6.4.2}$$

where $\mathbf{Y}_k$, $k = 1, 2, \ldots$, are some constant 2×2 matrices.

Our interest is in the normalizing constants h_{nn}, which are encoded in the expansion of $\mathbf{Y}_n(z)$ at $z = \infty$ as

$$h_{nn} = -2\pi \mathrm{i}[\mathbf{Y}_1]_{12}, \quad h_{n,n-1} = -\frac{2\pi \mathrm{i}}{[\mathbf{Y}_1]_{21}}. \tag{6.4.3}$$

6.4.1. The first transformation. The first transformation of the RHP (see Section 2.4.3) involves the g-function and the Lagrange multiplier. Namely we define the matrix $\mathbf{T}_n(z)$ by the equation

$$\mathbf{Y}_n(z) = \mathrm{e}^{(nl_n/2)\sigma_3} \mathbf{T}_n(z) \mathrm{e}^{n(g_n(z) - l_n/2)\sigma_3}. \tag{6.4.4}$$

Recall then that $\mathbf{T}_n(z)$ approaches the identity uniformly for large z. That is, as $z \to \infty$,

$$\mathbf{T}_n(z) = I + \frac{\mathbf{T}_1}{z} + \frac{\mathbf{T}_2}{z^2} + \cdots, \tag{6.4.5}$$

for some constant matrices $\mathbf{T}_j$. Notice that, expanding (6.4.4) at $z = \infty$ using (6.3.93) and (6.4.2), we find that the off-diagonal entries of $\mathbf{T}_1$ differ from the off-diagonal entries of $\mathbf{Y}_1$ only by a factor of e^{nl_n}. Indeed, we can write (6.4.3) in terms of $\mathbf{T}_n$ instead of $\mathbf{Y}_n$, giving

$$h_{nn} = -2\pi \mathrm{i} \mathrm{e}^{-nl_n}[\mathbf{T}_1]_{12}, \quad h_{n,n-1} = -\frac{2\pi \mathrm{i} \mathrm{e}^{nl_n}}{[\mathbf{T}_1]_{21}}. \tag{6.4.6}$$

6.4.2. The second transformation. The second transformation of the RHP is the so-called "opening of the lenses" described in Section 2.4.4. However, there is an extra difficulty which comes from the fact that we are dealing with a sequence of potentials $V_n(x)$ which converges to a potential $V(x)$ which is not analytic at the origin. In particular, the function

$$\mathrm{e}^{nV_n(z)} = \mathrm{e}^{-n\zeta z} \frac{\sinh\big(n\pi z/(2\gamma)\big)}{\sinh\big(nz(\pi/(2\gamma) - 1)\big)} \tag{6.4.7}$$

has poles at the points

$$z_k = \frac{\mathrm{i}k\pi}{n(\pi/(2\gamma) - 1)}, \quad k = \pm 1, \pm 2, \ldots, \tag{6.4.8}$$

which accumulate on the imaginary axis as $n \to \infty$. In order to avoid these poles, we open the lenses as shown in Figure 6.1. That is, let Σ_n^+ be a contour which goes from α_n to β_n and which crosses the imaginary axis between z_1 and the origin, say at $z_1/2$, so that it comes closer and closer to the origin as $n \to \infty$. Outside of some fixed neighborhood of the points α_n, β_n, and 0, however, the contour should be uniformly (say at a distance ε) far from the real axis. For the boundary of the lower lens, we can take $\Sigma_n^- = \overline{\Sigma_n^+}$. The *upper* (*lower*) *lens* is then the region bounded by Σ_n^+ (Σ_n^-) and the real axis.

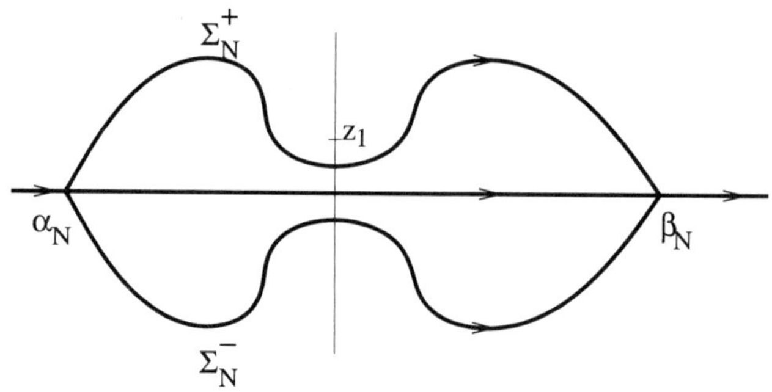

FIGURE 6.1. The opening of the lenses.

We are now ready to make the second transformation of the RHP. For $x \in [\alpha_n, \beta_n]$ we can define the function

$$G_n(x) = g_{n+}(x) - g_{n-}(x) = 2\pi \mathrm{i} \int_x^{\beta_n} \rho_n(t)\, \mathrm{d}t, \tag{6.4.9}$$

which extends to a complex neighborhood of $[\alpha_n, \beta_n]$. Define the matrices

$$j_+(z) = \begin{pmatrix} 1 & 0 \\ \mathrm{e}^{-nG_n(z)} & 1 \end{pmatrix}, \quad j_-(z) = \begin{pmatrix} 1 & 0 \\ \mathrm{e}^{nG_n(z)} & 1 \end{pmatrix}. \tag{6.4.10}$$

We then define the matrix $\mathbf{S}_n(z)$ as

$$\mathbf{S}_n(z) = \begin{cases} \mathbf{T}_n(z) j_+(x)^{-1} & \text{for } z \text{ in upper lens} \\ \mathbf{T}_n(z) j_-(x) & \text{for } z \text{ in lower lens} \\ \mathbf{T}_n(z) & \text{otherwise.} \end{cases} \tag{6.4.11}$$

We then have that $\mathbf{S}_n(z)$ is analytic for $z \in \mathbb{C} \setminus \Sigma$, where $\Sigma = \Sigma_n^+ \cup \Sigma_n^- \cup \mathbb{R}$. On the contour Σ, $\mathbf{S}_n(z)$ takes limiting values from the plus- and minus-sides, which are related by the jump condition

$$\mathbf{S}_{n+}(z) = \mathbf{S}_{n-}(z) j_S(z), \tag{6.4.12}$$

where

$$j_S(z) = \begin{cases} \begin{pmatrix} 1 & \mathrm{e}^{n(g_{n+}(z)+g_{n-}(z)-V_n(z)-l_n)} \\ 0 & 1 \end{pmatrix} & \text{for } z \in \mathbb{R} \setminus [\alpha_n, \beta_n], \\ \begin{pmatrix} 0 & 1 \\ -1 & 0 \end{pmatrix} & \text{for } z \in [\alpha_n, \beta_n], \\ \begin{pmatrix} 1 & 0 \\ \mathrm{e}^{-nG_n(z)} & 1 \end{pmatrix} & \text{for } z \in \Sigma_n^+, \\ \begin{pmatrix} 1 & 0 \\ \mathrm{e}^{nG_n(z)} & 1 \end{pmatrix} & \text{for } z \in \Sigma_n^-. \end{cases} \tag{6.4.13}$$

The transformation from $\mathbf{T}_n$ to $\mathbf{S}_n$ is local and therefore does not change the behavior at $z = \infty$. We therefore still have as $z \to \infty$,

$$\mathbf{S}_n(z) = I + \frac{\mathbf{S}_1}{z} + \frac{\mathbf{S}_2}{z^2} + \cdots, \tag{6.4.14}$$

and

$$h_{nn} = -2\pi \mathrm{i} e^{-nl_n}[\mathbf{S}_1]_{12}, \qquad h_{n,n-1} = -\frac{2\pi \mathrm{i} e^{nl_n}}{[\mathbf{S}_1]_{21}}. \tag{6.4.15}$$

6.4.3. The model RHP. Following Section 2.4.5 we now construct the solution to the model RHP. That is, we seek a 2×2 matrix function $\mathbf{M}_n(z)$ which satisfies the following conditions.

(1) $\mathbf{M}_n(z)$ is analytic in $\mathbb{C} \setminus [\alpha_n, \beta_q]$,

(2) For $z \in [\alpha_n, \beta_n]$, $\mathbf{M}_n(z)$ satisfies the jump condition

$$\mathbf{M}_{n+}(z) = \mathbf{M}_{n-}(z) \begin{pmatrix} 0 & 1 \\ -1 & 0 \end{pmatrix}, \tag{6.4.16}$$

where $\mathbf{M}_{n+}$ ($\mathbf{M}_{n-}$) refers to the limiting value from above (below) the real axis.

(3) As $z \to \infty$,

$$\mathbf{M}_n(z) = I + \frac{\mathbf{M}_1}{z} + \frac{\mathbf{M}_2}{z^2} + \cdots, \tag{6.4.17}$$

We are in the one-cut case, and the solution is given in Section 2.4.5. The solution is

$$\mathbf{M}_n(z) = \begin{pmatrix} \dfrac{\gamma_n(z) + \gamma_n(z)^{-1}}{2} & \dfrac{\gamma_n(z) - \gamma_n(z)^{-1}}{-2i} \\ \dfrac{\gamma_n(z) - \gamma_n(z)^{-1}}{2i} & \dfrac{\gamma_n(z) + \gamma_n(z)^{-1}}{2} \end{pmatrix}, \tag{6.4.18}$$

where

$$\gamma_n(z) = \left(\frac{z - \alpha_n}{z - \beta_n} \right)^{1/4}, \tag{6.4.19}$$

with the cut on $[\alpha_n, \beta_n]$, taking the branch such that $\gamma_n(\infty) = 1$. In particular notice that in the expansion of $\mathbf{M}_n(z)$ at $z = \infty$ we have

$$\mathbf{M}_1 = \begin{pmatrix} 0 & \dfrac{\beta_n - \alpha_n}{-4\mathrm{i}} \\ \dfrac{\beta_n - \alpha_n}{4\mathrm{i}} & 0 \end{pmatrix}, \quad \mathbf{M}_2 = \begin{pmatrix} \dfrac{(\beta_n - \alpha_n)^2}{32} & \dfrac{\beta_n^2 - \alpha_n^2}{-8\mathrm{i}} \\ \dfrac{\beta_n^2 - \alpha_n^2}{8\mathrm{i}} & \frac{(\beta_n - \alpha_n)^2}{32} \end{pmatrix}. \tag{6.4.20}$$

At the origin we have

$$\gamma_{n+}(0) = \left(\frac{-\alpha_n + \mathrm{i}0}{-\beta_n + \mathrm{i}0} \right)^{1/4} = e^{-\mathrm{i}\pi/4} \sqrt{\tan\left(\frac{\pi(1-\zeta)}{4} \right)} + \mathcal{O}(n^{-2}), \tag{6.4.21}$$

and we can therefore write

$$\mathbf{M}_{n+}(0) = \begin{pmatrix} p + \mathrm{i}q & p - \mathrm{i}q \\ -p + \mathrm{i}q & p + \mathrm{i}q \end{pmatrix}, \tag{6.4.22}$$

where

$$p, q = \frac{\sqrt{2}}{4} \left[\tan\left(\frac{\pi(1+\zeta)}{4} \right) \pm \tan\left(\frac{\pi(1-\zeta)}{4} \right) \right]. \tag{6.4.23}$$

From the conjugate symmetry

$$\mathbf{M}_n(\bar{z}) = \sigma_3 \overline{\mathbf{M}_n(z)} \sigma_3, \tag{6.4.24}$$

we also find

$$\mathbf{M}_{n-}(0) = \begin{pmatrix} p - \mathrm{i}q & -p - \mathrm{i}q \\ p + \mathrm{i}q & p - \mathrm{i}q \end{pmatrix}. \tag{6.4.25}$$

6.4.4. Parametrix at the edge points. We now seek a local parametrix in a neighborhood of the points α_n and β_n. For a small but fixed $r > 0$, we consider the disks of radius r around these points, denoting

$$D_n \equiv \{z : |z - \alpha_n| \le r\} \cup \{z : |z - \beta_n| \le r\}, \tag{6.4.26}$$

As in Section 2.4.6, we seek a 2×2 matrix $\mathbf{U}_n(z)$ defined on the union of these disks, such that

- $\mathbf{U}_n(z)$ is analytic on $D \setminus \Sigma$.
- For $z \in D \cap \Sigma$, $\mathbf{U}_n(z)$ satisfies the jump condition

$$\mathbf{U}_{n+}(z) = \mathbf{U}_{n-}(z) j_S(z). \tag{6.4.27}$$

- As $n \to \infty$,

$$\mathbf{U}_n(z) = \left(I + \mathcal{O}\left(\frac{1}{n}\right)\right)\mathbf{M}_n(z) \quad \text{uniformly for } z \in \partial D. \tag{6.4.28}$$

Let us summarize the solution, which is given in Section 2.4.6. For z close to the edge points, we define the functions

$$\begin{aligned} \psi_{\alpha_n}(z) &= -\left\{\frac{3\pi}{2}\int_{\alpha_n}^{z} \rho_n(t)\,\mathrm{d}t\right\}^{2/3}, & z \in D(\alpha_n, r), \\ \psi_{\beta_n}(z) &= -\left\{\frac{3\pi}{2}\int_{z}^{\beta_n} \rho_n(t)\,\mathrm{d}t\right\}^{2/3}, & z \in D(\beta_n, r). \end{aligned} \tag{6.4.29}$$

Define the matrix-valued functions $\mathbf{E}_n^{\alpha_n}(z)$ and $\mathbf{E}_n^{\beta_n}(z)$ as

$$\begin{aligned} \mathbf{E}_n^{\alpha_n}(z) &= 2\sqrt{\pi}\,\mathbf{M}_n(z)\begin{pmatrix} 1 & -\mathrm{i} \\ -1 & -\mathrm{i} \end{pmatrix}^{-1}\begin{pmatrix} n^{1/6}\psi_{\beta_n}^{1/4}(z) & 0 \\ 0 & n^{-1/6}\psi_{\beta_n}^{-1/4}(z) \end{pmatrix}, \\ \mathbf{E}_n^{\beta_n}(z) &= 2\sqrt{\pi}\,\mathbf{M}_n(z)\begin{pmatrix} 1 & \mathrm{i} \\ -1 & \mathrm{i} \end{pmatrix}^{-1}\begin{pmatrix} n^{1/6}\psi_{\beta_n}^{1/4}(z) & 0 \\ 0 & n^{-1/6}\psi_{\beta_n}^{-1/4}(z) \end{pmatrix}. \end{aligned} \tag{6.4.30}$$

Then the local solution $\mathbf{U}_n$ is given as

$$\mathbf{U}_n(z) = \begin{cases} \mathbf{E}_n^{\alpha_n}(z)\Phi_l\big(n^{2/3}\psi_{\alpha_n}(z)\big)\mathrm{e}^{-n(g_n(z)-V_n(z)/2-l_n/2)}, & z \in D(\alpha_n, r) \\ \mathbf{E}_n^{\beta_n}(z)\Phi_r\big(n^{2/3}\psi_{\beta_n}(z)\big)\mathrm{e}^{-n(g_n(z)-V_n(z)/2-l_n/2)}, & z \in D(\alpha_n, r), \end{cases} \tag{6.4.31}$$

where the functions Φ_l and Φ_r are explicitly defined in terms of Airy functions in (2.4.98) and (2.4.123). According to (2.4.119) and (2.4.130), as $n \to \infty$,

$$\mathbf{U}_n(z) = \begin{cases} \mathbf{M}_n(z)\left[I + \dfrac{\psi_{\alpha_n}(z)^{-3/2}}{48n}\begin{pmatrix} 1 & -6i \\ -6i & -1 \end{pmatrix} + \mathcal{O}(n^{-2})\right] & \text{for } z \in \partial D(\alpha_n, r), \\ \mathbf{M}_n(z)\left[I + \dfrac{\psi_{\beta_n}(z)^{-3/2}}{48n}\begin{pmatrix} 1 & 6i \\ 6i & -1 \end{pmatrix} + \mathcal{O}(n^{-2})\right] & \text{for } z \in \partial D(\beta_n, r). \end{cases} \tag{6.4.32}$$

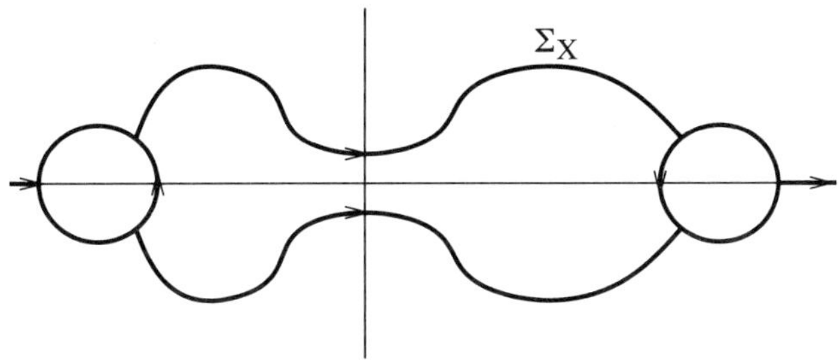

FIGURE 6.2. The contour Σ_X.

6.4.5. The third and final transformation of the RHP. The final transformation of the RHP brings us to the matrix $\mathbf{X}_n(z)$, defined as

$$\mathbf{X}_n(z) = \begin{cases} \mathbf{S}_n(z)\mathbf{U}_n(z)^{-1} & \text{for } z \in D(\alpha_n, r) \cup D(\beta_n, r) \\ \mathbf{S}_n(z)\mathbf{M}_n(z)^{-1} & \text{otherwise.} \end{cases} \tag{6.4.33}$$

Then $\mathbf{X}_n(z)$ satisfies the following RHP.

- $\mathbf{X}_n(z)$ is analytic on the contour Σ_X, where

$$\Sigma_X = \big\{\{\Sigma_n^+ \cup \Sigma_n^-\} \setminus D\big\} \cup \partial D(\alpha_n, r) \cup \partial D(\beta_n, r) \cup \Sigma_n^\infty, \tag{6.4.34}$$

where the circles $\partial D(\alpha_n, r)$ and $\partial D(\beta_n, r)$ are oriented in the counterclockwise direction, see Figure 6.2, and

$$\Sigma_n^\infty = (-\infty, \alpha_n - r) \cup (\beta_n + r, \infty). \tag{6.4.35}$$

- For $z \in \Sigma_X$, $\mathbf{X}_n(z)$ satisfies the jump conditions

$$\mathbf{X}_{n+}(z) = \mathbf{X}_{n-}(z) j_X(z), \tag{6.4.36}$$

where

(6.4.37)
$$j_X(z) = \begin{cases} \mathbf{M}_n(z) j_S(z) \mathbf{M}_n(z)^{-1} & \text{for } z \in \{\Sigma_n^+ \cup \Sigma_n^-\} \setminus \{D(\alpha_n, r) \cup D(\beta_n, r)\} \\ \mathbf{M}_n(z)\mathbf{U}_n(z)^{-1} & \text{for } z \in \partial D(\alpha_n, r) \cup \partial D(\beta_n, r). \end{cases}$$

- As $z \to \infty$,

$$\mathbf{X}_n(z) = I + \frac{\mathbf{X}_1}{z} + \frac{\mathbf{X}_2}{z^2} + \cdots. \tag{6.4.38}$$

Expanding (6.4.33) at $z = \infty$ gives

$$\mathbf{S}_1 = \mathbf{X}_1 + \mathbf{M}_1, \tag{6.4.39}$$

and thus by (6.4.15),

$$h_{nn} = -2\pi i e^{nl_n}[\mathbf{X}_1 + \mathbf{M}_1]_{12}, \quad h_{n,n-1} = -\frac{2\pi i e^{nl_n}}{[\mathbf{X}_1 + \mathbf{M}_1]_{21}}. \tag{6.4.40}$$

6.5. Estimates on the jumps for $\mathbf{X}_n$

The RHP for $\mathbf{X}_n$ can be solved by a series of perturbation theory as in Section 2.4.7, provided that the jump matrix $j_X(z)$ is uniformly close to the identity. This is nontrivial due to the fact that the lens boundaries $\Sigma_n^\pm$ pass close to the origin. We have the following proposition.

Proposition 6.5.1. *The jump matrix $j_X(z)$ is of the form*

$$j_X(z) = I + j_X^0(z), \tag{6.5.1}$$

where $j_X^0(z)$ satisfies the estimates

$$j_X^0(z) = \begin{cases} \mathcal{O}(n^{-1}) & \text{for } z \in \partial D(\alpha_n, r) \cup \partial D(\beta_n, r) \\ \mathcal{O}(\mathrm{e}^{-c(z)n}) & \text{for } z \in \Sigma_n^\infty. \end{cases} \tag{6.5.2}$$

For $z \in \Sigma_n^\pm \setminus \{D(\alpha_n, r) \cup D(\beta_n, r)\}$, we have the estimate

$$\|j_X^0(z)\| \le Cn^{-\delta}\mathrm{e}^{-cn|\mathrm{Im}\, z|}, \tag{6.5.3}$$

for some positive constants C, c, and δ.

Proof. The estimates (6.5.2) follow from (6.4.13), (6.3.20), and (6.4.32). We need only to prove the estimate (6.5.3). From (6.4.13) and (6.4.37), we have, for $z \in \Sigma_n^+$,

$$j_X^0(z) = \mathrm{e}^{-nG_n(z)}\mathbf{M}_n(z)\begin{pmatrix} 0 & 0 \\ 1 & 0 \end{pmatrix}\mathbf{M}_n(z)^{-1}. \tag{6.5.4}$$

Since $\mathbf{M}_n(z)$ and $\mathbf{M}_n(z)^{-1}$ are uniformly bounded for $z \in \Sigma_n^+ \setminus \{D(\alpha_n, r) \cup D(\beta_n, r)\}$, we need only prove the estimate (6.5.3) for the function $\mathrm{e}^{-nG_n(z)}$ for $z \in \Sigma_n^+ \setminus \{D(\alpha_n, r) \cup D(\beta_n, r)\}$. From (6.3.23) and (6.3.38) we see that

$$G_n(0) = 2\pi\mathrm{i}\int_0^{\beta_n} \rho_n(t)\,\mathrm{d}t = \pi\mathrm{i}(1+\zeta) + \mathcal{O}(n^{-2}). \tag{6.5.5}$$

We can use the formula for G_n,

$$G_n(z) = 2g_{n+}(z) - V_n(z) - l_n, \tag{6.5.6}$$

and thus, for $z \in \Sigma_n^+$,

$$\mathrm{e}^{-nG_n(z)} = \mathrm{e}^{-2ng_n(z) + nV_n(z) + nl_n}. \tag{6.5.7}$$

Using the equation (6.4.7) for $\mathrm{e}^{nV_n(z)}$, we can write

$$\begin{aligned} \mathrm{e}^{-nG_n(z)} &= \left(\frac{\pi - 2\gamma}{\pi}\right)\mathrm{e}^{-nG_n(0)}\mathrm{e}^{-2n(g_n(z) - g_n(0)) + nV_n(z)} \\ &= \left(\frac{\pi - 2\gamma}{\pi}\right)\mathrm{e}^{-n\pi\mathrm{i}(1+\zeta)}(1 + \mathcal{O}(n^{-1}))\mathrm{e}^{-2n(g_n(z) - g_n(0))}\mathrm{e}^{-n\zeta z} \\ &\qquad\qquad \times \frac{\sinh\big(n\pi z/(2\gamma)\big)}{\sinh\big(nz(\pi/(2\gamma) - 1)\big)}. \end{aligned} \tag{6.5.8}$$

Let us estimate the difference $g_n(z) - g_n(0)$. For $\Im z > 0$ we have

$$g_n(z) - g_n(0) = \int_0^z \omega_n(t)\,\mathrm{d}t. \tag{6.5.9}$$

Using (6.3.83), we have

$$g_n(z) - g_n(0) = -\frac{\mathrm{i}z\log n}{\pi} + \int_0^z b(t)\,\mathrm{d}t - \frac{1}{2\pi^2}\int_0^z m_0(nt)\,\mathrm{d}t + \mathcal{O}(n^{-2}|z|). \tag{6.5.10}$$

In particular, for $z = \mathrm{i}y/n$ for some positive y which is bounded away from zero, we have

$$g_n(\mathrm{i}y/n) - g_n(0) = \frac{y\log n}{n\pi} + \frac{\mathrm{i}yb_+(0)}{n}\,\mathrm{d}t - \frac{1}{2\pi^2 n}M_0(-y) + \mathcal{O}(n^{-1}), \tag{6.5.11}$$

where

$$M_0(z) = \int_0^z m_0(t)\,\mathrm{d}t. \tag{6.5.12}$$

Then from (6.5.8) we have

$$\mathrm{e}^{-nG_n(\mathrm{i}y/n)} = \mathrm{e}^{-in\omega} k_n(y) \frac{\sin\big(\pi y/(2\gamma)\big)}{\sin\big(y(\pi/(2\gamma)-1)\big)}\big(1+\mathcal{O}(n^{-1})\big), \tag{6.5.13}$$

where

$$\omega = -\pi(1+\zeta), \quad k_n(y) = \frac{\pi - 2\gamma}{\pi} e^{\varphi(y)} n^{-2\gamma/\pi}, \tag{6.5.14}$$

and

$$\begin{aligned} \varphi(y) &= -\mathrm{i}b_+(0)y + \frac{1}{\pi^2} M_0(iy) - \mathrm{i}\zeta y \\ &= -\mathrm{i}y - \frac{2\gamma}{\pi}\log\left(2\pi\cos\left(\frac{\pi\zeta}{2}\right)\right) + \frac{1}{\pi^2} M_0(\mathrm{i}y), \end{aligned} \tag{6.5.15}$$

where we have used (6.3.87). Using (6.3.85) and (6.5.12), we have

$$\frac{1}{\pi^2} M_0(\mathrm{i}y) = Q(\mathrm{i}y) - Q_+(0), \tag{6.5.16}$$

where

$$Q(z) := \frac{\mathrm{i}}{\pi}\left[\int_{-\infty}^{\infty} \log(z-x) f(x)\,\mathrm{d}x - 2z\log z + 2z\right]. \tag{6.5.17}$$

Since $f(x)$ is an odd function, we have that

$$Q(iy) - Q_+(0) = \frac{2}{\pi}\left[\int_0^{\infty} \arg(\mathrm{i}y + x) f(x)\,\mathrm{d}x + y\log y - y\right] + \mathrm{i}y, \tag{6.5.18}$$

from which we obtain

$$\varphi(y) = -\frac{2\gamma}{\pi}\log\left(2\pi\cos\left(\frac{\pi\zeta}{2}\right)\right) + \frac{2}{\pi}\left[\int_0^{\infty} \arg(iy+x) f(x)\,\mathrm{d}x + y\log y - y\right]. \tag{6.5.19}$$

Similarly, in the case $\operatorname{Im} z < 0$, we have

$$\mathrm{e}^{nG_n(-\mathrm{i}y/n)} = \mathrm{e}^{-in\omega} k_n(y) \frac{\sin\big(\pi y/(2\gamma)\big)}{\sin\big(y(\pi/(2\gamma)-1)\big)}\big(1+\mathcal{O}(n^{-1})\big), \tag{6.5.20}$$

□

6.6. Evaluation of $\mathbf{X}_1$

We have that

$$\begin{aligned} \mathbf{X}_1 &= -\frac{1}{2\pi\mathrm{i}}\int_{\Sigma_X} j_X^0(x)\,\mathrm{d}x + \mathcal{O}(n^{-2}) \\ &= I_\alpha + I_\beta + I_+ + I_- + \mathcal{O}(n^{-2}), \end{aligned} \tag{6.6.1}$$

where

$$\begin{aligned} I_\alpha &= -\frac{1}{2\pi\mathrm{i}}\int_{\partial D(\alpha_n, r)} j_X^0(x)\,\mathrm{d}x, \quad I_\beta = -\frac{1}{2\pi\mathrm{i}}\int_{\partial D(\beta_n, r)} j_X^0(x)\,\mathrm{d}x, \\ I_\pm &= -\frac{1}{2\pi\mathrm{i}}\int_{\Sigma_n^\pm \setminus D} j_X(x)\,\mathrm{d}x. \end{aligned} \tag{6.6.2}$$

Integrals around the circles. From (6.4.32) and (6.4.37) we have that

(6.6.3)
$$j_X^0(z) = \begin{cases} -\dfrac{\psi_{\alpha_n}(z)^{-3/2}}{48n}\mathbf{M}_n(z)\begin{pmatrix}1 & -6\mathrm{i}\\ -6\mathrm{i} & -1\end{pmatrix}\mathbf{M}_n(z)^{-1} + \mathcal{O}(n^{-2}) & \text{for } z \in \partial D(\alpha_n, r)\\ -\dfrac{\psi_{\beta_n}(z)^{-3/2}}{48n}\mathbf{M}_n(z)\begin{pmatrix}1 & 6\mathrm{i}\\ 6\mathrm{i} & -1\end{pmatrix}\mathbf{M}_n(z)^{-1} + \mathcal{O}(n^{-2}) & \text{for } z \in \partial D(\beta_n, r).\end{cases}$$

We are particularly interested in the off-diagonal entries. Multiplying out (6.6.3) we find

$$(6.6.4)\quad \begin{aligned}[j_X^0(z)]_{12} &= [j_X^0(z)]_{21}\\ &= \begin{cases} \dfrac{\mathrm{i}\psi_{\alpha_n}(z)^{-3/2}}{96n}(7\gamma_n(z)^2 + 5\gamma_n(z)^{-2}) + \mathcal{O}(n^{-2}) & \text{for } z \in \partial D(\alpha_n, r)\\ -\dfrac{\mathrm{i}\psi_{\beta_n}(z)^{-3/2}}{96n}(5\gamma_n(z)^2 + 7\gamma_n(z)^{-2}) + \mathcal{O}(n^{-2}) & \text{for } z \in \partial D(\beta_n, r).\end{cases}\end{aligned}$$

In the first-order approximation, we may use the limit as $n \to \infty$ in the above expressions. That is, if we denote

$$(6.6.5)\quad \begin{gathered}\psi_\alpha(z) = -\left\{\frac{3\pi}{2}\int_\alpha^z \rho(t)\,\mathrm{d}t\right\}^{2/3}, \quad \psi_\beta(z) = -\left\{\frac{3\pi}{2}\int_z^\beta \rho(t)\,\mathrm{d}t\right\}^{2/3},\\ \gamma(z) = \left(\frac{z-\alpha}{z-\beta}\right)^{1/4},\end{gathered}$$

then we can write (6.6.4) as

$$(6.6.6)\quad \begin{aligned}[j_X^0(z)]_{12} &= [j_X^0(z)]_{21}\\ &= \begin{cases} \dfrac{\mathrm{i}\psi_{\alpha}(z)^{-3/2}}{96n}(7\gamma(z)^2 + 5\gamma(z)^{-2}) + \mathcal{O}(n^{-2}) & \text{for } z \in \partial D(\alpha_n, r)\\ -\dfrac{\mathrm{i}\psi_{\beta}(z)^{-3/2}}{96n}(5\gamma(z)^2 + 7\gamma(z)^{-2}) + \mathcal{O}(n^{-2}) & \text{for } z \in \partial D(\beta_n, r).\end{cases}\end{aligned}$$

Since $\gamma^{\pm 2}(z)$ has square root behavior at the points α and β, these functions extend to meromorphic functions inside the discs $D(\alpha_n, r)$ and $D(\beta_n, r)$, and we can therefore evaluate the integrals I_α and I_β via a residue calculation. Using formula (6.3.9) for $\rho(x)$, we find as $x \to \alpha$,

$$(6.6.7)\quad \rho(x) = \frac{2\sqrt{\beta}\sqrt{x-\alpha}}{\pi^2\sqrt{-\alpha(\beta-\alpha)}}\left(1 + \frac{2\beta - 3\alpha}{6(-\alpha)(\beta-\alpha)}(x-\alpha) + \mathcal{O}((x-\alpha)^2)\right),$$

and as $x \to \beta$,

$$(6.6.8)\quad \rho(x) = \frac{2\sqrt{-\alpha}\sqrt{\beta-x}}{\pi^2\sqrt{\beta(\beta-\alpha)}}\left(1 + \frac{3\beta - 2\alpha}{6\beta(\beta-\alpha)}(\beta-x) + \mathcal{O}((\beta-x)^2)\right),$$

Thus as $z \to \alpha$,

$$(6.6.9)\quad \begin{aligned}\psi_\alpha^{-3/2}(z) &= \frac{\pi\sqrt{-\alpha(\beta-\alpha)}}{2\sqrt{\beta}}(\alpha - z)^{-3/2}\\ &\quad\times\left(1 + \frac{2\beta - 3\alpha}{10(-\alpha)(\beta-\alpha)}(\alpha - z) + \mathcal{O}((\alpha-z)^2)\right),\end{aligned}$$

and as $z \to \beta$,

$$\psi_\beta^{-3/2}(z) = \frac{\pi\sqrt{\beta(\beta-\alpha)}}{2\sqrt{-\alpha}}(z-\beta)^{-3/2} \times \left(1 + \frac{3\beta - 2\alpha}{10\beta(\beta-\alpha)}(z-\beta) + \mathcal{O}((z-\beta)^2)\right). \tag{6.6.10}$$

Using the above formulas, along with the relationship $(-\alpha\beta) = \pi^2$ given in (6.3.7), we find that

$$\begin{aligned} [I_\alpha]_{12} &= [I_\alpha]_{21} + \mathcal{O}(n^{-2}) \\ &= -\frac{\mathrm{i}}{96n}\operatorname*{Res}_{z=\alpha}\{\psi_\alpha(z)^{-3/2}(7\gamma(z)^2 + 5\gamma(z)^{-2})\} + \mathcal{O}(n^{-2}) \\ &= -\frac{\mathrm{i}(11\alpha - \beta)}{192n} + \mathcal{O}(n^{-2}). \end{aligned} \tag{6.6.11}$$

and

$$\begin{aligned} [I_\beta]_{12} &= [I_\beta]_{21} + \mathcal{O}(n^{-2}) \\ &= \frac{\mathrm{i}}{96n}\operatorname*{Res}_{z=\beta}\{\psi_\beta(z)^{-3/2}(5\gamma(z)^2 + 7\gamma(z)^{-2})\} + \mathcal{O}(n^{-2}). \\ &= \frac{\mathrm{i}(11\beta - \alpha)}{192n} + \mathcal{O}(n^{-2}). \end{aligned} \tag{6.6.12}$$

Notice that

$$[I_\alpha + I_\beta]_{12} = [I_\alpha + I_\beta]_{21} = -\frac{1}{16\mathrm{i}n}(\beta - \alpha) + \mathcal{O}(n^{-2}), \tag{6.6.13}$$

where we have used (6.3.7).

Integrals over the lens boundaries. We now turn to the evaluation of the integrals $I_\pm$. Consider first the upper lens boundary. We have

$$I_+ = -\frac{1}{2\pi\mathrm{i}}\int_{\Sigma_n^+\setminus D} \mathrm{e}^{-nG_n(z)}\mathbf{M}_n(z)\begin{pmatrix}0 & 0\\ 1 & 0\end{pmatrix}\mathbf{M}_n(z)^{-1}\,\mathrm{d}z. \tag{6.6.14}$$

From (6.5.8) we see that the function $\mathrm{e}^{-nG_n(z)}$ has simple poles at the points

$$z_j = \frac{\mathrm{i}j\pi}{n(\pi/(2\gamma) - 1)}, \quad j = 1, 2, \ldots, \tag{6.6.15}$$

with residues

$$\operatorname*{Res}_{z=z_j}\left[\mathrm{e}^{-nG_n(z)}\right] = \mathrm{e}^{\mathrm{i}n\omega}k_n(z_j)\frac{\mathrm{i}(-1)^j\sin(z_j\pi/(2\gamma))}{n(\pi/(2\gamma) - 1)}(1 + \mathcal{O}(n^{-1})). \tag{6.6.16}$$

Using 6.5.14 we can write this as

$$\operatorname*{Res}_{z=z_j}\left[\mathrm{e}^{-nG_n(z)}\right] = iC_j\mathrm{e}^{\mathrm{i}n\omega}n^{-\kappa_j}(1 + \mathcal{O}(n^{-1})), \tag{6.6.17}$$

where

$$C_j = \frac{2\gamma}{\pi}\mathrm{e}^{\varphi(y_j)}(-1)^j\sin\left(\frac{\pi j}{1 - 2\gamma/\pi}\right), \quad y_j = \frac{j\pi}{\pi/(2\gamma) - 1}, \tag{6.6.18}$$

and

$$\kappa_j = 1 + \frac{2j}{\pi/(2\gamma) - 1} > 1. \tag{6.6.19}$$

Similarly,

$$\operatorname*{Res}_{z=-z_j}\left[\mathrm{e}^{nG_n(z)}\right] = -\mathrm{i}C_j\mathrm{e}^{-\mathrm{i}n\omega}n^{-\kappa_j}\left(1+\mathcal{O}(n^{-1})\right), \tag{6.6.20}$$

We can deform the contour Σ_n^+ away from the origin, crossing poles at the points z_j. Each time we cross a pole, we must add the residue to the integral (6.6.14), and the integral becomes smaller than the contribution from the pole. This gives the asymptotic expansion as $n \to \infty$,

$$I_+ = -\sum_{j=1}^{\infty} \operatorname*{Res}_{z=z_j}\left[\mathrm{e}^{-nG_n(z)}\right]\mathbf{M}_n(z_j)\begin{pmatrix}0 & 0\\ 1 & 0\end{pmatrix}\mathbf{M}_n(z_j)^{-1}. \tag{6.6.21}$$

We are interested in an asymptotic series up to the order n^{-2}. We therefore consider only terms in the above expansion for which $\kappa_j \leq 2$. Also, since $\kappa_j > 1$ and $z_j = \mathcal{O}(n^{-1})$ for these terms, we may replace $\mathbf{M}_n(z_j)$ with $\mathbf{M}_{n+}(0)$, giving

$$I_+ = -\mathbf{M}_{n+}(0)\begin{pmatrix}0 & 0\\ 1 & 0\end{pmatrix}\mathbf{M}_{n+}(0)^{-1}\sum_{j:\kappa_j\leq 2}\operatorname*{Res}_{z=z_j}\left[\mathrm{e}^{-nG_n(z)}\right] + \mathcal{O}(n^{-2-\varepsilon}), \tag{6.6.22}$$

for some $\varepsilon > 0$. We are particularly interested in the off-diagonal entries of I_+. Using (6.4.22) and (6.4.25) we find that

$$\begin{aligned}\left[\mathbf{M}_{n+}(0)\begin{pmatrix}0 & 0\\ 1 & 0\end{pmatrix}\mathbf{M}_{n+}(0)^{-1}\right]_{12} &= -(p-\mathrm{i}q)^2,\\ \left[\mathbf{M}_{n+}(0)\begin{pmatrix}0 & 0\\ 1 & 0\end{pmatrix}\mathbf{M}_{n+}(0)^{-1}\right]_{21} &= (p+\mathrm{i}q)^2.\end{aligned} \tag{6.6.23}$$

Inserting this along with (6.6.20) into (6.6.21) gives

$$\begin{aligned}[I_+]_{12} &= \mathrm{i}\mathrm{e}^{\mathrm{i}n\omega}(p-\mathrm{i}q)^2\sum_{j:\kappa_j\leq 2} C_j n^{-\kappa_j} + \mathcal{O}(n^{-2-\varepsilon}),\\ [I_+]_{21} &= -\mathrm{i}\mathrm{e}^{\mathrm{i}n\omega}(p+\mathrm{i}q)^2\sum_{j:\kappa_j\leq 2} C_j n^{-\kappa_j} + \mathcal{O}(n^{-2-\varepsilon}).\end{aligned} \tag{6.6.24}$$

We can evaluate I_- is an analogous fashion, using (6.6.20) and (6.4.25) to find

$$\begin{aligned}[I_+]_{12} &= \mathrm{i}\mathrm{e}^{-\mathrm{i}n\omega}(p+\mathrm{i}q)^2\sum_{j:\kappa_j\leq 2} C_j n^{-\kappa_j} + \mathcal{O}(n^{-2-\varepsilon}),\\ [I_+]_{21} &= -\mathrm{i}\mathrm{e}^{-\mathrm{i}n\omega}(p-\mathrm{i}q)^2\sum_{j:\kappa_j\leq 2} C_j n^{-\kappa_j} + \mathcal{O}(n^{-2-\varepsilon}).\end{aligned} \tag{6.6.25}$$

Adding (6.6.24) and (6.6.25) gives

$$\begin{aligned}[I_+ + I_-]_{12} &= 2\mathrm{i}[(p^2-q^2)\cos(n\omega) + 2pq\sin(n\omega)]\sum_{j:\kappa_j\leq 2} C_j n^{-\kappa_j} + \mathcal{O}(n^{-2-\varepsilon}),\\ [I_+ + I_-]_{21} &= -2\mathrm{i}[(p^2-q^2)\cos(n\omega) - 2pq\sin(n\omega)]\sum_{j:\kappa_j\leq 2} C_j n^{-\kappa_j} + \mathcal{O}(n^{-2-\varepsilon}).\end{aligned} \tag{6.6.26}$$

Notice that

$$p^2 - q^2 = \frac{1}{2}, \quad 2pq = \frac{1}{2}\tan\left(\frac{\pi\zeta}{2}\right), \tag{6.6.27}$$

and therefore (6.6.26) is

$$
\begin{aligned}
[I_+ + I_-]_{12} &= \mathrm{i}\left[(\cos(n\omega) + \tan\left(\frac{\pi\zeta}{2}\right)\sin(n\omega)\right]\sum_{j:\kappa_j\le 2} C_j n^{-\kappa_j} + \mathcal{O}(n^{-2-\varepsilon}),\\
[I_+ + I_-]_{21} &= -\mathrm{i}\left[\cos(n\omega) - \tan\left(\frac{\pi\zeta}{2}\right)\sin(n\omega)\right]\sum_{j:\kappa_j\le 2} C_j n^{-\kappa_j} + \mathcal{O}(n^{-2-\varepsilon}).
\end{aligned}
\tag{6.6.28}
$$

6.7. Proof of Proposition 6.1.2

From (6.4.40), (6.6.1), and (6.4.20), we have

$$
h_{nn} = -2\pi \mathrm{i}\mathrm{e}^{nl_n}\left(\frac{\beta-\alpha}{-4\mathrm{i}} + [I_\alpha + I_\beta + I_+ + I_-]_{12} + \mathcal{O}(n^{-2})\right). \tag{6.7.1}
$$

Using (6.6.13), and (6.6.28), this is

$$
h_{nn} = \mathrm{e}^{nl_n}\frac{\pi(\beta\ \ \alpha)}{2}\left(1 + \frac{1}{4n} + \varepsilon_n + \mathcal{O}(n^{-2})\right), \tag{6.7.2}
$$

where

$$
\varepsilon_n = \left[\cos(n\omega) + \tan\left(\frac{\pi\zeta}{2}\right)\sin(n\omega)\right]\sum_{j:\kappa_j\le 2} c_j n^{-\kappa_j}, \quad c_j = \frac{4C_j}{\beta-\alpha}. \tag{6.7.3}
$$

Using (6.3.138), this is

$$
h_{nn} = \frac{\pi(\beta-\alpha)^{2n+1}}{2^{4n+1}\mathrm{e}^{2n}}\left(1 + \frac{1}{4n} - \frac{2\gamma^2}{3n\pi(\pi-2\gamma)} + \varepsilon_n + \mathcal{O}(n^{-2})\right). \tag{6.7.4}
$$

From (6.2.4) we then have

$$
h_n = \frac{\pi(n(\beta-\alpha)/\gamma)^{2n+1}}{2^{4n+1}\mathrm{e}^{2n}}\left(1 + \frac{1}{4n} - \frac{2\gamma^2}{3n\pi(\pi-2\gamma)} + \varepsilon_n + \mathcal{O}(n^{-2})\right). \tag{6.7.5}
$$

Using Stirling's formula, we then have

$$
\begin{aligned}
\frac{h_n}{(n!)^2} &= \left(\frac{\beta-\alpha}{4\gamma}\right)^{2n+1}\left(1 + \frac{1}{n}\left(\frac{1}{4} - \frac{1}{6} - \frac{2\gamma^2}{3\pi(\pi-2\gamma)}\right) + \varepsilon_n + \mathcal{O}(n^{-2})\right)\\
&= \left(\frac{\beta-\alpha}{4\gamma}\right)^{2n+1}\left(1 + \frac{\kappa}{n} + \varepsilon_n + \mathcal{O}(n^{-2})\right).
\end{aligned}
\tag{6.7.6}
$$

Applying the formula (6.3.7) for $\beta-\alpha$, we get (6.1.21). □

6.8. The constant term

Let us now prove the formula (6.1.30) for the constant C in the asymptotic expansion for Z_n. From (6.1.18) we have

$$
\log Z_n = n^2\log(ab) + \log\tau_n - 2\sum_{k=0}^{n-1}\log k!, \quad \tau_n = \prod_{k=0}^{n-1} h_k. \tag{6.8.1}
$$

As described in (4.4.15), τ_n satisfies the Toda equation

$$
(\log\tau_n)'' = \frac{\tau_{n+1}\tau_{n-1}}{\tau_n^2} = \frac{h_n}{h_{n-1}}, \quad (') = \frac{\partial}{\partial t}, \tag{6.8.2}
$$

and we therefore have

$$(\log Z_n)'' = n^2(\log ab)'' + \frac{h_n}{h_{n-1}}. \tag{6.8.3}$$

We can use Proposition 6.1.2 to write an expansion of the ratio h_n/h_{n-1}. Notice that the coefficients in the subleading terms contained in ε_n are quasi-periodic in n, and therefore do not cancel in the ratio. Nonetheless, using some trigonometric identities we find that

$$\frac{h_n}{h_{n-1}} = n^2G^2\left(1 + 2(-1)^n\cos(n\pi\zeta)\sum_{j:\kappa_j\le 2} c_j n^{-\kappa_j} - \frac{\kappa}{n^2} + \mathcal{O}(n^{-2-\varepsilon})\right), \tag{6.8.4}$$

for some $\varepsilon > 0$. From (6.1.22), (6.3.26), (6.5.19), (6.6.18), and (6.7.3), it is easy to see that the product G^2c_j does not depend on t. We therefore have

$$\frac{h_n}{h_{n-1}} = n^2G^2 - G^2\kappa + \cos(n\pi\zeta)\sum_{j:\kappa_j\le 2} d_j n^{2-\kappa_j} + \mathcal{O}(n^{-\varepsilon}), \tag{6.8.5}$$

where d_j are some numbers which depend on γ but not on n or t. Notice that

$$G^2 = -\left(\log\left(\cos\left(\frac{\pi t}{2\gamma}\right)\right)\right)'', \qquad \cos(n\pi\zeta) = -\left(n^{-2}\left(\frac{\gamma}{\pi}\right)^2\cos(n\pi\zeta)\right)''. \tag{6.8.6}$$

We can therefore write (6.8.5) as

$$\frac{h_n}{h_{n-1}} = -\left(\log\left(\cos\left(\frac{\pi t}{2\gamma}\right)\right)\right)''(n^2-\kappa) - \left(\frac{\gamma}{\pi}\right)^2(\cos(n\pi\zeta))''\sum_{j:\kappa_j\le 2} d_j n^{-\kappa_j} + \mathcal{O}(n^{-\varepsilon}). \tag{6.8.7}$$

Inserting this into (6.8.3) gives

$$\begin{aligned}(\log Z_n)'' = n^2(\log ab)'' &- \left(\log\left(\cos\left(\frac{\pi t}{2\gamma}\right)\right)\right)''(n^2-\kappa)\\ &- \left(\frac{\gamma}{\pi}\right)^2(\cos(n\pi\zeta))''\sum_{j:\kappa_j\le 2} d_j n^{-\kappa_j} + \mathcal{O}(n^{-\varepsilon}).\end{aligned} \tag{6.8.8}$$

Now integrating twice with respect to t, we find

$$\begin{aligned}\log Z_n &= n^2(\log ab) - \log\left(\cos\left(\frac{\pi t}{2\gamma}\right)\right)(n^2-\kappa)\\ &\quad - \left(\frac{\gamma}{\pi}\right)^2\cos(n\pi\zeta)\sum_{j:\kappa_j\le 2} d_j n^{-\kappa_j} + d_1(\gamma,n)t + d_0(\gamma,n) + \mathcal{O}(n^{-\varepsilon})\\ &= n^2(\log ab) - \log\left(\cos\left(\frac{\pi t}{2\gamma}\right)\right)(n^2-\kappa) + d_1(\gamma,n)t + d_0(\gamma,n) + \mathcal{O}(n^{-\varepsilon}),\end{aligned} \tag{6.8.9}$$

where d_1 and d_0 may depend on γ and on n, but not on t.

On the other hand, from Theorem 6.1.3, we have

$$
\begin{aligned}
(6.8.10)\quad \log Z_n &= \log C + \kappa \log n + n^2 \log F + \mathcal{O}(n^{-1}) \\
&= \log C + \kappa \log n + n^2 \log\left(\frac{ab\pi}{2\gamma}\right) - n^2 \log\left(\cos\left(\frac{\pi t}{2\gamma}\right)\right) + \mathcal{O}(n^{-1}).
\end{aligned}
$$

Combining (6.8.9) and (6.8.10), we find that

$$
\log C = \kappa \log\left(\cos\left(\frac{\pi t}{2\gamma}\right)\right) + d_1(\gamma, n)t + d_2(\gamma, n) + \mathcal{O}(n^{-\varepsilon}), \tag{6.8.11}
$$

where

$$
d_2(\gamma, n) = d_0(\gamma, n) - \kappa \log n - n^2 \log\left(\frac{\pi}{2\pi}\right). \tag{6.8.12}
$$

Consider γ fixed and two distinct values of t, say $0 < t_1 < t_2 < \gamma$. From (6.8.11) we have that

$$
\log C(t_2) - \log C(t_1) = \kappa \log\left(\frac{\cos(\pi t_2/(2\gamma))}{\cos(\pi t_1/(2\gamma))}\right) + d_1(\gamma, n)(t_2 - t_1), \tag{6.8.13}
$$

from which we find that $d_1(\gamma, n)$ has a limit as $n \to \infty$, which we can denote $d_1(\gamma)$. It then follows that $d_2(\gamma, n)$ has a limit as $n \to \infty$ as well, which we can denote $d_2(\gamma)$. Takng limits as $n \to \infty$ in (6.8.11) and exponentiating, we then have

$$
C = \left(\cos\left(\frac{\pi t}{2\gamma}\right)\right)^{\kappa} \mathrm{e}^{d_1(\gamma)t + d_2(\gamma)}. \tag{6.8.14}
$$

From the Izergin – Korepin formula, Z_n is an even function of t, and therefore $d_1(\gamma) \equiv 0$. This proves (6.1.30). □

CHAPTER 7

Antiferroelectric Phase

7.1. Introduction

In this chapter we discuss the antiferroelectric phase region, and we will use parameterization (4.4.3):

$$a = \sinh(\gamma - t), \quad b = \sinh(\gamma + t), \quad c = \sinh(2\gamma), \qquad |t| < \gamma. \tag{7.1.1}$$

The parameter Δ in the antiferroelectric phase region reduces to

$$\Delta = -\cosh(2\gamma) < -1. \tag{7.1.2}$$

In this phase region, the function $\phi(t)$ appearing in the Hankel determinant τ_n is the Laplace transform of a discrete measure lying on $\mathbb{Z}$. Indeed, we have

$$\begin{aligned} \phi(t) &= \frac{\sinh(2\gamma)}{\sinh(\gamma + t)\sinh(\gamma - t)} \\ &= \frac{2(1 - e^{4\gamma})}{(1 - e^{-2(\gamma - t)})(1 - e^{-2(\gamma + t)})} \\ &= \frac{2}{1 - e^{-2(\gamma - t)}} + \frac{2}{1 - e^{-2(\gamma + t)}} - 2 \\ &= 2\sum_{l=0}^{\infty} e^{-2l(\gamma - t)} + 2\sum_{l=0}^{\infty} e^{-2l(\gamma + t)} - 2 \\ &= 2\sum_{l=0}^{\infty} e^{-2l(\gamma - t)} + 2\sum_{l=1}^{\infty} e^{-2l(\gamma + t)} = 2\sum_{l=-\infty}^{\infty} e^{2tl} e^{-2\gamma|l|}. \end{aligned} \tag{7.1.3}$$

We therefore introduce the discrete monic polynomials $P_j(x) = x^j + \dots$ orthogonal on the set $\mathbb{Z}$ with respect to the weight

$$w(l) = e^{2tl} e^{-2\gamma|l|}, \tag{7.1.4}$$

so that

$$\sum_{l=-\infty}^{\infty} P_j(l) P_k(l) w(l) = h_k \delta_{jk}. \tag{7.1.5}$$

Then it follows from (7.1.3) and (4.4.14) that

$$\tau_n = 2^{n^2} \prod_{k=0}^{n-1} h_k, \tag{7.1.6}$$

and thus by (4.4.5),

$$Z_n = (2ab)^{n^2} \prod_{k=0}^{n-1} \frac{h_k}{(k!)^2}. \tag{7.1.7}$$

We will evaluate the asymptotics of the orthogonal polynomials (7.1.5) via the Riemann – Hilbert approach of Chapter 3. To employ this method, we will have to rescale the polynomials. Set

$$\Delta_n = \frac{2\gamma}{n}, \quad x = l\Delta_n, \quad w_n(x) = \mathrm{e}^{-nV(x)}, \quad \zeta = \frac{t}{\gamma}, \quad V(x) = |x| - \zeta x, \tag{7.1.8}$$

and

$$P_{nk}(x) = \Delta_n^k P_k\left(\frac{x}{\Delta_n}\right). \tag{7.1.9}$$

Consider also the lattice

$$L_n = \left\{x = \frac{2\gamma k}{n}, k \in \mathbb{Z}\right\}. \tag{7.1.10}$$

Then from (7.1.5) we obtain that the monic polynomials $P_{nk}(x)$ satisfy the orthogonality condition,

$$\sum_{x\in L_n} P_{nj}(x)P_{nk}(x)w_n(x)\Delta_n = h_{nk}\delta_{jk}, \quad h_{nk} = h_k\Delta_n^{2k+1}. \tag{7.1.11}$$

To employ the Riemann–Hilbert method, we will need to extend the weight $w_n(x)$, which is singular at the origin, to the complex plane. We do so by defining $w_n(z)$ on the complex plane as

$$w_n(z) = \mathrm{e}^{-nV(z)}, \tag{7.1.12}$$

where

$$V(z) = \begin{cases} z - \zeta z & \text{when } \operatorname{Re} z \geq 0 \\ -z - \zeta z & \text{when } \operatorname{Re} z \leq 0, \end{cases} \tag{7.1.13}$$

so that $V(z)$, and thus $w_n(z)$, is two-valued on the imaginary axis.

7.2. Jacobi theta functions: Definitions and properties

The formulation and proofs of the large n asymptotic formula for Z_n in the antiferroelectric phase region involve the Jacobi theta functions, which are the genus one version of the Riemann theta functions described in Section 2.4.5. Let us review their definition and basic properties. For a more complete description, see, e.g., [76].

There are four Jacobi theta functions:

$$\begin{aligned} \vartheta_1(z) &= 2\sum_{n=0}^{\infty}(-1)^n q^{(n+1/2)^2}\sin\big((2n+1)z\big), \\ \vartheta_2(z) &= 2\sum_{n=0}^{\infty} q^{(n+1/2)^2}\cos\big((2n+1)z\big), \\ \vartheta_3(z) &= 1 + 2\sum_{n=1}^{\infty} q^{n^2}\cos(2nz), \\ \vartheta_4(z) &= 1 + 2\sum_{n=1}^{\infty}(-1)^n q^{n^2}\cos(2nz), \end{aligned} \tag{7.2.1}$$

where q is the *elliptic nome.* In general, q can be any complex number such that $|q| < 1$, but we will consider only real values $0 < q < 1$. Figure 7.1 shows the graphs

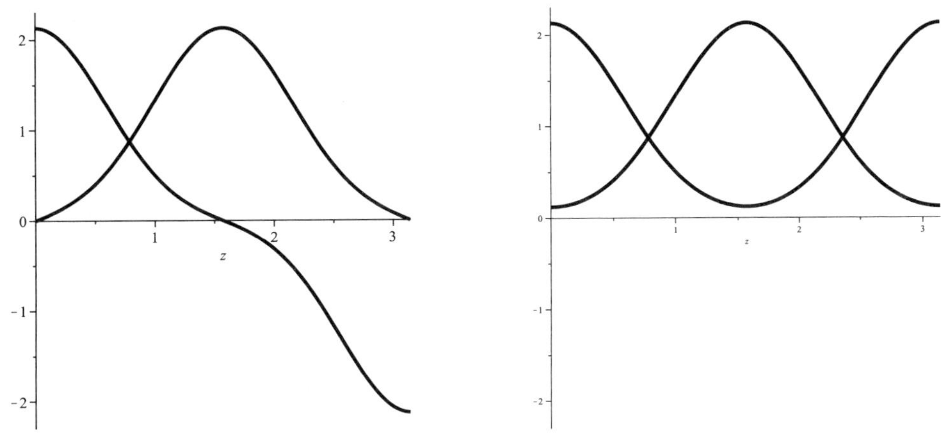

FIGURE 7.1. The graphs of ϑ_1, ϑ_2 (left figure) and ϑ_3, ϑ_4 (right figure) on the interval $[0, \pi]$ for $q = 0.5$.

of ϑ_1, ϑ_2 (left figure) and ϑ_3, ϑ_4 (right figure) on the interval $[0, \pi]$ for $q = 0.5$. Observe that ϑ_1, ϑ_4 are increasing on $[0, \pi/2]$ while ϑ_2, ϑ_3 are decreasing on this interval.

The Jacobi theta functions satisfy the following periodicity conditions:

$$\begin{aligned}
&\vartheta_1(z+\pi) = -\vartheta_1(z), && \vartheta_1(z+\pi\tau) = -\mathrm{e}^{-2\mathrm{i}z}q^{-1}\vartheta_1(z),\\
&\vartheta_2(z+\pi) = -\vartheta_2(z), && \vartheta_2(z+\pi\tau) = \mathrm{e}^{-2\mathrm{i}z}q^{-1}\vartheta_2(z),\\
&\vartheta_3(z+\pi) = \vartheta_3(z), && \vartheta_3(z+\pi\tau) = \mathrm{e}^{-2\mathrm{i}z}q^{-1}\vartheta_3(z),\\
&\vartheta_4(z+\pi) = \vartheta_4(z), && \vartheta_4(z+\pi\tau) = -\mathrm{e}^{-2\mathrm{i}z}q^{-1}\vartheta_4(z),
\end{aligned} \tag{7.2.2}$$

where τ is related to q by the equation,

$$q = \mathrm{e}^{\mathrm{i}\pi\tau}. \tag{7.2.3}$$

Notice that in general τ can be any complex number with positive imaginary part, but if q is to be real, τ must lie on the positive imaginary axis. The theta functions also satisfy the symmetry conditions

$$\begin{aligned}
&\vartheta_1(-z) = -\vartheta_1(z), && \vartheta_2(-z) = \vartheta_2(z),\\
&\vartheta_3(-z) = \vartheta_3(z), && \vartheta_4(-z) = \vartheta_4(z),
\end{aligned} \tag{7.2.4}$$

and the equations

$$\begin{gathered}
\vartheta_1(z) = \vartheta_2\left(z - \frac{\pi}{2}\right), \quad \vartheta_3(z) = \vartheta_4\left(z + \frac{\pi}{2}\right),\\
\vartheta_1(z) = -\mathrm{i}\mathrm{e}^{\mathrm{i}z+\mathrm{i}\pi\tau/4}\vartheta_4\left(z + \frac{\pi\tau}{2}\right).
\end{gathered} \tag{7.2.5}$$

The only zeroes of the theta functions are

$$\vartheta_1(0) = 0, \quad \vartheta_2\left(\frac{\pi}{2}\right) = 0, \quad \vartheta_3\left(\frac{\pi}{2} + \frac{\pi\tau}{2}\right) = 0, \quad \vartheta_4\left(\frac{\pi\tau}{2}\right) = 0, \tag{7.2.6}$$

and their shifts by $m\pi + n\pi\tau$; $m, n \in \mathbb{Z}$. There are many non-trivial identities satisfied by the theta functions. Let us summarize those identities which we will need. All of the following formulas can be found in or easily derived from formulas found in [76]. There are the identities involving derivatives of theta functions:

$$\vartheta_1'(0) = \vartheta_2(0)\vartheta_3(0)\vartheta_4(0), \tag{7.2.7}$$

$$\begin{aligned}
\vartheta_4'(z) &= \frac{\vartheta_1'(z)\vartheta_4(z) - \vartheta_4(0)^2\vartheta_2(z)\vartheta_3(z)}{\vartheta_1(z)},\\
\vartheta_2'(z) &= \frac{\vartheta_1'(z)\vartheta_2(z) - \vartheta_2(0)^3\vartheta_3(z)\vartheta_4(z)}{\vartheta_1(z)},\\
\vartheta_3'(z) &= \frac{\vartheta_1'(z)\vartheta_3(z) - \vartheta_3(0)^2\vartheta_2(z)\vartheta_4(z)}{\vartheta_1(z)},
\end{aligned} \tag{7.2.8}$$

$$\begin{aligned}
\vartheta_4'(z) &= \frac{\vartheta_2'(z)\vartheta_4(z) + \vartheta_1(z)\vartheta_3(z)\vartheta_3^2(0)}{\vartheta_2(z)},\\
\vartheta_1'(z) &= \frac{\vartheta_2'(z)\vartheta_1(z) + \vartheta_3(z)\vartheta_4(z)\vartheta_2^2(0)}{\vartheta_2(z)},\\
\vartheta_3'(z) &= \frac{\vartheta_2'(z)\vartheta_3(z) + \vartheta_1(z)\vartheta_4(z)\vartheta_4^2(0)}{\vartheta_2(z)},
\end{aligned} \tag{7.2.9}$$

$$\begin{aligned}
\vartheta_4''(z) &= \frac{\vartheta_2''(z)\vartheta_4(z)}{\vartheta_2(z)} + \frac{2\vartheta_2'(z)\vartheta_1(z)\vartheta_3(z)\vartheta_3^2(0)}{\vartheta_2^2(z)}\\
&\qquad + \frac{\vartheta_4(z)\vartheta_3^2(0)}{\vartheta_2^2(z)}\big(\vartheta_3^2(z)\vartheta_2^2(0) + \vartheta_1^2(z)\vartheta_4^2(0)\big),\\
\vartheta_1''(z) &= \frac{\vartheta_2''(z)\vartheta_1(z)}{\vartheta_2(z)} + \frac{2\vartheta_2'(z)\vartheta_3(z)\vartheta_4(z)\vartheta_2^2(0)}{\vartheta_2^2(z)}\\
&\qquad + \frac{\vartheta_1(z)\vartheta_2^2(0)}{\vartheta_2^2(z)}\big(\vartheta_3^2(z)\vartheta_3^2(0) + \vartheta_4^2(z)\vartheta_4^2(0)\big),\\
\vartheta_3''(z) &= \frac{\vartheta_2''(z)\vartheta_3(z)}{\vartheta_2(z)} + \frac{2\vartheta_2'(z)\vartheta_1(z)\vartheta_4(z)\vartheta_4^2(0)}{\vartheta_2^2(z)}\\
&\qquad + \frac{\vartheta_3(z)\vartheta_3^2(0)}{\vartheta_2^2(z)}\big(\vartheta_4^2(z)\vartheta_2^2(0) + \vartheta_1^2(z)\vartheta_4^2(0)\big),
\end{aligned} \tag{7.2.10}$$

$$\begin{aligned}
\vartheta_4''(z) &= \frac{\vartheta_1''(z)\vartheta_4(z)}{\vartheta_1(z)} - \frac{2\vartheta_1'(z)\vartheta_2(z)\vartheta_3(z)\vartheta_4^2(0)}{\vartheta_1^2(z)}\\
&\qquad + \frac{\vartheta_4^2(0)\vartheta_4^2(z)}{\vartheta_1^2(z)}\big(\vartheta_3^2(z)\vartheta_2^2(0) + \vartheta_2^2(z)\vartheta_3^2(0)\big),\\
\vartheta_2''(z) &= \frac{\vartheta_1''(z)\vartheta_2(z)}{\vartheta_1(z)} - \frac{2\vartheta_1'(z)\vartheta_3(z)\vartheta_4(z)\vartheta_2^2(0)}{\vartheta_1^2(z)}\\
&\qquad + \frac{\vartheta_2^2(0)\vartheta_2^2(z)}{\vartheta_1^2(z)}\big(\vartheta_4^2(z)\vartheta_3^2(0) + \vartheta_3^2(z)\vartheta_4^2(0)\big),\\
\vartheta_3''(z) &= \frac{\vartheta_1''(z)\vartheta_3(z)}{\vartheta_1(z)} - \frac{2\vartheta_1'(z)\vartheta_2(z)\vartheta_4(z)\vartheta_3^2(0)}{\vartheta_1^2(z)}\\
&\qquad + \frac{\vartheta_3^2(0)\vartheta_3^2(z)}{\vartheta_1^2(z)}\big(\vartheta_4^2(z)\vartheta_2^2(0) + \vartheta_2^2(z)\vartheta_4^2(0)\big),
\end{aligned} \tag{7.2.11}$$

the duplication formulas

$$\vartheta_1(2z) = \frac{2\vartheta_1(z)\vartheta_2(z)\vartheta_3(z)\vartheta_4(z)}{\vartheta_1'(0)}, \tag{7.2.12}$$

$$\vartheta_3(2z)\vartheta_3(0)\vartheta_2^2(0) = \vartheta_1(z)^2\vartheta_4(z)^2 + \vartheta_2(z)^2\vartheta_3(z)^2, \tag{7.2.13}$$

$$\vartheta_4(2z)\vartheta_4^3(0) = \vartheta_3^4(z) - \vartheta_2^4(z) = \vartheta_4^4(z) - \vartheta_1^4(z), \tag{7.2.14}$$

the addition formula

$$\begin{aligned}\vartheta_3(y+z)\vartheta_3(y-z)\vartheta_2^2(0) &= \vartheta_3^2(y)\vartheta_2^2(z) + \vartheta_4^2(y)\vartheta_1^2(z) \\ &= \vartheta_1^2(y)\vartheta_4^2(z) + \vartheta_2^2(y)\vartheta_3^2(z),\end{aligned} \tag{7.2.15}$$

and the identities relating squares of theta functions

$$\vartheta_1^2(z)\vartheta_4^2(0) = \vartheta_3^2(z)\vartheta_2^2(0) - \vartheta_2^2(z)\vartheta_3^2(0), \tag{7.2.16}$$

$$\vartheta_2^2(z)\vartheta_4^2(0) = \vartheta_4^2(z)\vartheta_2^2(0) - \vartheta_1^2(z)\vartheta_3^2(0), \tag{7.2.17}$$

$$\vartheta_3^2(z)\vartheta_4^2(0) = \vartheta_4^2(z)\vartheta_3^2(0) - \vartheta_1^2(z)\vartheta_2^2(0), \tag{7.2.18}$$

$$\vartheta_4^2(z)\vartheta_4^2(0) = \vartheta_3^2(z)\vartheta_3^2(0) - \vartheta_2^2(z)\vartheta_2^2(0). \tag{7.2.19}$$

We will also use the Jacobi elliptic functions sn, cn, and dn. The function sn can be defined from the equation

$$u = \int_0^{\mathrm{sn}(u,k)} \frac{\mathrm{d}x}{\sqrt{(1-x^2)(1-k^2x^2)}}, \tag{7.2.20}$$

where $0 < k < 1$ is a parameter known as the *elliptic modulus.* The *complete elliptic integrals of the first and second kind* depend on the elliptic modulus and are, respectively,

$$\begin{aligned}K &= \int_0^1 \frac{\mathrm{d}v'}{\sqrt{(1-v'^2)(1-k^2v'^2)}}, \\ K' &= \int_1^{\frac{1}{k}} \frac{\mathrm{d}v'}{\sqrt{(v'^2-1)(1-k^2v'^2)}}.\end{aligned} \tag{7.2.21}$$

The functions sn, cn, and dn are expressed in terms of Jacobi theta functions as,

$$\begin{aligned}\mathrm{sn}(u) &= \frac{\vartheta_3(0)}{\vartheta_2(0)}\frac{\vartheta_1\big(\pi u/(2K)\big)}{\vartheta_4\big(\pi u/(2K)\big)}, \quad \mathrm{cn}(u) = \frac{\vartheta_4(0)}{\vartheta_2(0)}\frac{\vartheta_2\big(\pi u/(2K)\big)}{\vartheta_4\big(\pi u/(2K)\big)}, \\ \mathrm{dn}(u) &= \frac{\vartheta_4(0)}{\vartheta_3(0)}\frac{\vartheta_3\big(\pi u/(2K)\big)}{\vartheta_4\big(\pi u/(2K)\big)},\end{aligned} \tag{7.2.22}$$

where the elliptic nome for the theta functions given in terms of the complete elliptic integrals:

$$q = \mathrm{e}^{\mathrm{i}\pi\tau}, \quad \tau = \frac{\mathrm{i}K'}{K}. \tag{7.2.23}$$

Furthermore, we have

$$K = \frac{\pi\vartheta_3(0)^2}{2}. \tag{7.2.24}$$

7.3. Main result: Asymptotics of the partition function

In what follows we will use the following parameters, which are simply related to the parameters t and γ:

$$\zeta = \frac{t}{\gamma} \in (-1,1), \quad \omega = \frac{\pi(1+\zeta)}{2} \in (0,\pi). \tag{7.3.1}$$

The elliptic nome for all Jacobi theta functions in this chapter will be equal to

$$q = \mathrm{e}^{-\pi^2/(2\gamma)}. \tag{7.3.2}$$

The main technical result of this chapter is the following asymptotic formula for the normalizing coefficients h_n.

Proposition 7.3.1. *As $n \to \infty$,*

$$\frac{h_n}{(n!)^2} = G^{2n+1}\,\frac{\vartheta_4\big((n+1)\omega\big)}{\vartheta_4(n\omega)}\big(1+\mathcal{O}(n^{-2})\big), \tag{7.3.3}$$

where

$$G = \frac{\pi\vartheta_1'(0)}{4\gamma\vartheta_1(\omega)}. \tag{7.3.4}$$

From this proposition we can easily obtain an asymptotic formula for Z_n.

Theorem 7.3.2. *As $n \to \infty$,*

$$Z_n = C\vartheta_4(n\omega)F^{n^2}\big(1+\mathcal{O}(n^{-1})\big), \tag{7.3.5}$$

where $C > 0$ is a constant, and

$$F = \frac{\pi\sinh(\gamma-t)\sinh(\gamma+t)\vartheta_1'(0)}{2\gamma\vartheta_1(\omega)}. \tag{7.3.6}$$

Furthermore, the constant C depends only on the parameter γ, and not on the parameter t.

Proof. By substituting (7.3.3) into (7.1.6) we obtain that

$$\begin{aligned}
\frac{\tau_n}{\prod_{k=0}^{n-1}(k!)^2} &= 2^{n^2}\prod_{k=0}^{n-1}\frac{h_k}{(k!)^2} \\
&= 2^{n^2}h_0\prod_{k=1}^{n-1}\left[G^{2k+1}\,\frac{\vartheta_4((k+1)\omega)}{\vartheta_4(k\omega)}\big(1+\mathcal{O}(k^{-2})\big)\right] \\
&= C\vartheta_4(n\omega)(2G)^{n^2}\big(1+\mathcal{O}(n^{-1})\big),
\end{aligned} \tag{7.3.7}$$

where $C > 0$ does not depend on n. Thus, by (4.4.5),

$$Z_n = \frac{[\sinh(\gamma-t)\sinh(\gamma+t)]^{n^2}\tau_n}{(\prod_{k=0}^{n-1}k!)^2} = C\vartheta_4(n\omega)F^{n^2}\big(1+\mathcal{O}(n^{-1})\big), \tag{7.3.8}$$

where

$$F = 2G\sinh(\gamma-t)\sinh(\gamma+t) = \frac{\pi\sinh(\gamma-t)\sinh(\gamma+t)\vartheta_1'(0)}{2\gamma\vartheta_1(\omega)}. \tag{7.3.9}$$

This proves the formula (7.3.5) up to the constant term. The fact that the constant C depends only on γ is a slight improvement on the result obtained in [15], and the proof is given in Section 7.7. □

The asymptotic formula (7.3.5) proves the conjecture of Zinn-Justin in [78]. Let us now consider the equilibrium measure associated with the potential $V(x)$.

7.4. Equilibrium measure

We can construct the equilibrium measure for this problem. The constrained equilibrium problem (see Section 3.3) for the potential

$$V(x) = |x| - \zeta x, \tag{7.4.1}$$

is the minimizer of the functional

$$H(\nu) = \iint_{x \neq y} \log \frac{1}{|x-y|} \, \mathrm{d}\nu(x) \, \mathrm{d}\nu(y) + \int (|x| - \zeta x) \, \mathrm{d}\nu(x), \tag{7.4.2}$$

where all integrals are over $\mathbb{R}$. According to (7.1.10), the mesh of the lattice of orthogonality is $2\gamma/n$, and therefore the upper constraint on the equilibrium measure is is $\nu \leq \sigma/2\gamma$, where σ is Lebesgue measure. We thus define

$$E_0 = \inf_{\nu} H(\nu), \tag{7.4.3}$$

where the infimum is taken over the set

$$\mathcal{M} = \left\{ \text{Positive probability measures } \nu \text{ on } \mathbb{R} \text{ such that } \nu \leq \frac{\sigma}{2\gamma} \right\}. \tag{7.4.4}$$

As discussed in Section 3.3, there exists a unique minimizer ν_0, so that

$$E_0 = H(\nu_0). \tag{7.4.5}$$

Recall that the equilibrium measure ν_0 has support on a finite number of intervals, and is absolutely continuous with respect to the Lebesgue measure.

Denote the density function of the equilibrium measure as $\rho(x)$, and its resolvent as ω, so we have

$$\frac{\mathrm{d}\nu_0}{\mathrm{d}x} = \rho(x), \quad \omega(z) = \int \frac{\rho(x) \, \mathrm{d}x}{z - x}, \tag{7.4.6}$$

and

$$\rho(x) = \frac{1}{2\pi \mathrm{i}} \big(\omega(x - \mathrm{i}0) - \omega(x + \mathrm{i}0) \big). \tag{7.4.7}$$

Because the potential $V(x)$ is convex and vanishes at the origin, we expect that the equilibrium measure is supported on a single interval, and that the upper constraint is active at the origin. That is, ν_0 has support on an interval $[\alpha, \beta]$, with a saturated region $[\alpha', \beta']$ in which

$$\rho(x) = \frac{1}{2\gamma}, \quad x \in [\alpha', \beta'], \tag{7.4.8}$$

and two unsaturated regions, $[\alpha, \alpha']$ and $[\beta', \beta]$, in which

$$0 < \rho(x) < \frac{1}{2\gamma}, \quad x \in (\alpha, \alpha') \cup (\beta', \beta), \tag{7.4.9}$$

with

$$\alpha < \alpha' < 0 < \beta' < \beta, \tag{7.4.10}$$

see Figure 7.2.

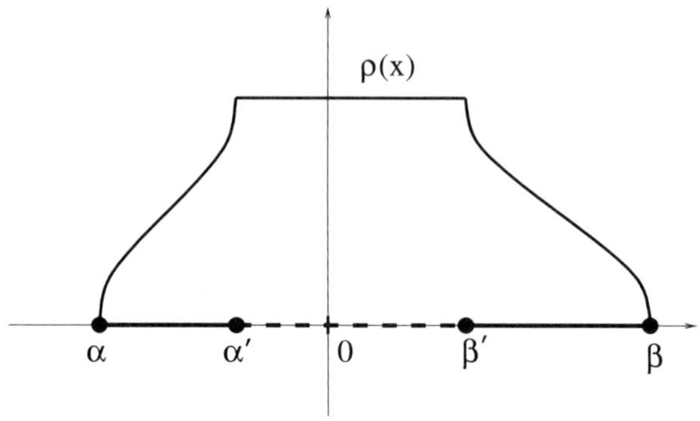

FIGURE 7.2. The equilibrium density function $\rho(x)$.

The Euler – Lagrange variational conditions (3.3.10) are

$$2\int \log|x-y|\,\mathrm{d}\nu_0(y) - (|x| - \zeta x) \begin{cases} = l & \text{for } x \in [\alpha,\alpha'] \cup [\beta',\beta] \\ \geq l & \text{for } x \in [\alpha',\beta'] \\ \leq l & \text{for } x \notin [\alpha,\beta], \end{cases} \tag{7.4.11}$$

where l is the Lagrange multiplier. The Euler – Lagrange variational conditions imply

$$\omega(x - \mathrm{i}0) + \omega(x + \mathrm{i}0) = -\zeta + \operatorname{sgn}(x) \quad \text{for } x \in [\alpha,\alpha'] \cup [\beta',\beta], \tag{7.4.12}$$

whereas in the saturated region, we have

$$\rho(x) = \frac{1}{2\pi\mathrm{i}}\big(\omega(x-\mathrm{i}0) - \omega(x+\mathrm{i}0)\big) = \frac{1}{2\gamma} \quad \text{for } x \in [\alpha',\beta']. \tag{7.4.13}$$

7.4.1. Formula for the equilibrium measure. Let us now give a detailed description of the equilibrium measure. These results appear in the papers [78] of Zinn-Justin and [15] of the current authors. We first describe the locations of the endpoints in Propostion 7.4.1, and then give an explicit formula for the equilibrium measure in Proposition 7.4.2. The proofs of these propositions are given in Section 7.4.3.

Proposition 7.4.1. *The end-points of the support of the equilibrium measure ν_0 are equal to*

$$\begin{aligned} \alpha &= -\pi\frac{\vartheta_1'(\omega/2)}{\vartheta_1(\omega/2)}, & \alpha' &= -\pi\frac{\vartheta_4'(\omega/2)}{\vartheta_4(\omega/2)}, \\ \beta' &= -\pi\frac{\vartheta_3'(\omega/2)}{\vartheta_3(\omega/2)}, & \beta &= -\pi\frac{\vartheta_2'(\omega/2)}{\vartheta_2(\omega/2)}. \end{aligned} \tag{7.4.14}$$

The differences between the end-points are equal to

$$\begin{aligned} \alpha' - \alpha &= \pi\vartheta_4^2(0)\frac{\vartheta_2(\omega/2)\vartheta_3(\omega/2)}{\vartheta_1(\omega/2)\vartheta_4(\omega/2)}, \quad \beta' - \alpha' = \pi\vartheta_2^2(0)\frac{\vartheta_1(\omega/2)\vartheta_2(\omega/2)}{\vartheta_3(\omega/2)\vartheta_4(\omega/2)}, \\ \beta - \beta' &= \pi\vartheta_4(0)^2\frac{\vartheta_1(\omega/2)\vartheta_4(\omega/2)}{\vartheta_2(\omega/2)\vartheta_3(\omega/2)}. \end{aligned} \tag{7.4.15}$$

and

$$\begin{aligned}\beta-\alpha&=\pi\vartheta_2^2(0)\frac{\vartheta_3(\omega/2)\vartheta_4(\omega/2)}{\vartheta_1(\omega/2)\vartheta_2(\omega/2)},\quad \beta-\alpha'=\pi\vartheta_3(0)^2\frac{\vartheta_1(\omega/2)\vartheta_3(\omega/2)}{\vartheta_2(\omega/2)\vartheta_4(\omega/2)},\\ \beta'-\alpha&=\pi\vartheta_3^2(0)\frac{\vartheta_2(\omega/2)\vartheta_4(\omega/2)}{\vartheta_1(\omega/2)\vartheta_3(\omega/2)},\end{aligned}\tag{7.4.16}$$

where the parameter ω *is as in* (7.3.1).

Proposition 7.4.2. *The equilibrium density function* $\rho(x)$ *is given by the formula*

$$\rho(x)=\begin{cases}\dfrac{1}{\pi}\displaystyle\int_\alpha^x\frac{\mathrm{d}x'}{\sqrt{(x'-\alpha)(\alpha'-x')(\beta'-x')(\beta-x')}}, & \alpha\le x\le\alpha'\\ \dfrac{1}{2\gamma}, & \alpha'\le x\le\beta'\\ \dfrac{1}{\pi}\displaystyle\int_x^\beta\frac{\mathrm{d}x'}{\sqrt{(x'-\alpha)(x'-\alpha')(x'-\beta')(\beta-x')}}, & \beta'\le x\le\beta.\end{cases}\tag{7.4.17}$$

Also,

$$\int_0^\beta\rho(x)\,\mathrm{d}x=\frac{1+\zeta}{2}.\tag{7.4.18}$$

The resolvent $\omega(z)$ *of the equilibrium measure is given as*

$$\omega(z)=\int_z^\infty\frac{\mathrm{d}z'}{\sqrt{(z'-\alpha)(z'-\alpha')(z'-\beta')(z'-\beta)}},\tag{7.4.19}$$

where integration takes place on the sheet of

$$\sqrt{R(z')}\equiv\sqrt{(z'-\alpha)(z'-\alpha')(z'-\beta')(z'-\beta)}$$

for which $\sqrt{R(z')}>0$ *for* $z'>\beta$, *with cuts on* $[\alpha,\alpha']$ *and* $[\beta',\beta]$.

The proof of this proposition is given in Section 7.4.3.

7.4.2. The g-function. Let us now describe the g-function and the Lagrange multiplier, which we will use in the first transformation of the Riemann–Hilbert problem. The g-function is defined for $z\in\mathbb{C}\setminus[-\infty,\beta]$ as

$$g(z)=\int_\alpha^\beta\log(z-x)\,\mathrm{d}\nu_0(x)\tag{7.4.20}$$

where we take the principal branch for logarithm. Let us review the properties of the g-function (see Section 3.4).

Properties of $g(z)$.

(1) $g(z)$ is analytic in $\mathbb{C}\setminus(-\infty,\beta]$.
(2) For large z,

$$g(z)=\log z-\sum_{j=1}^\infty\frac{g_j}{z^j},\quad g_j=\int_\alpha^\beta\frac{x^j}{j}\,\mathrm{d}\nu_0(x).\tag{7.4.21}$$

(3) $g'(z)=\omega(z)$.

(4) From the first relation in (7.4.11) we have that

$$g_+(x) + g_-(x) = |x| - \zeta x + l \quad \text{for } x \in [\alpha, \alpha'] \cup [\beta', \beta], \tag{7.4.22}$$

where g_+ and g_- refer to the limiting values of g from the upper and lower half-planes, respectively. By differentiating this equation we obtain that

$$\omega_+(x) + \omega_-(x) = g'_+(x) + g'_-(x) = \operatorname{sgn} x - \zeta \quad \text{for } x \in [\alpha, \alpha'] \cup [\beta', \beta]. \tag{7.4.23}$$

Consider the function

$$f(x) = g_+(x) + g_-(x) - (|x| - \zeta x + l). \tag{7.4.24}$$

We have from (7.4.22), (7.4.23) that

$$f(x) = f'(x) = 0 \quad \text{for } x = \alpha, \alpha', \beta', \beta, \tag{7.4.25}$$

and from (7.4.19) that

$$f''(x) = -\frac{1}{\sqrt{(x-\alpha)(x-\alpha')(x-\beta')(x-\beta)}} \qquad \text{for } x \in (-\infty, \alpha) \cup (\alpha', \beta') \cup (\beta, \infty). \tag{7.4.26}$$

Since

$$f''(x) < 0 \quad \text{for} x \in (-\infty, \alpha) \cup (\beta, \infty). \tag{7.4.27}$$

and

$$f''(x) > 0 \quad \text{for } x \in (\alpha', \beta'),\ x \neq 0, \tag{7.4.28}$$

we obtain that

$$g_+(x) + g_-(x) \begin{cases} = |x| - \zeta x + l & \text{for } x \in [\alpha, \alpha'] \cup [\beta', \beta], \\ > |x| - \zeta x + l & \text{for } x \in (\alpha', \beta'), \\ < |x| - \zeta x + l & \text{for } x \in \mathbb{R} \setminus [\alpha, \beta]. \end{cases} \tag{7.4.29}$$

In the language of Chapter 3, this means that the equilibrium measure is *regular*.

(5) Equation (7.4.20) implies that the function

$$G(x) \equiv g_+(x) - g_-(x), \tag{7.4.30}$$

is pure imaginary for all real x, and

$$G(x) = \begin{cases} 2\pi \mathrm{i} & \text{for } -\infty < x \le \alpha \\ 2\pi \mathrm{i} - 2\pi \mathrm{i} \displaystyle\int_\alpha^x \rho(s)\, \mathrm{d}s & \text{for } \alpha \le x \le \alpha' \\ 2\pi \mathrm{i} \left(\dfrac{1+\zeta}{2} - \dfrac{x}{2\gamma} \right) & \text{for } \alpha' \le x \le \beta' \\ 2\pi \mathrm{i} \displaystyle\int_x^\beta \rho(s)\, \mathrm{d}s & \text{for } \beta' \le x \le \beta \\ 0 & \text{for } \beta \le x < \infty, \end{cases} \tag{7.4.31}$$

where the formula in the saturated region follows from (7.4.18). From (7.4.29) and (7.4.31) we obtain that

$$2g_\pm(x) = \begin{cases} |x| - \zeta x + l \pm \left[2\pi\mathrm{i} - 2\pi\mathrm{i}\int_\alpha^x \rho(s)\,\mathrm{d}s\right] & \text{for } \alpha \le x \le \alpha' \\ |x| - \zeta x + l \pm 2\pi\mathrm{i}\int_x^\beta \rho(s)\,\mathrm{d}s & \text{for } \beta' \le x \le \beta. \end{cases} \tag{7.4.32}$$

(6) $G(x)$ is real analytic on each of the intervals $(-\infty,\alpha), (\alpha,\alpha'), (\alpha',\beta')$, (β',β), and (β,∞), and can therefore be extended to the complex plane in a neighborhood of any of these intervals. The Cauchy–Riemann equations imply that

$$\left.\frac{\mathrm{d}G(x+\mathrm{i}y)}{\mathrm{d}y}\right|_{y=0} = 2\pi\rho(x) > 0, \quad x \in (\alpha,\beta). \tag{7.4.33}$$

(7) From (7.4.22) we have that

$$G(x) = 2g_+(x) - V(x) - l = -[2g_-(x) - V(x) - l], \quad x \in [\alpha,\alpha'] \cup [\beta',\beta], \tag{7.4.34}$$

where $V(x) \equiv |x| - \zeta x$.

The value of the Lagrange multiplier l is described in the following proposition.

Proposition 7.4.3. *The Lagrange multiplier l solves the equation*

$$\mathrm{e}^{l/2} = \frac{\pi\vartheta_1'(0)}{2\mathrm{e}\vartheta_1(\omega)}. \tag{7.4.35}$$

The proof of this proposition is given in the next subsection.

7.4.3. Proof of Propositions 7.4.1, 7.4.2, and 7.4.3.

PROOF OF PROPOSITION 7.4.1. In order to calculate the equilibrium measure, we make the elliptic change of variables

$$u(z) = \frac{1}{2}\sqrt{(\beta'-\alpha)(\beta-\alpha')}\int_\beta^z \frac{\mathrm{d}z'}{\sqrt{(z'-\alpha)(z'-\alpha')(z'-\beta')(z'-\beta)}}, \tag{7.4.36}$$

where integration takes place on the sheet of $\sqrt{R(z')}$ specified in Proposition 7.4.2. Here, α, α', β', and β are as yet unknown numbers satisfying (7.4.10). To understand this integral in terms of the Jacobi elliptic functions, we first make the change of variables

$$v(z') = \frac{(\beta - z')(\beta'-\alpha)}{(\beta'-z')(\beta-\alpha)}, \tag{7.4.37}$$

so that

$$z' = \frac{\beta'(\beta-\alpha)v - \beta(\beta'-\alpha)}{(\beta-\alpha)v - (\beta'-\alpha)}. \tag{7.4.38}$$

Note that $v(\beta) = 0$, $v(\beta') = \infty$, and $v(\alpha) = 1$. When we substitute v into equation (7.4.36), we have

$$u(z) = \frac{1}{2k}\int_0^{v(z)} \frac{dv}{\sqrt{v(v-1)(v-\frac{1}{k^2})}}, \tag{7.4.39}$$

where

$$k = \sqrt{\frac{(\beta-\alpha)(\beta'-\alpha')}{(\beta'-\alpha)(\beta-\alpha')}}. \tag{7.4.40}$$

We next take $v' = \sqrt{v}$, obtaining

$$u(z) = \int_0^{\sqrt{v(z)}} \frac{\mathrm{d}v'}{\sqrt{(1-v'^2)(1-k^2v'^2)}}. \tag{7.4.41}$$

This corresponds to $\sqrt{v(z)} = \mathrm{sn}(u,k)$, see (7.2.20), so that

$$\frac{(\beta - z)(\beta' - \alpha)}{(\beta' - z)(\beta - \alpha)} = \mathrm{sn}^2(u), \quad \mathrm{sn}(u) = \mathrm{sn}(u,k). \tag{7.4.42}$$

Notice that u maps the upper z-plane conformally and bijectively onto the rectangle $[0,K]\times[0,\mathrm{i}K']$, and the lower z-plane conformally and bijectively onto the rectangle $[0,K]\times[-\mathrm{i}K',0]$, where K and K' are the usual complete integrals of the first and second kind, respectively, see (7.2.21). More specifically (see Figure 7.3),

(1) The upper (resp. lower) cusp of the interval $[\beta',\beta]$ is mapped onto the interval $[0,\mathrm{i}K']$ (resp. $[0,-\mathrm{i}K']$).
(2) The upper (resp. lower) cusp of the interval $[\alpha,\alpha']$ is mapped onto the interval $[K, K+\mathrm{i}K']$ (resp. $[K,K-\mathrm{i}K']$).
(3) The interval $[\alpha',\beta']$ is mapped onto the inteval $[\mathrm{i}K', K+\mathrm{i}K']$ or the inteval $[-\mathrm{i}K', K-\mathrm{i}K']$, depending on the path of integration.
(4) The remaining part of the real axis, $[-\infty,\alpha]\cup[\beta,\infty]$, is mapped onto the interval $[0,K]$, with $u(\infty) = u_* = u_\infty$.

We will denote the rectangle $[0,K]\times[-\mathrm{i}K',\mathrm{i}K']$ as R, the *fundamental domain* of the function $z(u)$. We can now define

$$\widetilde{\omega}(u) := \omega\big(z(u)\big) \quad \text{for } u \in R. \tag{7.4.43}$$

The Euler–Lagrange equation (7.4.12) and the equation (7.4.13) then become

$$\widetilde{\omega}(u) + \widetilde{\omega}(-u) = 1 - \zeta \quad \text{for } u \in [-\mathrm{i}K', \mathrm{i}K'], \tag{7.4.44}$$

$$\widetilde{\omega}(u) + \widetilde{\omega}(-u+2K) = -1 - \zeta \quad \text{for } u \in [K-\mathrm{i}K', K+\mathrm{i}K'], \tag{7.4.45}$$

$$\widetilde{\omega}(u + 2\mathrm{i}K') - \widetilde{\omega}(u) = -\frac{\mathrm{i}\pi}{\gamma} \quad \text{for } u \in [-\mathrm{i}K', K-\mathrm{i}K']. \tag{7.4.46}$$

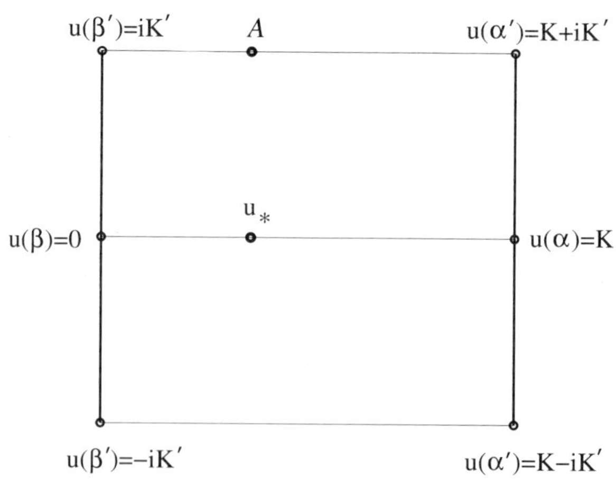

FIGURE 7.3. The u-plane. Here $u_* = u_\infty \equiv u(\infty)$ and $a = u_* + \mathrm{i}K'$.

The function $\omega(z)$ is analytic in $\mathbb{C} \setminus [\alpha, \beta]$, but can be analytically continued from either above or below through any of the cuts $[\alpha, \alpha']$, $[\alpha', \beta']$, and $[\beta', \beta]$. These analytic continuations in the z-plane give an analytic continuation of $\widetilde{\omega}$ in the u-plane into a neighborhood of R, which can then be continued by equations (7.4.44) – (7.4.46) to the entire u-plane. We therefore have that $\widetilde{\omega}$ is analytic and satisfies equations (7.4.44) – (7.4.46) throughout the u-plane. Equations (7.4.44) and (7.4.45) can be combined as

$$\widetilde{\omega}(u + 2K) = \widetilde{\omega}(u) - 2. \tag{7.4.47}$$

It therefore follows that $\widetilde{\omega}$ is a linear function of u, as its derivative is a doubly periodic entire function and therefore constant by the Liouville theorem. The slope is determined by (7.4.47), and we therefore have

$$\widetilde{\omega}(u) = -\frac{1}{K}u + C, \tag{7.4.48}$$

for some constant C. At the same time, from the fact that $\omega(z) \sim 1/z$ at infinity, we see that

$$\widetilde{\omega}(u) = -\frac{2}{\sqrt{(\beta' - \alpha)(\beta - \alpha')}}(u - u_\infty) + \mathcal{O}\big((u - u_\infty)^2\big), \tag{7.4.49}$$

in some neighborhood of u_∞, where u_∞ is the image of infinity under the map $u(z)$. We therefore have

$$\widetilde{\omega}(u) = -\frac{1}{K}(u - u_\infty), \tag{7.4.50}$$

and

$$\sqrt{(\beta' - \alpha)(\beta - \alpha')} = 2K. \tag{7.4.51}$$

The slope of $\widetilde{\omega}(u)$ can also be read from (7.4.46). Comparing with (7.4.48), we find

$$\frac{K'}{K} = \frac{\pi}{2\gamma}. \tag{7.4.52}$$

Also, from (7.4.44) and (7.4.50) we find that

$$\frac{u_\infty}{K} = \frac{1 - \zeta}{2}. \tag{7.4.53}$$

and from (7.4.42) we obtain that

$$\frac{\beta' - \alpha}{\beta - \alpha} = \operatorname{sn}^2(u_\infty). \tag{7.4.54}$$

This implies that

$$\begin{aligned} \operatorname{cn}^2(u_\infty) &= 1 - \operatorname{sn}^2(u_\infty) = 1 - \frac{\beta' - \alpha}{\beta - \alpha} = \frac{\beta - \beta'}{\beta - \alpha}, \\ \operatorname{dn}^2(u_\infty) &= 1 - k^2 \operatorname{sn}^2(u_\infty) = 1 - \frac{(\beta - \alpha)(\beta' - \alpha')}{(\beta' - \alpha)(\beta - \alpha')} \frac{(\beta' - \alpha)}{(\beta - \alpha)} = \frac{\beta - \beta'}{\beta - \alpha'}. \end{aligned} \tag{7.4.55}$$

From equations (7.4.51), (7.4.54), and (7.4.55), we obtain the distances between the turning points in terms of u_∞:

$$\begin{gathered} \beta - \alpha = 2K \frac{\operatorname{dn}(u_\infty)}{\operatorname{sn}(u_\infty) \operatorname{cn}(u_\infty)}, \quad \beta - \alpha' = 2K \frac{\operatorname{cn}(u_\infty)}{\operatorname{sn}(u_\infty) \operatorname{dn}(u_\infty)}, \\ \beta - \beta' = 2K \frac{\operatorname{cn}(u_\infty) \operatorname{dn}(u_\infty)}{\operatorname{sn}(u_\infty)}. \end{gathered} \tag{7.4.56}$$

By (7.2.23) and (7.4.52), the half-period ratio τ and the elliptic nome q of the theta functions are

$$\tau=\frac{\mathrm{i}K'}{K}=\frac{\mathrm{i}\pi}{2\gamma}\quad\text{and}\quad q=\mathrm{e}^{-\pi K'/K}=\mathrm{e}^{-\pi^2/(2\gamma)}. \tag{7.4.57}$$

Using (7.2.24) and (7.4.53), we can write equations for the distances between the turning points that involve the original parameters only:

$$\begin{gathered}\beta-\alpha=\pi\vartheta_2^2(0)\frac{\vartheta_3(\omega/2)\vartheta_4(\omega/2)}{\vartheta_1(\omega/2)\vartheta_2(\omega/2)},\quad \beta-\alpha'=\pi\vartheta_3(0)^2\frac{\vartheta_1(\omega/2)\vartheta_3(\omega/2)}{\vartheta_2(\omega/2)\vartheta_4(\omega/2)},\\ \beta'-\alpha=\pi\vartheta_3^2(0)\frac{\vartheta_2(\omega/2)\vartheta_4(\omega/2)}{\vartheta_1(\omega/2)\vartheta_3(\omega/2)},\end{gathered} \tag{7.4.58}$$

which is (7.4.16). These equations determine the end-points $\alpha,\alpha',\beta',\beta$ up to a shift. To fix the shift we use the equation (7.4.11) at the points α' and β' to obtain

$$\int_{\alpha'}^{\beta'}\big(\omega(z+\mathrm{i}0)+\omega(z-\mathrm{i}0)\big)\,\mathrm{d}z=(1-\zeta)\beta'+(1+\zeta)\alpha'. \tag{7.4.59}$$

Writing this integral in terms of u gives

$$\begin{aligned}\int_{\mathrm{i}K'}^{K+\mathrm{i}K'}\frac{1}{K}(u-u_\infty)r'(u)\,\mathrm{d}u+\int_{-\mathrm{i}K'}^{K-\mathrm{i}K'}\frac{1}{K}(u-u_\infty)r'(u)\,\mathrm{d}u&\\ =(1-\zeta)\beta'+(1+\zeta)\alpha',&\end{aligned} \tag{7.4.60}$$

where

$$\begin{aligned}r(u)&=\frac{\beta'(\beta-\alpha)\,\mathrm{sn}^2(u)-\beta(\beta'-\alpha)}{(\beta-\alpha)\,\mathrm{sn}^2(u)-(\beta'-\alpha)}=\frac{\beta-\beta'\,\mathrm{sn}^2(u)/\,\mathrm{sn}^2(u_\infty)}{1-\mathrm{sn}^2(u)/\,\mathrm{sn}^2(u_\infty)},\\ r'(u)&=\frac{\mathrm{d}}{\mathrm{d}u}r(u).\end{aligned} \tag{7.4.61}$$

Note that $r(\pm\mathrm{i}K')=\beta'$ and $r(K\pm\mathrm{i}K')=\alpha'$. Integrating by parts gives

$$\begin{aligned}\frac{2}{K}\big((K-u_\infty)\alpha'+\beta'u_\infty\big)-\int_{\mathrm{i}K'}^{K+\mathrm{i}K'}\frac{r(u)}{K}\,\mathrm{d}u-\int_{-\mathrm{i}K'}^{K-\mathrm{i}K'}\frac{r(u)}{K}\,\mathrm{d}u&\\ =(1-\zeta)\beta'+(1+\zeta)\alpha',&\end{aligned} \tag{7.4.62}$$

or equivalently

$$\int_{\mathrm{i}K'}^{K+\mathrm{i}K'}r(u)\,\mathrm{d}u+\int_{-\mathrm{i}K'}^{K-\mathrm{i}K'}r(u)\,\mathrm{d}u=0. \tag{7.4.63}$$

We can evaluate these integrals by first writing $r(u)$ in the form

$$r(u)=\beta+\left(\frac{\beta-\beta'}{\mathrm{sn}^2(u_\infty)}\right)\frac{\mathrm{sn}^2(u)}{1-\mathrm{sn}^2(u)/\,\mathrm{sn}^2(u_\infty)}, \tag{7.4.64}$$

and using the standard functions

$$\Theta(u)=\vartheta_4\left(\frac{\pi u}{2K}\right),\quad Z(u)=\frac{\Theta'(u)}{\Theta(u)}. \tag{7.4.65}$$

The addition formulas for the sn and Z functions are (see [76])

$$\begin{aligned}\mathrm{sn}(u\pm a)&=\frac{\mathrm{sn}(u)\,\mathrm{cn}(a)\,\mathrm{dn}(a)\pm\mathrm{sn}(a)\,\mathrm{cn}(u)\,\mathrm{dn}(u)}{1-k^2\,\mathrm{sn}^2(a)\,\mathrm{sn}^2(u)},\\ Z(u\pm a)&=Z(u)\pm Z(a)\mp k^2\,\mathrm{sn}(u)\,\mathrm{sn}(a)\,\mathrm{sn}(u\pm a).\end{aligned} \tag{7.4.66}$$

We thus have

$$Z(u-a)-Z(u+a)+2Z(a) = k^2\operatorname{sn}(u)\operatorname{sn}(a)\big(\operatorname{sn}(u+a)+\operatorname{sn}(u-a)\big) = \frac{k^2\operatorname{sn}(u)\operatorname{sn}(a)[2\operatorname{sn}(u)\operatorname{cn}(a)\operatorname{dn}(a)]}{1-k^2\operatorname{sn}^2(a)\operatorname{sn}^2(u)} = \frac{2k^2\operatorname{sn}(a)\operatorname{cn}(a)\operatorname{dn}(a)\operatorname{sn}^2(u)}{1-k^2\operatorname{sn}^2(a)\operatorname{sn}^2(u)}. \tag{7.4.67}$$

We also have the half- and quarter-period identities (see [76]),

$$\operatorname{sn}(u+\mathrm{i}K') = \frac{1}{k\operatorname{sn}(u)},\quad \operatorname{cn}(u+\mathrm{i}K') = \frac{-\mathrm{i}}{k}\frac{\operatorname{dn}(u)}{\operatorname{sn}(u)},\quad \operatorname{dn}(u+\mathrm{i}K') = -\mathrm{i}\frac{\operatorname{cn}(u)}{\operatorname{sn}(u)}. \tag{7.4.68}$$

In particular, notice that $1/\operatorname{sn}(u_\infty) = k\operatorname{sn}(u_\infty+\mathrm{i}K')$. Using the addition formula (7.4.67) we can write $r(u)$ as

$$r(u) = \beta+\left(\frac{\beta-\beta'}{2k^2\operatorname{sn}(a)\operatorname{cn}(a)\operatorname{dn}(a)\operatorname{sn}^2(u_\infty)}\right)\big(Z(u-a)-Z(u+a)+2Z(a)\big), \tag{7.4.69}$$

where $a = u_\infty+\mathrm{i}K'$, see Figure 7.3. From (7.4.68) and (7.4.56), it follows that

$$\frac{\beta-\beta'}{2k^2\operatorname{sn}(u_\infty+\mathrm{i}K')\operatorname{cn}(u_\infty+\mathrm{i}K')\operatorname{dn}(u_\infty+\mathrm{i}K')\operatorname{sn}^2(u_\infty)} = -K. \tag{7.4.70}$$

Thus we can write (7.4.69) as

$$r(u) = \beta - K[Z(u-u_\infty-\mathrm{i}K') - Z(u+u_\infty+\mathrm{i}K') + 2Z(u_\infty+\mathrm{i}K')]. \tag{7.4.71}$$

If we write $u = x+\mathrm{i}K'$ in the first integral of (7.4.63), and $u = x-\mathrm{i}K'$ in the second, we obtain

$$\int_0^K [2\beta-4KZ(u_\infty+\mathrm{i}K') - K[Z(x-u_\infty)-Z(x+u_\infty+2\mathrm{i}K')+Z(x-u_\infty-2\mathrm{i}K')-Z(x+u_\infty)]]\,\mathrm{d}x = 0. \tag{7.4.72}$$

From the periodic properties of ϑ_4, it follows that

$$Z(u\pm 2\mathrm{i}K') = Z(u) \mp \frac{\pi\mathrm{i}}{K}, \tag{7.4.73}$$

so we can write (7.4.72) as

$$\int_0^K \big(2\beta - 4KZ(u_\infty+\mathrm{i}K') - 2\pi\mathrm{i} + 2K\big[Z(x+u_\infty)-Z(x-u_\infty)\big]\big)\,\mathrm{d}x = 0 \tag{7.4.74}$$

This equation is readily integrated, as Z is the logarithmic derivative of the Θ function. Integrating gives

$$0 = \left[(2\beta-4KZ(u_\infty+\mathrm{i}K')-2\pi i)x + 2K\log\frac{\Theta(x+u_\infty)}{\Theta(x-u_\infty)}\right]_{x=0}^{K} = 2K\beta - 4K^2Z(u_\infty+\mathrm{i}K') - 2K\pi\mathrm{i} + 2K\log\left(\frac{\Theta(K+u_\infty)}{\Theta(K-u_\infty)}\frac{\Theta(-u_\infty)}{\Theta(u_\infty)}\right). \tag{7.4.75}$$

The logarithmic term in this equation is zero due to the evenness and periodicity (period $2K$) of the Θ function and the fact that the relevant term in the integration is real on the entire contour of integration. Thus we have that

$$\beta = 2KZ(u_\infty + \mathrm{i}K') + \pi\mathrm{i}. \tag{7.4.76}$$

From (7.2.5), we can deduce that

$$Z(u_\infty + \mathrm{i}K') = -\frac{\pi}{2K}\left(\frac{\vartheta_2'(\omega/2)}{\vartheta_2(\omega/2)} + \mathrm{i}\right), \tag{7.4.77}$$

and write (7.4.76) as

$$\beta = -\pi\,\frac{\vartheta_2'(\omega/2)}{\vartheta_2(\omega/2)}. \tag{7.4.78}$$

This equation, together with equations (7.4.58) and (7.4.57), determines the end-points $\alpha, \alpha', \beta', \beta$. Similar to (7.4.78) we have the following explicit formulas for the other end-points:

$$\alpha = -\pi\,\frac{\vartheta_1'(\omega/2)}{\vartheta_1(\omega/2)}, \quad \alpha' = -\pi\,\frac{\vartheta_4'(\omega/2)}{\vartheta_4(\omega/2)}, \quad \beta' = -\pi\,\frac{\vartheta_3'(\omega/2)}{\vartheta_3(\omega/2)}. \tag{7.4.79}$$

This follows from (7.4.58), (7.4.78), and the identities (7.2.8). Similarly, in addition to the formulas (7.4.58) for distances between turning points, we get (7.4.15). □

PROOF OF PROPOSITION 7.4.2. From equations (7.4.50), (7.4.36), and (7.4.51) we obtain formula (7.4.19), cf. [78]. From formula (7.4.19) and equations (7.4.7), (7.4.13) we obtain that the equilibrium density function $\rho(x)$ is given by formulas (7.4.17). We are left to prove formula (7.4.18).

By (7.4.36), (7.4.51), and (7.4.17), on the interval $[\beta', \beta]$,

$$\rho(x) = \frac{1}{\mathrm{i}K\pi}u_+(x) \quad \text{for } x \in [\beta', \beta]. \tag{7.4.80}$$

It follows that

$$\int_{\beta'}^{\beta} \rho(x)\,\mathrm{d}x = \frac{1}{\mathrm{i}K\pi}\int_{\beta'}^{\beta} u_+(x)\,\mathrm{d}x = \frac{1}{\mathrm{i}K\pi}\int_{\mathrm{i}K'}^{0} ur'(u)\,\mathrm{d}u, \tag{7.4.81}$$

where $r(u)$ is defined in (7.4.61). If we use equation (7.4.71), together with formula (7.4.76), we can write $r(u)$ as

$$r(u) = \mathrm{i}\pi - K[Z(u - u_\infty - \mathrm{i}K') - Z(u + u_\infty + \mathrm{i}K')]. \tag{7.4.82}$$

Integrating (7.4.81) by parts, we get

$$\begin{aligned} \int_{\mathrm{i}K'}^{0} &ur'(u)\,\mathrm{d}u \\ &= -\beta'\mathrm{i}K' - \pi K' - K\left[\log\left(\frac{\Theta(-u_\infty - \mathrm{i}K')\Theta(u_\infty + 2\mathrm{i}K')}{\Theta(-u_\infty)\Theta(u_\infty + \mathrm{i}K')}\right)\right]. \end{aligned} \tag{7.4.83}$$

Using the fact that Θ is an even function and the identity

$$\Theta(u + 2\mathrm{i}K') = \mathrm{e}^{\mathrm{i}\pi}\mathrm{e}^{-\mathrm{i}\pi\tau}\mathrm{e}^{-\mathrm{i}\pi u/K}\Theta(u), \tag{7.4.84}$$

we can write (7.4.83) as

$$\begin{aligned} \int_{\mathrm{i}K'}^{0} ur'(u)\,\mathrm{d}u &= -\beta'\mathrm{i}K' - \pi K' + K\left(\mathrm{i}\pi + \pi\frac{K'}{K} - \frac{\mathrm{i}\pi u_\infty}{K}\right) \\ &= \mathrm{i}(K\pi - \beta'K' - \pi u_\infty). \end{aligned} \tag{7.4.85}$$

Remark. There is perhaps some question here as to which branch of the logarithm to take, but it is clear that we have chosen the correct branch, as it is the only one that gives $0 < \int_{\beta'}^{\beta} \rho(x)\,dx < 1$.

Thus from (7.4.81) and (7.4.85), we have

$$\int_{\beta'}^{\beta} \rho(x)\,dx = \frac{1}{iK\pi} i(K\pi - \beta' K' - \pi u_\infty) = 1 - \frac{\beta' K'}{\pi K} - \frac{u_\infty}{K} = 1 - \frac{\beta'}{2\gamma} - \frac{1-\zeta}{2}, \tag{7.4.86}$$

hence by (7.4.17),

$$\int_0^{\beta} \rho(x)\,dx = \int_0^{\beta'} \rho(x)\,dx + \int_{\beta'}^{\beta} \rho(x)\,dx = \frac{\beta'}{2\gamma} + 1 - \frac{\beta'}{2\gamma} - \frac{1-\zeta}{2} = \frac{1+\zeta}{2}, \tag{7.4.87}$$

which proves formula (7.4.18). □

PROOF OF PROPOSITION 7.4.3. By taking $x = \beta$ we obtain from (7.4.29) that

$$l = 2g(\beta) - V(\beta) = 2g(\beta) - (1-\zeta)\beta. \tag{7.4.88}$$

We also have that

$$\lim_{A\to\infty} [g(A) - \log A] = 0, \tag{7.4.89}$$

hence

$$\begin{aligned} l &= -2 \lim_{A\to\infty} [g(A) - g(\beta) - \log A] - (1-\zeta)\beta \\ &= -2 \lim_{A\to\infty} \left[\int_\beta^A \omega(z)\,dz - \log A \right] - (1-\zeta)\beta. \end{aligned} \tag{7.4.90}$$

Writing this integral in terms of u (so $z = r(u)$) gives

$$\begin{aligned} l &= 2 \lim_{A\to\infty} \left[\int_0^B \frac{1}{K}(u - u_\infty) r'(u)\,du + \log A \right] - (1-\zeta)\beta \\ &= 2 \lim_{B\to u_\infty} \left[\int_0^B \frac{1}{K}(u - u_\infty) r'(u)\,du + \log r(B) \right] - (1-\zeta)\beta, \end{aligned} \tag{7.4.91}$$

where $A = r(B)$. Integrating by parts gives

$$l = 2 \lim_{B\to u_\infty} \left[\frac{1}{K}(u - u_\infty) r(u) \Big|_{u=0}^{B} - \frac{1}{K}\int_0^B r(u)\,du + \log r(B) \right] - (1-\zeta)\beta. \tag{7.4.92}$$

From (7.4.61) we obtain that $r(0) = \beta$ and

$$\lim_{B\to u_\infty} (B - u_\infty) r(B) = -\frac{(\beta - \beta')\,\mathrm{sn}(u_\infty)}{2\,\mathrm{sn}'(u_\infty)} = -\frac{(\beta - \beta')\,\mathrm{sn}(u_\infty)}{2\,\mathrm{cn}(u_\infty)\,\mathrm{dn}(u_\infty)} = -K, \tag{7.4.93}$$

hence

$$\begin{aligned} l &= 2\left[-1 + \frac{\beta(1-\zeta)}{2} \right] - 2 \lim_{B\to u_\infty} \left[\frac{1}{K}\int_0^B r(u)\,du - \log r(B) \right] - (1-\zeta)\beta \\ &= -2 - 2 \lim_{B\to u_\infty} \left[\frac{1}{K}\int_0^D r(u)\,du - \log r(B) \right]. \end{aligned} \tag{7.4.94}$$

Using equation (7.4.82) for $r(u)$, we immediately get that

$$\frac{1}{K}\int_0^B r(u)\,du = \frac{B\pi i}{K} + \log\left[\frac{\Theta(B + u_\infty + iK')}{\Theta(B - u_\infty - iK')} \right]. \tag{7.4.95}$$

Now using equation (7.4.61) for $r(u)$, we have

$$\begin{aligned}(7.4.96)\quad &\lim_{B\to u_\infty}\left[\frac{1}{K}\int_0^B r(u)\,\mathrm{d}u-\log r(B)\right]\\ &=\frac{u_\infty\pi\mathrm{i}}{K}+\lim_{B\to u_\infty}\log\left[\frac{\Theta(B+u_\infty+\mathrm{i}K')\big(\mathrm{sn}^2(u_\infty)-\mathrm{sn}^2(B)\big)}{\Theta(B-u_\infty-\mathrm{i}K')\big(\beta\,\mathrm{sn}^2(u_\infty)-\beta'\,\mathrm{sn}^2(B)\big)}\right]\\ &=\frac{u_\infty\pi\mathrm{i}}{K}+\log\left[\frac{\Theta(2u_\infty+\mathrm{i}K')2\,\mathrm{sn}(u_\infty)\,\mathrm{sn}'(u_\infty)}{\Theta'(\mathrm{i}K')(\beta-\beta')\,\mathrm{sn}^2(u_\infty)}\right]\\ &=\frac{u_\infty\pi\mathrm{i}}{K}+\log\left[\frac{2\mathrm{e}^{-\mathrm{i}\pi u_\infty/K}\vartheta_1(\pi u_\infty/K)}{\pi\vartheta_1'(0)}\right]\\ &=\log\left[\frac{2\vartheta_1(\omega)}{\pi\vartheta_1'(0)}\right].\end{aligned}$$

Plugging this into (7.4.94) gives

$$(7.4.97)\qquad l=-2+2\log\left(\frac{\pi\vartheta_1'(0)}{2\vartheta_1(\omega)}\right),$$

and thus we obtain that

$$(7.4.98)\qquad \mathrm{e}^{l/2}=\frac{\pi\vartheta_1'(0)}{2e\vartheta_1(\omega)}.$$

□

7.5. Riemann – Hilbert analysis

We can now begin the Riemann-Hilbert analysis of the orthogonal polynomials (7.1.9). The first step is the formulation of the interpolation problem.

7.5.1. Interpolation problem. We consider the lattice L_n defined in (7.1.10) and the weight $w_n(x)$ defined in (7.1.8).

Interpolation Problem. For a given $n=0,1,\dots$, find a 2×2 matrix-valued function $\mathbf{P}_n(z)=\big(\mathbf{P}_{nij}(z)\big)_{i,j=1}^2$ with the following properties:

Analyticity: $\mathbf{P}_n(z)$ is an analytic function of z for $z\in\mathbb{C}\setminus L_n$.

Residues at poles: At each node $x\in L_n$, the elements $\mathbf{P}_{n11}(z)$ and $\mathbf{P}_{n21}(z)$ of the matrix $\mathbf{P}_n(z)$ are analytic functions of z, and the elements $\mathbf{P}_{n12}(z)$ and $\mathbf{P}_{n22}(z)$ have a simple pole with the residues,

$$(7.5.1)\qquad \operatorname*{Res}_{z=x}\mathbf{P}_{nj2}(z)=\Delta_n w_n(x)\mathbf{P}_{nj1}(x),\quad j=1,2.$$

Asymptotics at infinity: There exists a function $r(x)>0$ on L_n such that

$$(7.5.2)\qquad \lim_{x\to\infty} r(x)=0,$$

and such that as $z\to\infty$, $\mathbf{P}_n(z)$ admits the asymptotic expansion,

$$(7.5.3)\quad \mathbf{P}_n(z)\sim\left(I+\frac{\mathbf{P}_1}{z}+\frac{\mathbf{P}_2}{z^2}+\cdots\right)\begin{pmatrix} z^n & 0\\ 0 & z^{-n}\end{pmatrix},\quad z\in\mathbb{C}\setminus\left[\bigcup_{x\in L_n}^{\infty} D\big(x,r(x)\big)\right],$$

where $D(x,r(x))$ denotes a disk of radius $r(x)>0$ centered at x and I is the identity matrix.

As described in Section 3.2, the IP has a unique solution, which is

$$\mathbf{P}_n(z) = \begin{pmatrix} P_{nn}(z) & C(w_n P_{nn})(z) \\ (h_{n,n-1})^{-1} P_{n,n-1}(z) & (h_{n,n-1})^{-1} C(w_n P_{n,n-1})(z) \end{pmatrix}, \tag{7.5.4}$$

where the discrete Cauchy transform C is defined by the formula

$$C(f)(z) = \Delta_n \sum_{x \in L_n} \frac{f(x)}{z-x}. \tag{7.5.5}$$

The normalizing constants for the orthogonal polynomials (7.1.11) are given as

$$h_{nn} = [\mathbf{P}_1]_{12}, \quad h_{n,n-1}^{-1} = [\mathbf{P}_1]_{21}. \tag{7.5.6}$$

7.5.2. Reduction of IP to RHP. As in Section 3.5, let us reduce the Interpolation Problem to a Riemann–Hilbert Problem (RHP). Introduce the function

$$\Pi(z) = \frac{2\gamma}{n\pi} \sin\left(\frac{n\pi z}{2\gamma}\right), \tag{7.5.7}$$

which has the properties

$$\Pi(x_k) = 0, \quad \Pi'(x_k) = (-1)^k, \quad \exp\left(\frac{in\pi x_k}{2\gamma}\right) = (-1)^k \quad \text{for} \quad x_k = \frac{2\gamma k}{n} \in L_n. \tag{7.5.8}$$

We will interpolate the poles in the IP using the upper triangular matrices,

$$\mathbf{D}^u_\pm(z) = \begin{pmatrix} 1 & -\frac{\Delta_n w_n(z)}{\Pi(z)} \mathrm{e}^{\pm in\pi z/(2\gamma)} \\ 0 & 1 \end{pmatrix}, \tag{7.5.9}$$

and the lower triangular matrices,

$$\begin{aligned} \mathbf{D}^l_\pm(z) &= \begin{pmatrix} \Pi(z)^{-1} & 0 \\ -\frac{1}{\Delta_n w_n(z)} \mathrm{e}^{\pm in\pi z/(2\gamma)} & \Pi(z) \end{pmatrix} \\ &= \begin{pmatrix} \Pi(z)^{-1} & 0 \\ 0 & \Pi(z) \end{pmatrix} \begin{pmatrix} 1 & 0 \\ -\frac{1}{\Pi(z)\Delta_n w_n(z)} \mathrm{e}^{\pm in\pi z/(2\gamma)} & 1 \end{pmatrix}. \end{aligned} \tag{7.5.10}$$

Define the matrix-valued functions,

$$\mathbf{R}^u_n(z) = \mathbf{P}_n(z) \times \begin{cases} \mathbf{D}^u_+(z) & \text{when } \operatorname{Im} z \geq 0 \\ \mathbf{D}^u_-(z) & \text{when } \operatorname{Im} z \leq 0, \end{cases} \tag{7.5.11}$$

and

$$\mathbf{R}^l_n(z) = \mathbf{P}_n(z) \times \begin{cases} \mathbf{D}^l_+(z), & \text{when } \operatorname{Im} z \geq 0 \\ \mathbf{D}^l_-(z), & \text{when } \operatorname{Im} z \leq 0. \end{cases} \tag{7.5.12}$$

Denote

$$\Delta = L_n \cap [\alpha', \beta'], \quad \nabla = L_n \setminus \Delta. \tag{7.5.13}$$

Consider the oriented contour Σ on the complex plane depicted on Figure 7.4, in which the horizontal lines are $\operatorname{Im} z = \varepsilon, 0, -\varepsilon$, where $\varepsilon > 0$ is a small positive constant which will be determined later, and the vertical segments pass through the points $z = \alpha'$ and $z = \beta'$. Consider the regions $\Omega^\Delta_\pm$ and $\Omega^\nabla_\pm$ bounded by the contour Σ, see Figure 7.4. Observe that the regions $\Omega^\nabla_\pm$ consist of two connected components, to the left and to the right of $\Omega^\Delta_\pm$.

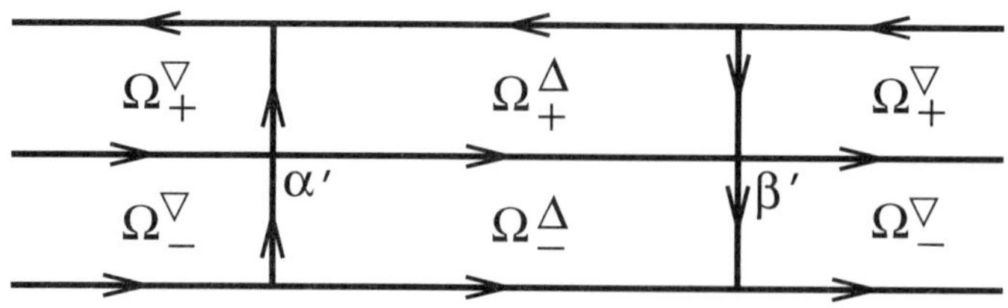

FIGURE 7.4. The contour Σ.

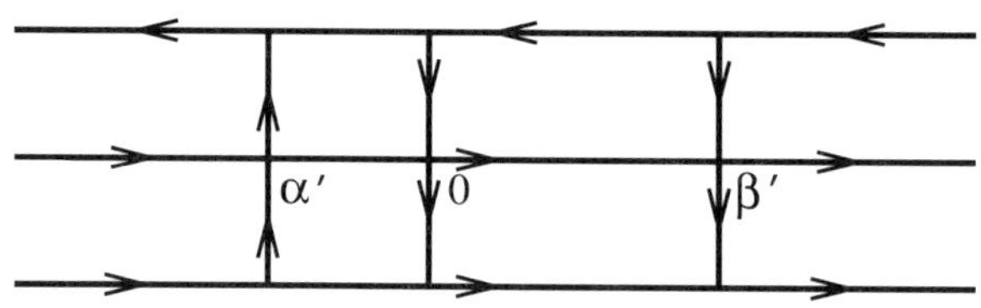

FIGURE 7.5. The contour Σ_R.

Define

$$\mathbf{R}_n(z) = \begin{cases} \mathbf{K}\mathbf{R}_n^u(z)\mathbf{K}^{-1} & \text{for } z \in \Omega_\pm^\nabla \\ \mathbf{K}\mathbf{R}_n^l(z)\mathbf{K}^{-1} & \text{for } z \in \Omega_\pm^\Delta \\ \mathbf{K}\mathbf{P}_n(z)\mathbf{K}^{-1} & \text{otherwise.} \end{cases} \tag{7.5.14}$$

where $\mathbf{K} = \begin{pmatrix} 1 & 0 \\ 0 & -2\pi\mathrm{i} \end{pmatrix}$.

Define a contour Σ_R by adding to the contour Σ a vertical segment $[\mathrm{i}\varepsilon, -\mathrm{i}\varepsilon]$, see Figure 7.5. The matrix-valued function $\mathbf{R}_n(z)$ has the following jumps on the contour Σ_R:

$$\mathbf{R}_{n+}(z) = \mathbf{R}_{n-}(z) j_R(z), \tag{7.5.15}$$

where
(7.5.16)

$$j_R(z) = \begin{cases} \begin{pmatrix} 1 & w_n(z) \\ 0 & 1 \end{pmatrix} & \text{when } z \in (-\infty, \alpha') \cup (\beta', \infty) \\ \begin{pmatrix} 1 & 0 \\ -\left(\frac{n\pi}{\gamma}\right)^2 w_n(z)^{-1} & 1 \end{pmatrix} & \text{when } z \in [\alpha', \beta'] \\ \mathbf{KD}^u_\pm(z)\mathbf{K}^{-1} = \begin{pmatrix} 1 & -\frac{i\gamma}{n\pi} \frac{w_n(z)e^{\pm in\pi z/(2\gamma)}}{\Pi(z)} \\ 0 & 1 \end{pmatrix} & \\ & \text{when } z \in (-\infty, \alpha') \cup (\beta', \infty) \pm i\varepsilon \\ \mathbf{KD}^l_\pm(z)\mathbf{K}^{-1} = \begin{pmatrix} \Pi(z)^{-1} & 0 \\ \frac{in\pi}{\gamma} \frac{e^{\pm in\pi z/(2\gamma)}}{w_n(z)} & \Pi(z) \end{pmatrix} & \text{when } z \in (\alpha', \beta') \pm i\varepsilon \\ \mathbf{KD}^l_\pm(z)^{-1}\mathbf{D}^u_\pm(z)\mathbf{K}^{-1} = \begin{pmatrix} \Pi(z) & \frac{\gamma}{n\pi i} w_n(z) e^{\pm in\pi z/(2\gamma)} \\ -\frac{n\pi i}{\gamma} w_n(z)^{-1} e^{\pm in\pi z/(2\gamma)} & \mp \frac{n\pi i}{\gamma} e^{\pm in\pi z/(2\gamma)} \end{pmatrix} & \\ & \text{when } z \in (0, \pm i\varepsilon) + \alpha' \text{ or } z \in (0, \pm i\varepsilon) + \beta' \\ \mathbf{KD}^0_\pm(z)\mathbf{K}^{-1} & \text{when } z \in (0, \pm i\varepsilon), \end{cases}$$

and

(7.5.17)
$$\mathbf{D}^0_\pm(z) = \begin{pmatrix} 1 & 0 \\ -\frac{2\sinh(nz)e^{-n\zeta z}e^{\pm in\pi z/(2\gamma)}}{\Delta_n \Pi(z)} & 1 \end{pmatrix}.$$

Up to a slightly different scaling, these jumps are almost identical to those described in (3.5.20). The difference is that there is an extra jump on vertical contours close to the origin, arising because the origin is a singular point of the potential $V(x)$. Notice, however, that this jump is exponentially close to the identity matrix, and we will see that it will not require any extra local analysis. This is a reflection of the fact that the origin is inside the saturated region.

7.5.3. The first transformation. As in Section 3.6, we define the matrix function $\mathbf{T}_n(z)$ as follows from the equation,

(7.5.18)
$$\mathbf{R}_n(z) = e^{(nl/2)\sigma_3}\mathbf{T}_n(z)e^{n(g(z)-l/2)\sigma_3},$$

where the Lagrange multiplier l and the function $g(z)$ are as described in Section 7.4 and $\sigma_3 = \left(\begin{smallmatrix} 1 & 0 \\ 0 & -1 \end{smallmatrix}\right)$ is the third Pauli matrix. We then have that as $z \to \infty$,

(7.5.19)
$$\mathbf{T}_n(z) \sim I + \frac{\mathbf{T}_1}{z} + \frac{\mathbf{T}_2}{z^2} + \cdots.$$

From (7.4.21) we have that

(7.5.20)
$$g(z) = \log z + \mathcal{O}(z^{-1}) \quad \text{as } z \to \infty,$$

which implies that

(7.5.21)
$$[\mathbf{T}_1]_{12} = e^{-nl}[\mathbf{R}_1]_{12}.$$

7.5.4. The second transformation. As in Section 3.7, let

$$(7.5.22)\quad \mathbf{S}_n(z) = \begin{cases} \mathbf{T}_n(z)\begin{pmatrix} 1 & 0 \\ -e^{-nG(z)} & 1 \end{pmatrix} & \text{for } z \in \{(\alpha,\alpha')\cup(\beta',\beta)\}\times(0,\mathrm{i}\varepsilon) \\ \mathbf{T}_n(z)\begin{pmatrix} 1 & 0 \\ e^{nG(z)} & 1 \end{pmatrix} & \text{for } z \in \{(\alpha,\alpha')\cup(\beta',\beta)\}\times(0,-\mathrm{i}\varepsilon) \\ \mathbf{T}_n(z)\begin{pmatrix} -\frac{\gamma}{n\pi \mathrm{i}}e^{-\mathrm{i}n\pi z/(2\gamma)} & 0 \\ 0 & -\frac{n\pi\mathrm{i}}{\gamma}e^{\mathrm{i}n\pi z/(2\gamma)} \end{pmatrix} & \text{for } z \in (\alpha',\beta')\times(0,\mathrm{i}\varepsilon) \\ \mathbf{T}_n(z)\begin{pmatrix} \frac{\gamma}{n\pi \mathrm{i}}e^{\mathrm{i}n\pi z/(2\gamma)} & 0 \\ 0 & \frac{n\pi\mathrm{i}}{\gamma}e^{-\mathrm{i}n\pi z/(2\gamma)} \end{pmatrix} & \text{for } z \in (\alpha',\beta')\times(0,-\mathrm{i}\varepsilon) \\ \mathbf{T}_n(z) & \text{otherwise.} \end{cases}$$

This function satisfies a similar RHP to $\mathbf{T}_n$, but jumps now occur on a new contour, Σ_S, which is obtained from Σ_R by adding the two segments $(\alpha-\mathrm{i}\varepsilon,\alpha+\mathrm{i}\varepsilon)$ and $(\beta-\mathrm{i}\varepsilon,\beta+\mathrm{i}\varepsilon)$, see Figure 7.6.

For $z\in\Sigma_S$, we have that

$$(7.5.23)\quad \mathbf{S}_{n+}(z) = \mathbf{S}_{n-}(z)j_S(z),$$

where

$$(7.5.24)\quad j_S(z) = \begin{cases} \begin{pmatrix} 0 & 1 \\ -1 & 0 \end{pmatrix} & \text{for } z\in(\alpha,\alpha')\cup(\beta',\beta) \\ \begin{pmatrix} (1-e^{\pm\mathrm{i}n\pi z/\gamma})^{-1} & \pm e^{\pm nG(z)}(1-e^{\mp\mathrm{i}n\pi z/\gamma})^{-1} \\ \mp e^{\mp nG(z)} & 1 \end{pmatrix} & \text{for } z\in\in(\alpha,\alpha')\cup(\beta',\beta)\pm\mathrm{i}\varepsilon \\ \begin{pmatrix} (1-e^{\pm\mathrm{i}n\pi z/\gamma})^{-1} & 0 \\ \mp e^{-n(2g(z)-l-V(z))} & 1-e^{\pm\mathrm{i}n\pi z/\gamma} \end{pmatrix} & \text{for } z\in[\alpha',\beta']\pm\mathrm{i}\varepsilon \\ \begin{pmatrix} -e^{-n\pi\mathrm{i}(1+\zeta)} & 0 \\ e^{-n(g_+(z)+g_-(z)-l-V(z))} & -e^{n\pi\mathrm{i}(1+\zeta)} \end{pmatrix} & \text{for} z\in[\alpha',\beta'] \\ \begin{pmatrix} 1 & 0 \\ \pm\frac{e^{-n(2g(z)-l-\zeta z+z)}-e^{-n(2g(z)-l-\zeta z-z)}}{1-e^{\pm\mathrm{i}n\pi z/\gamma}} & 1 \end{pmatrix} & \text{for } z\in[0,\pm\mathrm{i}\varepsilon]. \end{cases}$$

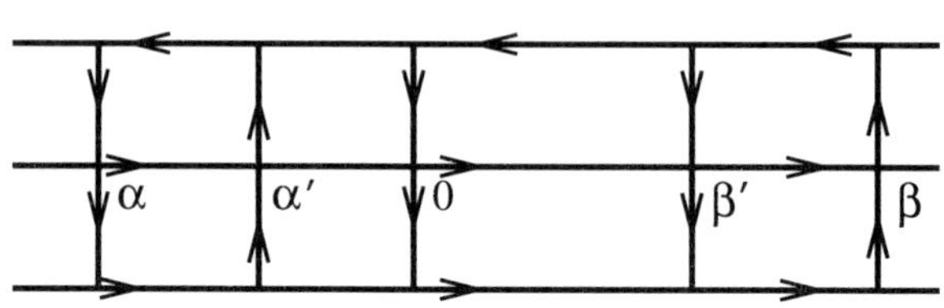

FIGURE 7.6. The contour Σ_S.

Notice that on the vertical segments near the origin, $[0, \pm i\varepsilon]$, the jump is uniformly exponentially close to the identity matrix. For the rest of the jumps, we use formula (7.4.31) for the G-function and the upper constraint (7.4.9) on the density ρ, to see that for sufficiently small $\varepsilon > 0$ and $x \in (\alpha, \alpha') \cup (\beta', \beta)$,

$$0 < \pm \operatorname{Re} G(x \pm i\varepsilon) = 2\pi\varepsilon\rho(x) + \mathcal{O}(\varepsilon^2) < \frac{\pi\varepsilon}{\gamma} + \mathcal{O}(\varepsilon^2). \tag{7.5.25}$$

Combined with property (7.4.29) of the g-function, this implies that all jumps on horizontal segments are exponentially close to the identity matrix, provided that they are bounded away from the interval $[\alpha, \beta]$. For what follows we denote

$$\Omega_n = \pi + n2\pi \int_0^\beta \rho(x)\, dx = \pi + n\pi(1 + \zeta), \tag{7.5.26}$$

so that

$$-e^{-n\pi i(1+\zeta)} = e^{-i\Omega_n}. \tag{7.5.27}$$

7.5.5. Model RHP. The model RHP, which appears when we drop in the jump matrix $j_S(z)$ the terms that vanish as $n \to \infty$, is the following.

(1) $\mathbf{M}(z)$ is analytic in $\mathbb{C} \setminus [\alpha, \beta]$.

(2) $\mathbf{M}_+(z) = \mathbf{M}_-(z) j_M(z)$ for $z \in [\alpha, \beta]$, where

$$j_M(z) = \begin{cases} \begin{pmatrix} 0 & 1 \\ -1 & 0 \end{pmatrix} & \text{for } z \in [\alpha, \alpha'] \cup [\beta, \beta'] \\ e^{-i\Omega_n \sigma_3} & \text{for } z \in [\alpha', \beta']. \end{cases} \tag{7.5.28}$$

(3) As $z \to \infty$,

$$\mathbf{M}(z) \sim I + \frac{\mathbf{M}_1}{z} + \frac{\mathbf{M}_2}{z^2} + \cdots. \tag{7.5.29}$$

As shown in Section 2.4.5, this problem is solved in two steps. In the first step, we solve the auxiliary RHP:

(1) $\mathbf{Q}(z)$ is analytic in $\mathbb{C} \setminus [\alpha, \alpha'] \cup [\beta', \beta]$.

(2) $\mathbf{Q}_+(z) = \mathbf{Q}_-(z) \begin{pmatrix} 0 & 1 \\ -1 & 0 \end{pmatrix}$ for $z \in [\alpha, \alpha'] \cup [\beta', \beta]$.

(3) $\mathbf{Q}(z) = I + \mathcal{O}(z^{-1})$ as $z \to \infty$.

This RHP has the unique solution

$$\mathbf{Q}(z) = \begin{pmatrix} \dfrac{\gamma(z) + \gamma^{-1}(z)}{2} & \dfrac{\gamma(z) - \gamma^{-1}(z)}{-2i} \\ \dfrac{\gamma(z) - \gamma^{-1}(z)}{2i} & \dfrac{\gamma(z) + \gamma^{-1}(z)}{2} \end{pmatrix}, \tag{7.5.30}$$

where

$$\gamma(z) = \left(\frac{(z - \alpha)(z - \beta')}{(z - \alpha')(z - \beta)} \right)^{1/4}, \tag{7.5.31}$$

with cuts on $[\alpha, \alpha'] \cup [\beta', \beta]$.

In this "two-cut" case, the Riemann theta functions used to solve the full model RHP in Section 2.4.5 reduce to the Jacobi elliptic functions defined in (7.2.2). Define the function

$$f(s) = \frac{\vartheta_3(s + d + c)}{\vartheta_3(s + d)}, \tag{7.5.32}$$

where ϑ_3 is as defined in (7.2.1) with elliptic nome $q = \mathrm{e}^{\mathrm{i}\pi\tau} = \mathrm{e}^{-\pi^2/(2\gamma)}$ ($\tau = \mathrm{i}\pi/(2\gamma)$), and d and c are arbitrary complex numbers. Notice that f has the periodic properties

$$f(s+\pi) = f(s), \quad f(s+\pi\tau) = \mathrm{e}^{-2\mathrm{i}c} f(s), \tag{7.5.33}$$

and that f is an even function of its argument. Now let

$$\tilde{u}(z) = \frac{\pi}{2K} u(z) = \frac{\pi}{2}\int_\beta^z \frac{\mathrm{d}z'}{\sqrt{R(z')}}, \tag{7.5.34}$$

where u is as defined in (7.4.36). Then $\tilde{u}$ is two-valued on $[\alpha,\beta]$ and satisfies

$$\tilde{u}_+(x) - \tilde{u}_-(x) = \pi\tau \quad \text{for } x\in[\alpha',\beta']. \tag{7.5.35}$$

Also,

$$\tilde{u}_\pm(\alpha) = \frac{\pi}{2}, \quad \tilde{u}_\pm(\alpha') = \frac{\pi}{2} \pm \frac{\pi\tau}{2}, \quad \tilde{u}_\pm(\beta') = \pm\frac{\pi\tau}{2}, \quad \tilde{u}_\pm(\beta) = 0, \tag{7.5.36}$$

see Figure 7.3. Because $\sqrt{R(x)}_+ = -\sqrt{R(x)}_-$ for $x\in[\alpha,\alpha']\cup[\beta',\beta]$, it immediately follows that

$$\tilde{u}_+(x) + \tilde{u}_-(x) = 0 \quad \text{for } x\in[\beta',\beta], \tag{7.5.37}$$

and that

$$\tilde{u}_+(x) + \tilde{u}_-(x) = \tilde{u}_+(\alpha') - \tilde{u}_+(\beta') + \tilde{u}_-(\alpha') - \tilde{u}_-(\beta') = \pi \quad \text{for } x\in[\alpha,\alpha']. \tag{7.5.38}$$

We now define

$$f_1(z) = \frac{\vartheta_3(\tilde{u}(z) + d + \Omega_n/2)}{\vartheta_3(\tilde{u}(z)+d)}, \quad f_2(z) = \frac{\vartheta_3(-\tilde{u}(z) + d + \Omega_n/2)}{\vartheta_3(-\tilde{u}(z)+d)} \quad \text{for } z\in\mathbb{C}\setminus[\alpha,\beta], \tag{7.5.39}$$

where d is an arbitrary complex number. It then follows from (7.5.33) and (7.5.35) that

$$f_{1+}(x) = \mathrm{e}^{-\mathrm{i}\Omega_n} f_{1-}(x) \quad \text{and} \quad f_{2+}(x) = \mathrm{e}^{\mathrm{i}\Omega_n} f_{2-}(x) \qquad \text{for } x\in[\alpha',\beta'], \tag{7.5.40}$$

and from (7.5.33), (7.5.37), and (7.5.38) that

$$f_{1+}(x) = f_{2-}(x) \quad \text{and} \quad f_{2+}(x) = f_{1-}(x) \qquad \text{for } x\in[\alpha,\alpha']\cup[\beta'\beta]. \tag{7.5.41}$$

Define the matrix valued function

$$\mathbf{F}(z) = \begin{pmatrix} \dfrac{\vartheta_3(\tilde{u}(z)+d_1+\Omega_n/2)}{\vartheta_3(\tilde{u}(z)+d_1)} & \dfrac{\vartheta_3(-\tilde{u}(z)+d_1+\Omega_n/2)}{\vartheta_3(-\tilde{u}(z)+d_1)} \\ \dfrac{\vartheta_3(\tilde{u}(z)+d_2+\Omega_n/2)}{\vartheta_3(\tilde{u}(z)+d_2)} & \dfrac{\vartheta_3(-\tilde{u}(z)+d_2+\Omega_n/2)}{\vartheta_3(-\tilde{u}(z)+d_2)} \end{pmatrix}, \tag{7.5.42}$$

where d_1 and d_2 are yet undetermined complex constants. Then from (7.5.40) and (7.5.41) we have that

$$\begin{aligned} \mathbf{F}_+(x) &= \mathbf{F}_-(x)\begin{pmatrix}\mathrm{e}^{-\mathrm{i}\Omega_n} & 0\\ 0 & \mathrm{e}^{\mathrm{i}\Omega_n}\end{pmatrix} && \text{for } x\in[\alpha',\beta'],\\ \mathbf{F}_+(x) &= \mathbf{F}_-(x)\begin{pmatrix}0 & 1\\ 1 & 0\end{pmatrix} && \text{for } x\in[\alpha,\alpha']\cup[\beta',\beta]. \end{aligned} \tag{7.5.43}$$

We can now combine (7.5.30) and (7.5.42) to obtain

$$\mathbf{M}(z) = \mathbf{F}(\infty)^{-1} \times \begin{pmatrix} \frac{\gamma(z)+\gamma^{-1}(z)}{2}\frac{\vartheta_3(\tilde{u}(z)+d_1+\Omega_n/2)}{\vartheta_3(\tilde{u}(z)+d_1)} & \frac{\gamma(z)-\gamma^{-1}(z)}{-2\mathrm{i}}\frac{\vartheta_3(-\tilde{u}(z)+d_1\Omega_n/2)}{\vartheta_3(-\tilde{u}(z)+d_1)} \\ \frac{\gamma(z)-\gamma^{-1}(z)}{2\mathrm{i}}\frac{\vartheta_3(\tilde{u}(z)+d_2+\Omega_n/2)}{\vartheta_3(\tilde{u}(z)+d_2)} & \frac{\gamma(z)+\gamma^{-1}(z)}{2}\frac{\vartheta_3(-\tilde{u}(z)+d_2+\Omega_n/2)}{\vartheta_3(-\tilde{u}(z)+d_2)} \end{pmatrix}, \tag{7.5.44}$$

where

$$\mathbf{F}(\infty) = \begin{pmatrix} \frac{\vartheta_3(\tilde{u}_\infty+d_1+\Omega_n/2)}{\vartheta_3(\tilde{u}_\infty+d_1)} & 0 \\ 0 & \frac{\vartheta_3(-\tilde{u}_\infty+d_2+\Omega_n/2)}{\vartheta_3(-\tilde{u}_\infty+d_2)} \end{pmatrix}. \tag{7.5.45}$$

and $\tilde{u}_\infty \equiv \tilde{u}(\infty)$. This matrix satisfies conditions (7.5.28) and (7.5.29) of the model RHP. We are left to choose the constants d_1 and d_2 so that (7.5.44) is analytic on $\mathbb{C}\setminus[\alpha,\beta]$. That is, we can choose the constants d_1 and d_2 such that the zeroes of $\vartheta_3(\pm\tilde{u}(z)+d_{1,2})$ coincide with the zeroes of $\gamma(z)\pm\gamma^{-1}(z)$ and are thus cancelled in the product.

First consider the zeroes of $\gamma(z)\pm\gamma^{-1}(z)$. These are the zeroes of $\gamma^2(z)\pm1$ and thus of $\gamma^4(z)-1$. Thus there is only one zero, which uniquely solves the equation

$$p(z) \equiv \frac{(z-\alpha)(z-\beta')}{(z-\alpha')(z-\beta)} = 1, \tag{7.5.46}$$

which is

$$x_0 = \frac{\beta\alpha'-\alpha\beta'}{(\alpha'-\alpha)+(\beta-\beta')} \in (\alpha',\beta'). \tag{7.5.47}$$

It is easy to check that $\gamma(x_0) = 1$, thus x_0 is the unique zero of $\gamma(z)-\gamma^{-1}(z)$, whereas there are no zeroes of $\gamma(z)+\gamma^{-1}(z)$ on the specified sheet. We use here the change of variables v defined in (7.4.37). Notice that, by (7.4.56),

$$v(x_0) = \frac{\beta'-\alpha}{\beta'-\alpha'} = \frac{\mathrm{dn}^2(u_\infty)}{k^2\,\mathrm{cn}^2(u_\infty)}, \tag{7.5.48}$$

implying that

$$\mathrm{sn}^2\left(u(x_0)\right) = \frac{\mathrm{dn}^2(u_\infty)}{k^2\,\mathrm{cn}^2(u_\infty)}. \tag{7.5.49}$$

Since $x_0 \in (\alpha',\beta')$, we must have $u(x_0) \in (\mathrm{i}K', K+\mathrm{i}K')$ (if we choose to take u_+). Since sn^2 is a one-to-one function on this interval, there is a unique point $u_0 \in (\mathrm{i}K', K+\mathrm{i}K')$ such that $\mathrm{sn}^2(u_0) = \mathrm{dn}^2(u_\infty)/\left(k^2\,\mathrm{cn}^2(u_\infty)\right)$. The simple period identity

$$\mathrm{sn}(u+K+\mathrm{i}K') = \frac{\mathrm{dn}(u)}{k\,\mathrm{cn}(u)}, \tag{7.5.50}$$

along with (7.5.49), gives that we must have

$$u_0 = u(x_0) = K - u_\infty + \mathrm{i}K'. \tag{7.5.51}$$

Thus,

$$\tilde{u}(x_0) = \frac{\pi}{2K}(K-u_\infty+\mathrm{i}K') = \frac{\tau\pi}{2}+\frac{\pi}{2}-\tilde{u}_\infty. \tag{7.5.52}$$

We now consider zeroes of the function $\vartheta_3(\tilde{u}(z)-d) \equiv \vartheta_3(-\tilde{u}(z)+d)$. The zeroes of this function are the solutions to the equation

$$\tilde{u}(z)-d=(2m+1)\frac{\pi}{2}+(2k+1)\frac{\tau\pi}{2}, \tag{7.5.53}$$

for any $m,k\in\mathbb{Z}$. Because $\tilde{u}$ maps the first sheet of X to the rectangular domain $[0,\pi/2]\times[-\tau\pi/2,\tau\pi/2]$, it is clear that this equation can have at most one solution, and without any loss of generality we may take $m=k=0$. Then, if we want the solution of this equation to be x_0, we need to let

$$d=\tilde{u}(x_0)-\frac{\pi}{2}(1+\tau)=-\tilde{u}_\infty. \tag{7.5.54}$$

This choice of d also ensures that $\vartheta_3(\tilde{u}(z)+d)\equiv\vartheta_3(-\tilde{u}(z)-d)$ has no zeroes on the first sheet of X. We can then let

$$d_1=d,\quad d_2=-d, \tag{7.5.55}$$

so that (7.5.44) and (7.5.45) become

(7.5.56)

$$\mathbf{M}(z)$$
$$=\mathbf{F}(\infty)^{-1}\begin{pmatrix}\frac{\gamma(z)+\gamma^{-1}(z)}{2}\frac{\vartheta_3(\tilde{u}(z)+d+\Omega_n/2)}{\vartheta_3(\tilde{u}(z)+d)} & \frac{\gamma(z)-\gamma^{-1}(z)}{-2i}\frac{\vartheta_3(-\tilde{u}(z)+d+\Omega_n/2)}{\vartheta_3(-\tilde{u}(z)+d)}\\ \frac{\gamma(z)-\gamma^{-1}(z)}{2i}\frac{\vartheta_3(\tilde{u}(z)-d+\Omega_n/2)}{\vartheta_3(\tilde{u}(z)-d)} & \frac{\gamma(z)+\gamma^{-1}(z)}{2}\frac{\vartheta_3(-\tilde{u}(z)-d+\Omega_n/2)}{\vartheta_3(-\tilde{u}(z)-d)}\end{pmatrix}$$
$$=\mathbf{F}(\infty)^{-1}\begin{pmatrix}\frac{\gamma(z)+\gamma^{-1}(z)}{2}\frac{\vartheta_3(\tilde{u}(z)+(n+\frac12)\omega)}{\vartheta_4(\tilde{u}(z)+\omega/2)} & \frac{\gamma(z)-\gamma^{-1}(z)}{-2i}\frac{\vartheta_3(\tilde{u}(z)-(n+\frac12)\omega)}{\vartheta_4(\tilde{u}(z)-\omega/2)}\\ \frac{\gamma(z)-\gamma^{-1}(z)}{2i}\frac{\vartheta_3(\tilde{u}(z)+(n-\frac12)\omega)}{\vartheta_4(\tilde{u}(z)-\omega/2)} & \frac{\gamma(z)+\gamma^{-1}(z)}{2}\frac{\vartheta_3(\tilde{u}(z)-(n-\frac12)\omega)}{\vartheta_4(\tilde{u}(z)+\omega/2)}\end{pmatrix},$$

where

$$\mathbf{F}(\infty)=\begin{pmatrix}\frac{\vartheta_3(\Omega_n/2)}{\vartheta_3(0)} & 0\\ 0 & \frac{\vartheta_3(\Omega_n/2)}{\vartheta_3(0)}\end{pmatrix}=\begin{pmatrix}\frac{\vartheta_4(n\omega)}{\vartheta_3(0)} & 0\\ 0 & \frac{\vartheta_4(n\omega)}{\vartheta_3(0)}\end{pmatrix}, \tag{7.5.57}$$

solving the model RHP. The asymptotics at infinity are

$$\mathbf{M}(z)=I+\frac{\mathbf{M}_1}{z}+\mathcal{O}(z^{-2}), \tag{7.5.58}$$

where the matrix $\mathbf{M}_1$ has the form

(7.5.59)

$$\mathbf{M}_1=\begin{pmatrix} * & \frac{\vartheta_3(-\tilde{u}_\infty+d+\Omega_n/2)\vartheta_3(\tilde{u}_\infty+d)}{\vartheta_3(\tilde{u}_\infty+d+\Omega_n/2)\vartheta_3(-\tilde{u}_\infty+d)}\frac{(\beta-\beta')+(\alpha'-\alpha)}{-4i}\\ \frac{\vartheta_3(\tilde{u}_\infty-d+\Omega_n/2)\vartheta_3(-\tilde{u}_\infty-d)}{\vartheta_3(-\tilde{u}_\infty-d+\Omega_n/2)\vartheta_3(\tilde{u}_\infty-d)}\frac{(\beta-\beta')+(\alpha'-\alpha)}{4i} & *\end{pmatrix}.$$

The matrix $\mathbf{M}_1$ can be written in a cleaner fashion and in terms of the original parameters as follows.

Proposition 7.5.1. *We have that*

$$[\mathbf{M}_1]_{12}=\frac{iA(\omega)\vartheta_4\big((n+1)\omega\big)}{\vartheta_4(n\omega)},\qquad [\mathbf{M}_1]_{21}=\frac{A(\omega)\vartheta_4(n\omega)}{i\vartheta_4\big((n-1)\omega\big)}, \tag{7.5.60}$$

where

$$\omega = \frac{\pi(1+\zeta)}{2}, \quad A(\omega) = \frac{\pi\vartheta_1'(0)}{2\vartheta_1(\omega)}. \tag{7.5.61}$$

PROOF. From (7.4.53)) and (7.5.34)), we have

$$\tilde{u}_\infty = \frac{\pi}{4}(1-\zeta). \tag{7.5.62}$$

Using the formula (7.5.26) for Ω_n, this immediately gives

$$\frac{\vartheta_3(\tilde{u}_\infty + d)\vartheta_3(-\tilde{u}_\infty + d + \Omega_n/2)}{\vartheta_3(-\tilde{u}_\infty + d)\vartheta_3(\tilde{u}_\infty + d + \Omega_n/2)} = \frac{\vartheta_3(0)\vartheta_4((n+1)\omega)}{\vartheta_3(\omega)\vartheta_4(n\omega)}. \tag{7.5.63}$$

Formulas (7.4.15) give that

$$\frac{(\beta-\beta')+(\alpha'-\alpha)}{4\mathrm{i}} = \frac{\pi\vartheta_4^2(0)}{4\mathrm{i}}\left[\frac{\vartheta_1^2(\omega/2)\vartheta_4^2(\omega/2)+\vartheta_2^2(\omega/2)\vartheta_3^2(\omega/2)}{\vartheta_1(\omega/2)\vartheta_2(\omega/2)\vartheta_3(\omega/2)\vartheta_4(\omega/2)}\right]. \tag{7.5.64}$$

Plugging the duplication formulas (7.2.12) and (7.2.13) into (7.5.64) yields

$$\frac{(\beta-\beta')+(\alpha'-\alpha)}{4\mathrm{i}} = \frac{\pi}{2\mathrm{i}}\frac{\vartheta_2(0)\vartheta_4(0)\vartheta_3(\omega)}{\vartheta_1(\omega)}. \tag{7.5.65}$$

Combining (7.5.63) and (7.5.65), we can write the [12] entry of (7.5.59) as

$$\begin{aligned}[\mathbf{M}_1]_{12} &= \frac{\mathrm{i}\pi}{2}\frac{\vartheta_4\big((n+1)\omega\big)}{\vartheta_4(n\omega)}\frac{\vartheta_3(0)}{\vartheta_3(\omega)}\frac{\vartheta_2(0)\vartheta_4(0)\vartheta_3(\omega)}{\vartheta_1(\omega)} \\ &= \frac{\mathrm{i}\pi}{2}\frac{\vartheta_4\big((n+1)\omega\big)}{\vartheta_4(n\omega)}\frac{\vartheta_1'(0)}{\vartheta_2(\pi\zeta/2)} = \frac{\mathrm{i}A\vartheta_4\big((n+1)\omega\big)}{\vartheta_4(n\omega)}.\end{aligned} \tag{7.5.66}$$

Similarly, we can write the [21] entry of (7.5.59) as

$$[\mathbf{M}_1]_{21} = \frac{A\vartheta_4(n\omega)}{\mathrm{i}\vartheta_4\big((n-1)\omega\big)}. \tag{7.5.67}$$

□

Notice that since $\mathbf{M}$ solves the model RHP, we have that

$$\det \mathbf{M}(z) = 1, \quad z \in \mathbb{C}. \tag{7.5.68}$$

7.5.6. Parametrix at outer turning points. We now consider small disks $D(\alpha,\varepsilon)$ and $D(\beta,\varepsilon)$ centered at the outer turning points. Denote $D = D(\alpha,\varepsilon) \cup D(\beta,\varepsilon)$. We will seek a local parametrix $\mathbf{U}_n(z)$ defined on D such that

(1) $\mathbf{U}_n(z)$ is analytic on $D \setminus \Sigma_S$.
(2) $\mathbf{U}_{n+}(z) = \mathbf{U}_{n-}(z)j_S(z)$ for $z \in D \cap \Sigma_S$.
(3) $\mathbf{U}_n(z) = \mathbf{M}(z)\big(I + \mathcal{O}(n^{-1})\big)$ uniformly for $z \in \partial D$.

The solution is as described in Section 2.4.6. For $z \in D(\beta,\varepsilon)$, the solution is

$$\mathbf{U}_n(z) = \mathbf{E}_n^\beta(z)\Phi_\beta\big(n^{2/3}\psi_\beta(z)\big)\mathrm{e}^{-n(g(z)-V(z)/2-l/2)\sigma_3}, \tag{7.5.69}$$

where

$$\begin{aligned}&\mathbf{E}_n^\beta(z) = \mathbf{M}(z)\mathbf{L}_n^\beta(z)^{-1}, \\ &\mathbf{L}_n^\beta(z) = \frac{1}{2\sqrt{\pi}}\begin{pmatrix} n^{-1/6}\psi_\beta^{-1/4}(z) & 0 \\ 0 & n^{1/6}\psi_\beta^{1/4}(z)\end{pmatrix}\begin{pmatrix} 1 & \mathrm{i} \\ -1 & \mathrm{i}\end{pmatrix},\end{aligned} \tag{7.5.70}$$

$$\psi_\beta(z) = -\left\{\frac{3\pi}{2}\int_z^\beta \rho(t)\,\mathrm{d}t\right\}^{2/3}, \tag{7.5.71}$$

$$\Phi_\beta(z) = \begin{cases} \begin{pmatrix} y_0(z) & -y_2(z) \\ y_0'(z) & -y_2'(z) \end{pmatrix} & \text{for } \arg z \in (0, \pi/2) \\ \begin{pmatrix} -y_1(z) & -y_2(z) \\ -y_1'(z) & -y_2'(z) \end{pmatrix} & \text{for } \arg z \in (\pi/2, \pi) \\ \begin{pmatrix} -y_2(z) & y_1(z) \\ -y_2'(z) & y_1'(z) \end{pmatrix} & \text{for } \arg z \in (-\pi, -\pi/2) \\ \begin{pmatrix} y_0(z) & y_1(z) \\ y_0'(z) & y_1'(z) \end{pmatrix} & \text{for } \arg z \in (-\pi/2, 0), \end{cases} \tag{7.5.72}$$

$$y_0(z) = \mathrm{Ai}(z), \quad y_1(z) = \omega\,\mathrm{Ai}(\omega z), \quad y_2(z) = \omega^2\,\mathrm{Ai}(\omega^2 z), \tag{7.5.73}$$

and we take the branch of $\psi_\beta^{1/4}$ which is positive on $(\beta, \beta+\varepsilon)$ and has a cut on $(\beta-\varepsilon, \beta)$.

For $z \in D(\alpha, \varepsilon)$, the solution is given by

$$\mathbf{U}_n(z) = \mathbf{E}_n^\alpha(z)\Phi_\alpha\big(n^{2/3}\psi_\alpha(z)\big)\mathrm{e}^{-n(g(z)-V(z)/2-l/2)\sigma_3}, \tag{7.5.74}$$

for $z \in D(\alpha, \varepsilon)$, where

$$\begin{aligned} \mathbf{E}_n^\alpha(z) &= \mathbf{M}(z)\mathbf{L}_n^\alpha(z)^{-1}, \\ \mathbf{L}_n^\alpha(z) &= \frac{1}{2\sqrt{\pi}} \begin{pmatrix} n^{-1/6}\psi_\alpha^{-1/4}(z) & 0 \\ 0 & n^{1/6}\psi_\alpha^{1/4}(z) \end{pmatrix} \begin{pmatrix} -1 & i \\ 1 & i \end{pmatrix}, \end{aligned} \tag{7.5.75}$$

where

$$\psi_\alpha(z) = -\left\{\frac{3\pi}{2}\int_\alpha^z \rho(t)\,\mathrm{d}t\right\}^{2/3}, \tag{7.5.76}$$

and

$$\Phi_\alpha(z) = \Phi_\beta(z)\begin{pmatrix} 1 & 0 \\ 0 & -1 \end{pmatrix}. \tag{7.5.77}$$

The large n behavior is as follows. For $z \in D(\beta, \varepsilon)$, as $n \to \infty$,

$$\mathbf{U}_n(z) = \mathbf{M}(z)\left[I + \frac{\psi_\beta(z)^{-3/2}}{48n}\begin{pmatrix} 1 & 6\mathrm{i} \\ 6\mathrm{i} & -1 \end{pmatrix} + \mathcal{O}(n^{-2})\right]. \tag{7.5.78}$$

For $z \in D(\alpha, \varepsilon)$, as $n \to \infty$,

$$\mathbf{U}_n(z) = \mathbf{M}(z)\left[I + \frac{\psi_\alpha(z)^{-3/2}}{48n}\begin{pmatrix} 1 & -6\mathrm{i} \\ -6\mathrm{i} & -1 \end{pmatrix} + \mathcal{O}(n^{-2})\right]. \tag{7.5.79}$$

7.5.7. Parametrix at the inner turning points. We now consider small disks $D(\alpha', \varepsilon)$ and $D(\beta', \varepsilon)$ centered at the inner turning points. Denote $\widetilde{D} = D(\alpha', \varepsilon) \cup D(\beta', \varepsilon)$. We will seek a local parametrix $\mathbf{U}_n(z)$ defined on $\widetilde{D}$ such that

(1) $\mathbf{U}_n(z)$ is analytic on $\widetilde{D} \setminus \Sigma_S$.
(2) $\mathbf{U}_{n+}(z) = \mathbf{U}_{n-}(z) j_S(z)$ for $z \in \widetilde{D} \cap \Sigma_S$.
(3) $\mathbf{U}_n(z) = \mathbf{M}(z)\big(I + \mathcal{O}(n^{-1})\big)$ uniformly for $z \in \partial\widetilde{D}$.

The solution is as described in Section 3.10. For $z \in D(\alpha', \varepsilon)$, the solution is

$$\begin{aligned} \mathbf{U}_n(z) = \mathbf{M}(z)&\mathrm{e}^{\pm(\mathrm{i}\Omega_n/2)\sigma_3}\mathbf{L}_n^{\alpha'}(z)^{-1}\Phi_{\alpha'}\big(n^{2/3}\psi_{\alpha'}(z)\big) \\ &\times \mathrm{e}^{\mp(\mathrm{i}n\pi z/(2\gamma))\sigma_3}\mathrm{e}^{-n(g(z)-V(z)/2-l/2)\sigma_3} \quad \text{for } \pm\operatorname{Im} z > 0, \end{aligned} \tag{7.5.80}$$

where

$$\Phi_{\alpha'}(z) = \begin{cases} \begin{pmatrix} y_2(z) & -y_0(z) \\ y_2'(z) & -y_0'(z) \end{pmatrix} & \text{for } \arg z \in (0, \pi/2) \\ \begin{pmatrix} y_2(z) & y_1(z) \\ y_2'(z) & y_1'(z) \end{pmatrix} & \text{for } \arg z \in (\pi/2, \pi) \\ \begin{pmatrix} y_1(z) & -y_2(z) \\ y_1'(z) & -y_2'(z) \end{pmatrix} & \text{for } \arg z \in (-\pi, -\pi/2) \\ \begin{pmatrix} y_1(z) & y_0(z) \\ y_1'(z) & y_0'(z) \end{pmatrix} & \text{for } \arg z \in (-\pi/2, 0), \end{cases} \tag{7.5.81}$$

$$\mathbf{E}_n^{\alpha'}(z) = \mathbf{M}(z) \mathrm{e}^{\pm(\mathrm{i}\Omega_n/2)\sigma_3} \tilde{\mathbf{L}}_n(z)^{-1} \quad \text{for } \pm \operatorname{Im} z \geq 0,$$

$$\mathbf{L}_n^{\alpha'}(z) = \frac{1}{2\sqrt{\pi}} \begin{pmatrix} n^{-1/6} \psi_{\alpha'}^{-1/4}(z) & 0 \\ 0 & n^{1/6} \psi_{\alpha'}^{1/4}(z) \end{pmatrix} \begin{pmatrix} 1 & \mathrm{i} \\ 1 & -\mathrm{i} \end{pmatrix}, \tag{7.5.82}$$

$$\psi_{\alpha'}(z) = -\left\{ \frac{3\pi}{2} \int_z^{\alpha'} \left(\frac{1}{2\gamma} - \rho(t) \right) \mathrm{d}t \right\}^{2/3}, \tag{7.5.83}$$

and we take the branch of $\psi_{\alpha'}^{1/4}$ which is positive on $(\alpha', \alpha' + \varepsilon)$ and has a cut on $(\alpha' - \varepsilon, \alpha')$.

For $z \in (\varepsilon, \beta')$ the solution is given by

$$\begin{aligned} \mathbf{U}_n(z) = \mathbf{M}(z) \mathrm{e}^{\pm(\mathrm{i}\Omega_n/2)\sigma_3} \mathbf{L}_n^{\beta'}(z)^{-1} \Phi_{\beta'}(n^{2/3} \psi_{\beta'}(z)) \\ \times \mathrm{e}^{\mp(\mathrm{i}n\pi z/(2\gamma))\sigma_3} \mathrm{e}^{-n(g(z) - V(z)/2 - l/2)\sigma_3} \quad \text{for } \pm \operatorname{Im} z > 0, \end{aligned} \tag{7.5.84}$$

where

$$\mathbf{L}_n^{\beta'}(z) = \frac{1}{2\sqrt{\pi}} \begin{pmatrix} n^{-1/6} \psi_{\beta'}^{-1/4}(z) & 0 \\ 0 & n^{1/6} \psi_{\beta'}^{1/4}(z) \end{pmatrix} \begin{pmatrix} -1 & \mathrm{i} \\ -1 & -\mathrm{i} \end{pmatrix}, \tag{7.5.85}$$

$$\psi_{\beta'}(z) = -\left\{ \frac{3\pi}{2} \int_{\beta'}^{z} \left(\frac{1}{2\gamma} - \rho(t) \, \mathrm{d}t \right) \right\}^{2/3}, \tag{7.5.86}$$

and

$$\Phi_{\beta'}(z) = \Phi_{\alpha'}(z) \begin{pmatrix} 1 & 0 \\ 0 & -1 \end{pmatrix}. \tag{7.5.87}$$

The large n behavior is as follows. For $z \in D(\alpha', \varepsilon)$, as $n \to \infty$,

$$\mathbf{U}_n(z) = \mathbf{M}(z) \left[I + \frac{\psi_{\alpha'}(z)^{-3/2}}{48n} \begin{pmatrix} -1 & -6\mathrm{i}\mathrm{e}^{\pm\mathrm{i}\Omega_n} \\ -6\mathrm{i}\mathrm{e}^{\mp\mathrm{i}\Omega_n} & 1 \end{pmatrix} + \mathcal{O}(n^{-2}) \right] \quad \text{for } \pm \operatorname{Im}(z) > 0. \tag{7.5.88}$$

For $z \in D(\beta', \varepsilon)$, as $n \to \infty$,

$$\mathbf{U}_n(z) = \mathbf{M}(z) \left[I + \frac{\psi_{\beta'}(z)^{-3/2}}{48n} \begin{pmatrix} -1 & 6\mathrm{i}\mathrm{e}^{\pm\mathrm{i}\Omega_n} \\ 6\mathrm{i}\mathrm{e}^{\mp\mathrm{i}\Omega_n} & 1 \end{pmatrix} + \mathcal{O}(n^{-2}) \right] \quad \text{for } \operatorname{Im} z > 0. \tag{7.5.89}$$

7.5.8. The third and final transformation of the RHP. We now consider the contour Σ_X, which consists of the circles $\partial D(\alpha,\varepsilon), \partial D(\alpha',\varepsilon), \partial D(\beta',\varepsilon)$, and $\partial D(\beta,\varepsilon)$, all oriented counterclockwise, together with the parts of

$$\Sigma_S \setminus \big([\alpha,\alpha'] \cup [\beta',\beta]\big)$$

which lie outside of the disks $D(\alpha,\varepsilon), D(\alpha',\varepsilon), D(\beta',\varepsilon)$, and $D(\beta,\varepsilon)$, see Figure 7.7.

We let

$$\mathbf{X}_n(z) = \begin{cases} \mathbf{S}_n(z)\mathbf{M}(z)^{-1} & \text{for } z \text{ outside the disks } D(\alpha,\varepsilon), D(\alpha',\varepsilon), \\ & D(\beta',\varepsilon), D(\beta,\varepsilon) \\ \mathbf{S}_n(z)\mathbf{U}_n(z)^{-1} & \text{for } z \text{ inside the disks } D(\alpha,\varepsilon), D(\alpha',\varepsilon), \\ & D(\beta',\varepsilon), D(\beta,\varepsilon). \end{cases} \tag{7.5.90}$$

Then $\mathbf{X}_n(z)$ solves the following RHP:

(1) $\mathbf{X}_n(z)$ is analytic on $\mathbb{C} \setminus \Sigma_X$.

(2) $\mathbf{X}_n(z)$ has the jump properties

$$\mathbf{X}_{n+}(x) = \mathbf{X}_{n-}(z) j_X(z), \tag{7.5.91}$$

where

$$j_X(z) = \begin{cases} \mathbf{M}(z)\mathbf{U}_n(z)^{-1} & \text{for } z \text{ on the circles} \\ \mathbf{M}(z) j_S(z)\mathbf{M}(z)^{-1} & \text{otherwise.} \end{cases} \tag{7.5.92}$$

(3) As $z \to \infty$,

$$\mathbf{X}_n(z) \sim I + \frac{\mathbf{X}_1}{z} + \frac{\mathbf{X}_2}{z^2} + \cdots . \tag{7.5.93}$$

Additionally, we have that $j_X(z)$ is uniformly close to the identity in the following sense:

$$j_X(z) = \begin{cases} I + \mathcal{O}(n^{-1}) & \text{uniformly on the circles} \\ I + \mathcal{O}(\mathrm{e}^{-C(z)n}) & \text{on the rest of } \Sigma_X, \end{cases} \tag{7.5.94}$$

where $C(z)$ is a positive, continuous function satisfying (1.1.7). If we set

$$j_X^0(z) = j_X(z) - I, \tag{7.5.95}$$

then (7.5.94) becomes

$$j_X^0(z) = \begin{cases} \mathcal{O}(n^{-1}) & \text{uniformly on the circles} \\ \mathcal{O}(\mathrm{e}^{-C(z)n}) & \text{on the rest of } \Sigma_X. \end{cases} \tag{7.5.96}$$

As shown in Sections 2.4.7 and 3.11, the solution is given as

$$\mathbf{X}_n(z) = I + \sum_{k=1}^{\infty} \mathbf{X}_{n,k}(z), \tag{7.5.97}$$

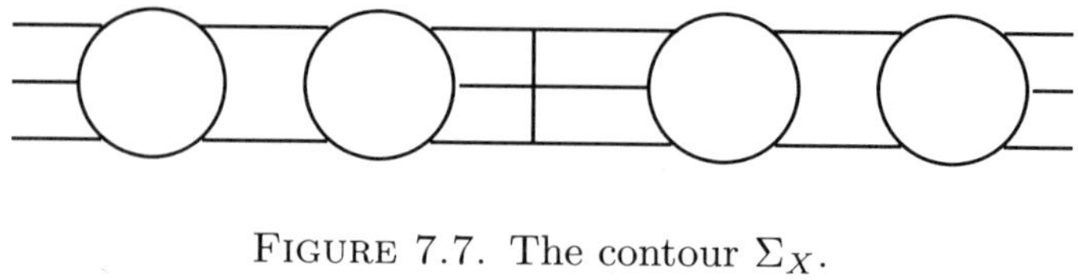

FIGURE 7.7. The contour Σ_X.

where

$$(7.5.98)\qquad \mathbf{X}_{n,k}(z) = -\frac{1}{2\pi \mathrm{i}}\int_{\Sigma_X} \frac{v_{k-1}(u) j_X^0(u)}{z-u}\,\mathrm{d}u,$$

and

$$(7.5.99)\qquad v_k(z) = -\frac{1}{2\pi \mathrm{i}}\int_{\Sigma_X} \frac{v_{k-1}(u) j_X^0(u)}{z_- - u}\,\mathrm{d}u, \quad v_0(z) = I.$$

In the expansion (7.5.97), the term $\mathbf{X}_{n,k}(z)$ is of the order $\mathcal{O}(n^{-k})$. Proposition 7.3.1 gives an expansion of h_n up to an error of the order n^{-2}. Thus in order to prove Proposition 7.3.1 using (7.5.6), we will need to compute the $1/z$-term in the expansion of $\mathbf{X}_{n,1}(z)$ at $z=\infty$. This computation is quite tedious, and we carry it out in the remaining sections of this chapter.

7.6. Evaluation of $\mathbf{X}_1$

We are interested in the matrix $\mathbf{X}_1$, which gives the $1/z$-term of $\mathbf{X}_n(z)$ at infinity, (see (7.5.93)) up to an error of the order $\mathcal{O}(n^{-2})$. It is therefore enough to consider the expansion of $\mathbf{X}_{n,1}(z)$, and we have that

$$(7.6.1)\qquad \mathbf{X}_1 = -\frac{1}{2\pi \mathrm{i}}\int_{\Sigma_X} j_X^0(u)\,\mathrm{d}u + \mathcal{O}(n^{-2}).$$

By (7.5.96), to compute the integral (7.6.1) up to an error of the order $\mathcal{O}(n^{-2})$, it is enough to evaluate this integral over the circles $\partial D(\alpha,\varepsilon), \partial D(\alpha',\varepsilon), \partial D(\beta',\varepsilon)$, and $\partial D(\beta,\varepsilon)$. As we will see in the next section, the matrix-valued function $j_X^0(z)$ is analytic in the punctured disks, hence

$$(7.6.2)\qquad \mathbf{X}_1 = -\Big(\operatorname*{Res}_{z=\alpha} + \operatorname*{Res}_{z=\alpha'} + \operatorname*{Res}_{z=\beta'} + \operatorname*{Res}_{z=\beta}\Big) j_X^0(z) + \mathcal{O}(n^{-2}).$$

We are especially interested in evaluation of the [12] element of the matrix $\mathbf{X}_1$, and we will prove the following asymptotic formula. Introduce the numbers

$$(7.6.3)\qquad \eta_\alpha = \bigg[5\frac{\vartheta_4''(n\omega+\omega/2)}{\vartheta_4(n\omega+\omega/2)} - 5\frac{\vartheta_3''(\omega/2)}{\vartheta_3(\omega/2)} + 7\bigg(\frac{\vartheta_4'(n\omega+\omega/2)}{\vartheta_4(n\omega+\omega/2)}\bigg)^2 + 17\bigg(\frac{\vartheta_3'(\omega/2)}{\vartheta_3(\omega/2)}\bigg)^2 - 24\frac{\vartheta_4'(n\omega+\omega/2)\vartheta_3'(\omega/2)}{\vartheta_4(n\omega+\omega/2)\vartheta_3(\omega/2)}\bigg],$$

$$(7.6.4)\qquad \eta_{\alpha'} = -\bigg[5\frac{\vartheta_1''(n\omega+\omega/2)}{\vartheta_1(n\omega+\omega/2)} - 5\frac{\vartheta_2''(\omega/2)}{\vartheta_2(\omega/2)} + 7\bigg(\frac{\vartheta_1'(n\omega+\omega/2)}{\vartheta_1(n\omega+\omega/2)}\bigg)^2 + 17\bigg(\frac{\vartheta_2'(\omega/2)}{\vartheta_2(\omega/2)}\bigg)^2 - 24\frac{\vartheta_1'(n\omega+\omega/2)\vartheta_2'(\omega/2)}{\vartheta_1(n\omega+\omega/2)\vartheta_2(\omega/2)}\bigg],$$

$$(7.6.5)\qquad \eta_{\beta'} = -\bigg[5\frac{\vartheta_2''(n\omega+\omega/2)}{\vartheta_2(n\omega+\omega/2)} - 5\frac{\vartheta_1''(\omega/2)}{\vartheta_1(\omega/2)} + 7\bigg(\frac{\vartheta_2'(n\omega+\omega/2)}{\vartheta_2(n\omega+\omega/2)}\bigg)^2 + 17\bigg(\frac{\vartheta_1'(\omega/2)}{\vartheta_1(\omega/2)}\bigg)^2 - 24\frac{\vartheta_2'(n\omega+\omega/2)\vartheta_1'(\omega/2)}{\vartheta_2(n\omega+\omega/2)\vartheta_1(\omega/2)}\bigg],$$

$$(7.6.6)\qquad \eta_\beta = \bigg[5\frac{\vartheta_3''(n\omega+\omega/2)}{\vartheta_3(n\omega+\omega/2)} - 5\frac{\vartheta_4''(\omega/2)}{\vartheta_4(\omega/2)} + 7\bigg(\frac{\vartheta_3'(n\omega+\omega/2)}{\vartheta_3(n\omega+\omega/2)}\bigg)^2 + 17\bigg(\frac{\vartheta_4'(\omega/2)}{\vartheta_4(\omega/2)}\bigg)^2 - 24\frac{\vartheta_3'(n\omega+\omega/2)\vartheta_4'(\omega/2)}{\vartheta_3(n\omega+\omega/2)\vartheta_4(\omega/2)}\bigg],$$

and

$$
\begin{aligned}
C_\alpha &= \frac{7}{2}(\beta'-\alpha)+\frac{3}{2}(\beta-\alpha)+\frac{3}{2}(\alpha'-\alpha)-\frac{(\alpha'-\alpha)(\beta-\alpha)}{(\beta'-\alpha)},\\
C_{\alpha'} &= -\frac{7}{2}(\beta-\alpha')-\frac{3}{2}(\beta'-\alpha')+\frac{3}{2}(\alpha'-\alpha)-\frac{(\alpha'-\alpha)(\beta'-\alpha')}{(\beta-\alpha')},\\
C_{\beta'} &= -\frac{7}{2}(\beta'-\alpha)-\frac{3}{2}(\beta'-\alpha')+\frac{3}{2}(\beta-\beta')-\frac{(\beta-\beta')(\beta'-\alpha')}{(\beta'-\alpha)},\\
C_\beta &= \frac{7}{2}(\beta-\alpha')+\frac{3}{2}(\beta-\alpha)+\frac{3}{2}(\beta-\beta')-\frac{(\beta-\beta')(\beta-\alpha)}{(\beta-\alpha')}.
\end{aligned}
\tag{7.6.7}
$$

Introduce also the numbers,

$$
\begin{aligned}
\Xi_\alpha &= \frac{\vartheta_3^2(0)\vartheta_4^2(n\omega+\omega/2)}{\vartheta_3^2(\omega/2)\vartheta_4^2(n\omega)}, &\qquad \Xi_{\alpha'} &= \frac{\vartheta_3^2(0)\vartheta_1^2(n\omega+\omega/2)}{\vartheta_2^2(\omega/2)\vartheta_4^2(n\omega)},\\
\Xi_{\beta'} &= \frac{\vartheta_3^2(0)\vartheta_2^2(n\omega+\omega/2)}{\vartheta_1^2(\omega/2)\vartheta_4^2(n\omega)}, &\qquad \Xi_\beta &= \frac{\vartheta_3^2(0)\vartheta_3^2(n\omega+\omega/2)}{\vartheta_4^2(\omega/2)\vartheta_4^2(n\omega)},
\end{aligned}
\tag{7.6.8}
$$

and

$$
\begin{aligned}
\xi_\alpha &= \frac{\vartheta_3'(\omega/2)}{\vartheta_3(\omega/2)}-\frac{\vartheta_4'(n\omega+\omega/2)}{\vartheta_4(n\omega+\omega/2)}, &\quad \xi_{\alpha'} &= -\left(\frac{\vartheta_2'(\omega/2)}{\vartheta_2(\omega/2)}-\frac{\vartheta_1'(n\omega+\omega/2)}{\vartheta_1(n\omega+\omega/2)}\right),\\
\xi_{\beta'} &= \frac{\vartheta_1'(\omega/2)}{\vartheta_1(\omega/2)}-\frac{\vartheta_2'(n\omega+\omega/2)}{\vartheta_2(n\omega+\omega/2)}, &\quad \xi_\beta &= -\left(\frac{\vartheta_4'(\omega/2)}{\vartheta_4(\omega/2)}-\frac{\vartheta_3'(n\omega+\omega/2)}{\vartheta_3(n\omega+\omega/2)}\right).
\end{aligned}
\tag{7.6.9}
$$

We have the following lemma.

Lemma 7.6.1. *As $n\to\infty$,*

$$
[\mathbf{X}_1]_{12} = \frac{1}{n}(X_\alpha+X_{\alpha'}+X_{\beta'}+X_\beta)+\mathcal{O}(n^{-2}),
\tag{7.6.10}
$$

where

$$
\begin{aligned}
X_\alpha &= \frac{\mathrm{i}\Xi_\alpha}{96}\left(C_\alpha+12\pi\xi_\alpha+\frac{\pi^2\eta_\alpha}{2(\beta'-\alpha)}\right),\\
X_{\alpha'} &= \frac{\mathrm{i}\Xi_\alpha'}{96}\left(C_{\alpha'}+12\pi\xi_{\alpha'}+\frac{\pi^2\eta_{\alpha'}}{2(\beta-\alpha')}\right),\\
X_{\beta'} &= \frac{\mathrm{i}\Xi_{\beta'}}{96}\left(C_{\beta'}+12\pi\xi_{\beta'}+\frac{\pi^2\eta_{\beta'}}{2(\beta'-\alpha)}\right),\\
X_\beta &= \frac{\mathrm{i}\Xi_\beta}{96}\left(C_\beta+12\pi\xi_\beta+\frac{\pi^2\eta_\beta}{2(\beta-\alpha')}\right).
\end{aligned}
\tag{7.6.11}
$$

Proof of this lemma is given in the next subsection.

7.6.1. Proof of Lemma 7.6.1. We have that

(7.6.12)
$$j_X(z) = \begin{cases} I - \dfrac{\psi_\alpha(z)^{-3/2}}{48n}\mathbf{M}(z)\begin{pmatrix} 1 & -6\mathrm{i} \\ -6\mathrm{i} & -1 \end{pmatrix}\mathbf{M}^{-1}(z) + \mathcal{O}(n^{-2}) \\ \qquad\qquad \text{for } z \in \partial D(\alpha,\varepsilon) \\ I - \dfrac{\psi_\beta(z)^{-3/2}}{48n}\mathbf{M}(z)\begin{pmatrix} 1 & 6\mathrm{i} \\ 6\mathrm{i} & -1 \end{pmatrix}\mathbf{M}^{-1}(z) + \mathcal{O}(n^{-2}) \\ \qquad\qquad \text{for } z \in \partial D(\beta,\varepsilon) \\ I - \dfrac{\psi_{\alpha'}(z)^{-3/2}}{48n}\mathbf{M}(z)\begin{pmatrix} -1 & -6\mathrm{i}e^{\pm \mathrm{i}\Omega_n} \\ -6\mathrm{i}e^{\mp \mathrm{i}\Omega_n} & 1 \end{pmatrix}\mathbf{M}^{-1}(z) + \mathcal{O}(n^{-2}) \\ \qquad\qquad \text{for } z \in \partial D(\alpha',\varepsilon),\ \pm\operatorname{Im} z > 0 \\ I - \dfrac{\psi_{\beta'}(z)^{-3/2}}{48n}\mathbf{M}(z)\begin{pmatrix} -1 & 6\mathrm{i}e^{\pm \mathrm{i}\Omega_n} \\ 6\mathrm{i}e^{\mp \mathrm{i}\Omega_n} & 1 \end{pmatrix}\mathbf{M}^{-1}(z) + \mathcal{O}(n^{-2}) \\ \qquad\qquad \text{for } z \in \partial D(\beta',\varepsilon),\ \pm\operatorname{Im} z > 0. \end{cases}$$

Thus

(7.6.13)
$$j_X^0(z) = \begin{cases} -\dfrac{\psi_\alpha(z)^{-3/2}}{48n}\mathbf{M}(z)\begin{pmatrix} 1 & -6\mathrm{i} \\ -6\mathrm{i} & -1 \end{pmatrix}\mathbf{M}^{-1}(z) + \mathcal{O}(n^{-2}) \\ \qquad\qquad \text{for } z \in \partial D(\alpha,\varepsilon) \\ -\dfrac{\psi_\beta(z)^{-3/2}}{48n}\mathbf{M}(z)\begin{pmatrix} 1 & 6\mathrm{i} \\ 6\mathrm{i} & -1 \end{pmatrix}\mathbf{M}^{-1}(z) + \mathcal{O}(n^{-2}) \\ \qquad\qquad \text{for } z \in \partial D(\beta,\varepsilon) \\ -\dfrac{\psi_{\alpha'}(z)^{-3/2}}{48n}\mathbf{M}(z)\begin{pmatrix} -1 & -6\mathrm{i}e^{\pm \mathrm{i}\Omega_n} \\ -6\mathrm{i}e^{\mp \mathrm{i}\Omega_n} & 1 \end{pmatrix}\mathbf{M}^{-1}(z) + \mathcal{O}(n^{-2}) \\ \qquad\qquad \text{for } z \in \partial D(\alpha',\varepsilon) \\ -\dfrac{\psi_{\beta'}(z)^{-3/2}}{48n}\mathbf{M}(z)\begin{pmatrix} -1 & 6\mathrm{i}e^{\pm \mathrm{i}\Omega_n} \\ 6\mathrm{i}e^{\mp \mathrm{i}\Omega_n} & 1 \end{pmatrix}\mathbf{M}^{-1}(z) + \mathcal{O}(n^{-2}) \\ \qquad\qquad \text{for } z \in \partial D(\beta',\varepsilon), \end{cases}$$

for $\pm\operatorname{Im} z > 0$. To simplify notation, we will write the model solution given in (7.5.56) and (7.5.57) as

$$\mathbf{M}(z) = \frac{1}{2}\begin{pmatrix} \big(\gamma(z)+\gamma^{-1}(z)\big)\vartheta_{11}(z) & \mathrm{i}\big(\gamma(z)-\gamma^{-1}(z)\big)\vartheta_{12}(z) \\ -\mathrm{i}\big(\gamma(z)-\gamma^{-1}(z)\big)\vartheta_{21}(z) & \big(\gamma(z)+\gamma^{-1}(z)\big)\vartheta_{22}(z) \end{pmatrix}, \tag{7.6.14}$$

where

$$\begin{aligned} \vartheta_{11}(z) &= \frac{\vartheta_3(\tilde u(z)-\tilde u_\infty+\Omega_n/2)\vartheta_3(0)}{\vartheta_3(\tilde u(z)-\tilde u_\infty)\vartheta_3(\Omega_n/2)}, \\ \vartheta_{12}(z) &= \frac{\vartheta_3(\tilde u(z)+\tilde u_\infty-\Omega_n/2)\vartheta_3(0)}{\vartheta_3(\tilde u(z)+\tilde u_\infty)\vartheta_3(\Omega_n/2)}, \\ \vartheta_{21}(z) &= \frac{\vartheta_3(\tilde u(z)+\tilde u_\infty+\Omega_n/2)\vartheta_3(0)}{\vartheta_3(\tilde u(z)+\tilde u_\infty)\vartheta_3(\Omega_n/2)}, \\ \vartheta_{22}(z) &= \frac{\vartheta_3(\tilde u(z)-\tilde u_\infty-\Omega_n/2)\vartheta_3(0)}{\vartheta_3(\tilde u(z)-\tilde u_\infty)\vartheta_3(\Omega_n/2)}. \end{aligned} \tag{7.6.15}$$

Notice that each of the functions ϑ_{ij} is analytic throughout the complex plane, except on the intervals (α,α') and (β',β), where they satisfy the relations

$$[\vartheta_{11}]_{\pm}=[\vartheta_{12}]_{\mp},\quad [\vartheta_{21}]_{\pm}=[\vartheta_{22}]_{\mp}, \tag{7.6.16}$$

and on the interval (α',β'), where they satisfy

$$\begin{aligned}[\vartheta_{11}]_+&=\mathrm{e}^{-\mathrm{i}\Omega_n}[\vartheta_{11}]_-, & [\vartheta_{12}]_+&=\mathrm{e}^{\mathrm{i}\Omega_n}[\vartheta_{12}]_-,\\ [\vartheta_{21}]_+&=\mathrm{e}^{-\mathrm{i}\Omega_n}[\vartheta_{21}]_-, & [\vartheta_{22}]_+&=\mathrm{e}^{\mathrm{i}\Omega_n}[\vartheta_{22}]_-.\end{aligned} \tag{7.6.17}$$

Multiplying out equation (7.6.13) gives

$$j_X^0(z)=\frac{\psi_\xi(z)^{-3/2}}{48n}\begin{pmatrix} j_{11}^\xi & j_{12}^\xi\\ j_{21}^\xi & j_{22}^\xi\end{pmatrix}+\mathcal{O}(n^{-2}) \qquad \text{for } z\in\partial D(\xi,\varepsilon),\ \xi=\alpha,\alpha',\beta',\beta, \tag{7.6.18}$$

where

$$\begin{aligned} j_{12}^{\alpha}&=\frac{\mathrm{i}}{2}\big[3((\gamma^2(z)+\gamma^{-2}(z))(\vartheta_{11}^2-\vartheta_{12}^2)\\ &\qquad\qquad +(\gamma^2(z)-\gamma^{-2}(z))\vartheta_{11}\vartheta_{12}+6(\vartheta_{11}^2-\vartheta_{12}^2)\big],\\ j_{12}^{\alpha'}&=\frac{\mathrm{i}}{2}\big[3(\gamma^2(z)+\gamma^{-2}(z))J(z)\\ &\qquad\qquad -(\gamma^2(z)-\gamma^{-2}(z))\vartheta_{11}\vartheta_{12}+6K(z)\big],\\ j_{12}^{\beta'}&=-\frac{\mathrm{i}}{2}\big[3(\gamma^2(z)+\gamma^{-2}(z))J(z)\\ &\qquad\qquad +(\gamma^2(z)-\gamma^{-2}(z))\vartheta_{11}\vartheta_{12}+6K(z)\big],\\ j_{12}^{\beta}&=-\frac{\mathrm{i}}{2}\big[3((\gamma^2(z)+\gamma^{-2}(z))(\vartheta_{11}^2-\vartheta_{12}^2)\\ &\qquad\qquad -(\gamma^2(z)-\gamma^{-2}(z))\vartheta_{11}\vartheta_{12}+6(\vartheta_{11}^2-\vartheta_{12}^2)\big],\end{aligned} \tag{7.6.19}$$

and

$$J(z)=\begin{cases}\vartheta_{11}^2\mathrm{e}^{\mathrm{i}\Omega_n}+\vartheta_{12}^2\mathrm{e}^{-\mathrm{i}\Omega_n} & \text{for } \operatorname{Im} z>0\\ \vartheta_{11}^2\mathrm{e}^{-\mathrm{i}\Omega_n}+\vartheta_{12}^2\mathrm{e}^{\mathrm{i}\Omega_n} & \text{for } \operatorname{Im} z<0,\end{cases} \tag{7.6.20}$$

$$K(z)=\begin{cases}\vartheta_{11}^2\mathrm{e}^{\mathrm{i}\Omega_n}-\vartheta_{12}^2\mathrm{e}^{-\mathrm{i}\Omega_n} & \text{for } \operatorname{Im} z>0\\ \vartheta_{11}^2\mathrm{e}^{-\mathrm{i}\Omega_n}-\vartheta_{12}^2\mathrm{e}^{\mathrm{i}\Omega_n} & \text{for } \operatorname{Im} z<0.\end{cases} \tag{7.6.21}$$

In order to integrate $j_X^0(z)$, let us examine the behavior of the various functions described above near each of the turning points. Introduce the numbers

$$\begin{aligned} A_\alpha&=\sqrt{(\alpha'-\alpha)(\beta'-\alpha)(\beta-\alpha)}, & A_{\alpha'}&=\sqrt{(\alpha'-\alpha)(\beta'-\alpha')(\beta-\alpha')},\\ A_{\beta'}&=\sqrt{(\beta'-\alpha)(\beta'-\alpha')(\beta-\beta')}, & A_\beta&=\sqrt{(\beta-\alpha)(\beta-\alpha')(\beta-\beta')},\end{aligned} \tag{7.6.22}$$

and

$$\begin{aligned} B_\alpha&=\frac{1}{\alpha'-\alpha}+\frac{1}{\beta'-\alpha}+\frac{1}{\beta-\alpha}, & B_{\alpha'}&=-\frac{1}{\alpha'-\alpha}+\frac{1}{\beta'-\alpha'}+\frac{1}{\beta-\alpha'},\\ B_{\beta'}&=\frac{1}{\beta'-\alpha}+\frac{1}{\beta'-\alpha'}-\frac{1}{\beta-\beta'}, & B_\beta&=\frac{1}{\beta-\alpha}+\frac{1}{\beta-\alpha'}+\frac{1}{\beta-\beta'}.\end{aligned} \tag{7.6.23}$$

For $x\in(\beta,\beta+\varepsilon)$, we have

$$\tilde u(x)=\frac{\pi}{2K}u(x)=\frac{\pi}{A_\beta}\sqrt{x-\beta}+\mathcal{O}\big((x-\beta)^{3/2}\big); \tag{7.6.24}$$

for $x \in (\alpha - \varepsilon, \alpha)$,

$$\tilde{u}(x) = \frac{\pi}{2K} u(x) = \frac{\pi}{2} - \frac{\pi}{A_\alpha}\sqrt{\alpha - x} + \mathcal{O}((\alpha - x)^{3/2}); \tag{7.6.25}$$

for $x \in (\alpha', \alpha' + \varepsilon)$,

$$\tilde{u}_\pm(x) = \frac{\pi}{2K} u_\pm(x) = \frac{\pi}{2} \pm \frac{\tau\pi}{2} - \frac{\pi}{A_{\alpha'}}\sqrt{x - \alpha'} + \mathcal{O}\big((x - \alpha')^{3/2}\big); \tag{7.6.26}$$

and for $x \in (\beta' - \varepsilon, \beta')$,

$$\tilde{u}_\pm(x) = \frac{\pi}{2K} u_\pm(x) = \pm\frac{\tau\pi}{2} + \frac{\pi}{A_{\beta'}}\sqrt{\beta' - x} + \mathcal{O}\big((\beta' - x)^{-3/2}\big). \tag{7.6.27}$$

Also, from (7.5.31) we have that

$$\gamma^2(z) \pm \gamma^{-2}(z) = \sqrt{\frac{(z-\alpha)(z-\beta')}{(z-\alpha')(z-\beta)}} \pm \sqrt{\frac{(z-\alpha')(z-\beta)}{(z-\alpha)(z-\beta')}}, \tag{7.6.28}$$

and from (7.5.71), (7.5.76), (7.5.83), (7.5.86), and (7.4.17), we have

$$\begin{aligned}
\psi_\alpha^{-3/2}(z) &= (\alpha - z)^{-3/2}\left[\frac{A_\alpha}{2} + \frac{1}{20}A_\alpha B_\alpha(\alpha - z) + \mathcal{O}\big((\alpha - z)^2\big)\right],\\
\psi_{\alpha'}^{-3/2}(z) &= (z - \alpha')^{-3/2}\left[\frac{A_{\alpha'}}{2} - \frac{1}{20}A_{\alpha'} B_{\alpha'}(z - \alpha') + \mathcal{O}\big((z - \alpha')^2\big)\right],\\
\psi_{\beta'}^{-3/2}(z) &= (\beta' - z)^{-3/2}\left[\frac{A_{\beta'}}{2} - \frac{1}{20}A_{\beta'} B_{\beta'}(\beta' - z) + \mathcal{O}\big((\beta' - z)^2\big)\right],\\
\psi_\beta^{-3/2}(z) &= (z - \beta)^{-3/2}\left[\frac{A_\beta}{2} + \frac{1}{20}A_\beta B_\beta(z - \beta) + \mathcal{O}\big((z - \beta)^2\big)\right].
\end{aligned} \tag{7.6.29}$$

It follows that the functions $(\gamma^2 \pm \gamma^{-2})\psi^{-3/2}(z)$ are meromorphic in a neighborhood of each of the turning points. In particular, at $z = \alpha$, we have

$$\begin{aligned}
(\gamma^2 \pm \gamma^{-2})\psi_\alpha^{-3/2}(z) &= \pm\frac{(\alpha'-\alpha)(\beta-\alpha)}{2(\alpha - z)^2}\\
&\quad + \frac{1}{(\alpha-z)}\left[\frac{(\beta'-\alpha)}{2} \pm \frac{3}{10}(\beta-\alpha) \pm \frac{3}{10}(\alpha'-\alpha) \mp \frac{1}{5}\frac{(\alpha'-\alpha)(\beta-\alpha)}{(\beta'-\alpha)}\right]\\
&\quad + \mathcal{O}(1);
\end{aligned} \tag{7.6.30}$$

at $z = \beta$, we have

$$\begin{aligned}
(\gamma^2 \pm \gamma^{-2})\psi_\beta^{-3/2}(z) &= \frac{(\beta-\alpha)(\beta-\beta')}{2(z-\beta)^2}\\
&\quad + \frac{1}{(z-\beta)}\left[\pm\frac{1}{2}(\beta-\alpha') + \frac{3}{10}(\beta-\alpha) + \frac{3}{10}(\beta-\beta') - \frac{1}{5}\frac{(\beta-\alpha)(\beta-\beta')}{(\beta-\alpha')}\right]\\
&\quad + \mathcal{O}(1);
\end{aligned} \tag{7.6.31}$$

at $z = \alpha'$, we have

$$\begin{aligned}
(\gamma^2 \pm \gamma^{-2})\psi_{\alpha'}^{-3/2}(z) &= \frac{(\alpha'-\alpha)(\beta'-\alpha')}{2(z-\alpha')^2}\\
&\quad + \frac{1}{(z-\alpha')}\left[\pm\frac{(\beta-\alpha')}{2} + \frac{3}{10}(\beta'-\alpha') - \frac{3}{10}(\alpha'-\alpha) + \frac{1}{5}\frac{(\alpha'-\alpha)(\beta'-\alpha')}{(\beta-\alpha')}\right]\\
&\quad + \mathcal{O}(1);
\end{aligned} \tag{7.6.32}$$

and at $z=\beta'$, we have

$$(\gamma^2 \pm \gamma^{-2})\psi_{\beta'}^{-3/2}(z) = \pm\frac{(\beta'-\alpha')(\beta-\beta')}{2(\beta'-z)^2} + \frac{1}{(\beta'-z)}\left[\frac{(\beta'-\alpha)}{2} \pm \frac{3}{10}(\beta'-\alpha') \mp \frac{3}{10}(\beta-\beta') \pm \frac{1}{5}\frac{(\beta'-\alpha')(\beta-\beta')}{\beta'-\alpha}\right] + \mathcal{O}(1). \tag{7.6.33}$$

Notice also, from the relations (7.6.15), that the functions $\vartheta_{11}^2 + \vartheta_{12}^2$ and $\vartheta_{11}\vartheta_{12}$ have no jumps in neighborhoods of α or β, and take finite values at $z=\alpha$ and $z=\beta$, thus are analytic in neighborhoods of α and β. Using (7.6.24) and (7.6.25), we see that these functions have Taylor expansions about $z=\beta$,

$$\begin{aligned} \vartheta_{11}^2(z)+\vartheta_{12}^2(z) &= 2\frac{\vartheta_3^2(0)}{\vartheta_3^2(\Omega_n/2)}\frac{\vartheta_3^2(\tilde{u}_\infty-\Omega_n/2)}{\vartheta_3^2(\tilde{u}_\infty)} + \frac{\pi^2}{2A_\beta^2}\frac{\mathrm{d}^2}{\mathrm{d}\tilde{u}^2}(\vartheta_{11}^2+\vartheta_{12}^2)\Big|_{z=\beta}(z-\beta)+\cdots, \\ \vartheta_{11}(z)\vartheta_{12}(z) &= \frac{\vartheta_3^2(0)}{\vartheta_3^2(\Omega_n/2)}\frac{\vartheta_3^2(\tilde{u}_\infty-\Omega_n/2)}{\vartheta_3^2(\tilde{u}_\infty)} + \frac{\pi^2}{2A_\beta^2}\frac{\mathrm{d}^2}{\mathrm{d}\tilde{u}^2}(\vartheta_{11}\vartheta_{12})\Big|_{z=\beta}(z-\beta)+\cdots, \end{aligned} \tag{7.6.34}$$

and about $z=\alpha$,

$$\begin{aligned} \vartheta_{11}^2(z)+\vartheta_{12}^2(z) &= 2\frac{\vartheta_3^2(0)}{\vartheta_3^2(\Omega_n/2)}\frac{\vartheta_4^2(\tilde{u}_\infty-\Omega_n/2)}{\vartheta_4^2(\tilde{u}_\infty)} + \frac{\pi^2}{2A_\alpha^2}\frac{\mathrm{d}^2}{\mathrm{d}\tilde{u}^2}(\vartheta_{11}^2+\vartheta_{12}^2)\Big|_{z=\alpha}(\alpha-z)+\cdots, \\ \vartheta_{11}(z)\vartheta_{12}(z) &= \frac{\vartheta_3^2(0)}{\vartheta_3^2(\Omega_n/2)}\frac{\vartheta_4^2(\tilde{u}_\infty-\Omega_n/2)}{\vartheta_4^2(\tilde{u}_\infty)} + \frac{\pi^2}{2A_\alpha^2}\frac{\mathrm{d}^2}{\mathrm{d}\tilde{u}^2}(\vartheta_{11}\vartheta_{12})\Big|_{z=\alpha}(\alpha-z)+\cdots. \end{aligned} \tag{7.6.35}$$

By a similar argument, $J(z)$ and $\vartheta_{11}\vartheta_{12}$ are also analytic in neighborhoods of α' and β' and using (7.6.26) and (7.6.27) we can write their Taylor expansions about $z=\alpha'$,

$$\begin{aligned} J(z) &= 2\frac{\vartheta_3^2(0)}{\vartheta_3^2(\Omega_n/2)}\frac{\vartheta_1^2(\tilde{u}_\infty-\Omega_n/2)}{\vartheta_1^2(\tilde{u}_\infty)} + \frac{\pi^2}{2A_{\alpha'}^2}\frac{\mathrm{d}^2}{\mathrm{d}\tilde{u}^2}J(z)\Big|_{z=\alpha'}(z-\alpha')+\cdots, \\ \vartheta_{11}(z)\vartheta_{12}(z) &= \frac{\vartheta_3^2(0)}{\vartheta_3^2(\Omega_n/2)}\frac{\vartheta_1^2(\tilde{u}_\infty-\Omega_n/2)}{\vartheta_1^2(\tilde{u}_\infty)} + \frac{\pi^2}{2A_{\alpha'}^2}\frac{\mathrm{d}^2}{\mathrm{d}\tilde{u}^2}(\vartheta_{11}\vartheta_{12})\Big|_{z=\alpha'}(z-\alpha')+\cdots, \end{aligned} \tag{7.6.36}$$

and about $z = \beta'$,

(7.6.37)
$$\begin{aligned} J(z) &= 2\frac{\vartheta_3^2(0)}{\vartheta_3^2(\Omega_n/2)}\frac{\vartheta_2^2(\tilde{u}_\infty - \Omega_n/2)}{\vartheta_2^2(\tilde{u}_\infty)} \\ &\qquad + \frac{\pi^2}{2A_{\beta'}^2}\frac{\mathrm{d}^2}{\mathrm{d}\tilde{u}^2}J(z)\bigg|_{z=\beta'}(\beta' - z) + \cdots, \\ \vartheta_{11}(z)\vartheta_{12}(z) &= \frac{\vartheta_3^2(0)}{\vartheta_3^2(\Omega_n/2)}\frac{\vartheta_2^2(\tilde{u}_\infty - \Omega_n/2)}{\vartheta_2^2(\tilde{u}_\infty)} \\ &\qquad + \frac{\pi^2}{2A_{\beta'}^2}\frac{\mathrm{d}^2}{\mathrm{d}\tilde{u}^2}(\vartheta_{11}\vartheta_{12})\bigg|_{z=\beta'}(\beta' - z) + \cdots . \end{aligned}$$

Finally, notice that the function $\vartheta_{11}^2 - \vartheta_{12}^2$ is an odd function of $\tilde{u}$, and using (7.6.24) and (7.6.25), we can write for $x \in (\beta, \beta + \varepsilon)$,

(7.6.38)
$$\begin{aligned} \vartheta_{11}^2(x) - \vartheta_{12}^2(x) &= \sqrt{x - \beta} \\ &\times \left[\frac{4\pi}{A_\beta}\frac{\vartheta_3^2(0)}{\vartheta_3^2(\Omega_n/2)}\frac{\vartheta_3^2(\tilde{u}_\infty - \Omega_n/2)}{\vartheta_3^2(\tilde{u}_\infty)}\left(\frac{\vartheta_3'(\tilde{u}_\infty)}{\vartheta_3(\tilde{u}_\infty)} - \frac{\vartheta_3'(\tilde{u}_\infty - \Omega_n/2)}{\vartheta_3(\tilde{u}_\infty - \Omega_n/2)}\right)\right. \\ &\qquad \left. + \mathcal{O}(x - \beta)\right], \end{aligned}$$

and for $x \in (\alpha - \varepsilon, \alpha)$,

(7.6.39)
$$\begin{aligned} \vartheta_{11}^2(x) - \vartheta_{12}^2(x) &= -\sqrt{\alpha - x} \\ &\times \left[\frac{4\pi}{A_\alpha}\frac{\vartheta_3^2(0)}{\vartheta_3^2(\Omega_n/2)}\frac{\vartheta_4^2(\tilde{u}_\infty - \Omega_n/2)}{\vartheta_4^2(\tilde{u}_\infty)}\left(\frac{\vartheta_4'(\tilde{u}_\infty)}{\vartheta_4(\tilde{u}_\infty)} - \frac{\vartheta_4'(\tilde{u}_\infty - \Omega_n/2)}{\vartheta_4(\tilde{u}_\infty - \Omega_n/2)}\right)\right. \\ &\qquad \left. + \mathcal{O}(\alpha - x)\right]. \end{aligned}$$

Similarly, using (7.6.26) and (7.6.27), we can write for $x \in (\alpha', \alpha' + \varepsilon)$,

(7.6.40)
$$\begin{aligned} K(x) &= -\sqrt{x - \alpha'} \\ &\times \left[\frac{4\pi}{A_{\alpha'}}\frac{\vartheta_3^2(0)}{\vartheta_3^2(\Omega_n/2)}\frac{\vartheta_1^2(\tilde{u}_\infty - \Omega_n/2)}{\vartheta_1^2(\tilde{u}_\infty)}\left(\frac{\vartheta_1'(\tilde{u}_\infty)}{\vartheta_1(\tilde{u}_\infty)} - \frac{\vartheta_1'(\tilde{u}_\infty - \Omega_n/2)}{\vartheta_1'(\tilde{u}_\infty - \Omega_n/2)}\right)\right. \\ &\qquad \left. + \mathcal{O}(x - \alpha')\right], \end{aligned}$$

and for $x \in (\beta' - \varepsilon, \beta')$,

(7.6.41)
$$\begin{aligned} K(x) &= \sqrt{\beta' - x} \\ &\times \left[\frac{4\pi}{A_{\beta'}}\frac{\vartheta_3^2(0)}{\vartheta_3^2(\Omega_n/2)}\frac{\vartheta_2^2(\tilde{u}_\infty - \Omega_n/2)}{\vartheta_2^2(\tilde{u}_\infty)}\left(\frac{\vartheta_2'(\tilde{u}_\infty)}{\vartheta_2(\tilde{u}_\infty)} - \frac{\vartheta_2'(\tilde{u}_\infty - \Omega_n/2)}{\vartheta_2'(\tilde{u}_\infty - \Omega_n/2)}\right)\right. \\ &\qquad \left. + \mathcal{O}(\beta' - x)\right]. \end{aligned}$$

From equations (7.6.29), (7.6.38), (7.6.39), (7.6.40), and (7.6.41), it follows that the functions $\big(\vartheta_{11}^2(z) - \vartheta_{12}^2(z)\big)\psi_\alpha^{-3/2}(z)$ and $\big(\vartheta_{11}^2(z) - \vartheta_{12}^2(z)\big)\psi_\beta^{-3/2}(z)$ are meromorphic in neighborhoods of α and β, respectively, and have simple poles at $z = \alpha$ and $z = \beta$, respectively, and that the functions $K(z)\psi_{\alpha'}^{-3/2}(z)$ and $K(z)\psi_{\beta'}^{-3/2}(z)$

are meromorphic in neighborhoods of α' and β', respectively, and have simple poles at $z = \alpha'$ and $z = \beta'$, respectively.

Let us compute the residues of functions that appear in (7.6.19). Observe that

$$\frac{\Omega_n}{2} = n\omega + \frac{\pi}{2}, \quad \tilde{u}_\infty = -\frac{\omega}{2} + \frac{\pi}{2}, \quad \frac{\Omega_n}{2} - \tilde{u}_\infty = n\omega + \frac{\omega}{2}. \tag{7.6.42}$$

From (7.6.30), (7.6.31), (7.6.34), and (7.6.35), we obtain that

$$\begin{aligned}\operatorname*{Res}_{z=\alpha} 3\big(\vartheta_{11}^2(z)+\vartheta_{12}^2(z)\big)\big(\gamma^2(z)+\gamma^{-2}(z)\big)\psi_\alpha^{-3/2}(z) \\ = \frac{\vartheta_3^2(0)\vartheta_4^2(n\omega+\omega/2)}{\vartheta_3^2(\omega/2)\vartheta_4^2(n\omega)}\left[-3(\beta'-\alpha)-\frac{9}{5}(\beta-\alpha)-\frac{9}{5}(\alpha'-\alpha)+\frac{6}{5}\frac{(\alpha'-\alpha)(\beta-\alpha)}{(\beta'-\alpha)}\right] \\ -\frac{3\pi^2}{4(\beta'-\alpha)}\frac{\mathrm{d}^2}{\mathrm{d}\tilde{u}^2}\big(\vartheta_{11}^2+\vartheta_{12}^2\big)\Big|_{z=\alpha},\end{aligned} \tag{7.6.43}$$

and

$$\begin{aligned}\operatorname*{Res}_{z=\beta} 3\big(\vartheta_{11}^2(z)+\vartheta_{12}^2(z)\big)\big(\gamma^2(z)+\gamma^{-2}(z)\big)\psi_\beta^{-3/2}(z) \\ = \frac{\vartheta_3^2(0)\vartheta_3^2(n\omega+\omega/2)}{\vartheta_4^2(\omega/2)\vartheta_4^2(n\omega)}\left[3(\beta-\alpha')+\frac{9}{5}(\beta-\beta')+\frac{9}{5}(\beta-\alpha)-\frac{6}{5}\frac{(\beta-\alpha)(\beta-\beta')}{(\beta-\alpha')}\right] \\ +\frac{3\pi^2}{4(\beta-\alpha')}\frac{\mathrm{d}^2}{\mathrm{d}\tilde{u}^2}\big(\vartheta_{11}^2+\vartheta_{12}^2\big)\Big|_{z=\beta}.\end{aligned} \tag{7.6.44}$$

Also,

$$\begin{aligned}\operatorname*{Res}_{z=\alpha}\big(\vartheta_{11}(z)\vartheta_{12}(z)\big)\big(\gamma^2(z)-\gamma^{-2}(z)\big)\psi_\alpha^{-3/2}(z) \\ = \frac{\vartheta_3^2(0)\vartheta_4^2(n\omega+\omega/2)}{\vartheta_3^2(\omega/2)\vartheta_4^2(n\omega)}\left[-\frac{(\beta'-\alpha)}{2}+\frac{3}{10}(\beta-\alpha)+\frac{3}{10}(\alpha'-\alpha)-\frac{1}{5}\frac{(\alpha'-\alpha)(\beta-\alpha)}{(\beta'-\alpha)}\right] \\ +\frac{\pi^2}{4(\beta'-\alpha)}\frac{\mathrm{d}^2}{\mathrm{d}\tilde{u}^2}\big(\vartheta_{11}\vartheta_{12}\big)\Big|_{z=\alpha},\end{aligned} \tag{7.6.45}$$

and

$$\begin{aligned}\operatorname*{Res}_{z=\beta}\big(\vartheta_{11}(z)\vartheta_{12}(z)\big)\big(\gamma^2(z)-\gamma^{-2}(z)\big)\psi_\beta^{-3/2}(z) \\ = \frac{\vartheta_3^2(0)\vartheta_3^2(n\omega+\omega/2)}{\vartheta_4^2(\omega/2)\vartheta_4^2(n\omega)}\left[-\frac{(\beta-\alpha')}{2}+\frac{3}{10}(\beta-\beta')+\frac{3}{10}(\beta-\alpha)-\frac{1}{5}\frac{(\beta-\alpha)(\beta-\beta')}{(\beta-\alpha')}\right] \\ +\frac{\pi^2}{4(\beta-\alpha')}\frac{\mathrm{d}^2}{\mathrm{d}\tilde{u}^2}\big(\vartheta_{11}\vartheta_{12}\big)\Big|_{z=\beta}.\end{aligned} \tag{7.6.46}$$

From (7.6.29), (7.6.38), and (7.6.39), we obtain that

$$\begin{aligned}\operatorname*{Res}_{z=\alpha} 6\big(\vartheta_{11}^2(z)-\vartheta_{12}^2(z)\big)\psi_\alpha^{-3/2}(z) \\ = -12\pi\frac{\vartheta_3^2(0)\vartheta_4^2(n\omega+\omega/2)}{\vartheta_3^2(\omega/2)\vartheta_4^2(n\omega)}\left[\frac{\vartheta_3'(\omega/2)}{\vartheta_3(\omega/2)}-\frac{\vartheta_4'(n\omega+\omega/2)}{\vartheta_4(n\omega+\omega/2)}\right]\end{aligned} \tag{7.6.47}$$

and

$$\operatorname*{Res}_{z=\beta} 6\big(\vartheta_{11}^2(z) - \vartheta_{12}^2(z)\big)\psi_\beta^{-3/2}(z) = -12\pi\frac{\vartheta_3^2(0)\vartheta_3^2(n\omega+\omega/2)}{\vartheta_4^2(\omega/2)\vartheta_4^2(n\omega)}\left[\frac{\vartheta_4'(\omega/2)}{\vartheta_4(\omega/2)} - \frac{\vartheta_3'(n\omega+\omega/2)}{\vartheta_3(n\omega+\omega/2)}\right]. \tag{7.6.48}$$

We now turn our attention to the residues at the inner turning points. From (7.6.32), (7.6.33), (7.6.36), and (7.6.37), we have

$$\begin{aligned}\operatorname*{Res}_{z=\alpha'} & 3\psi_{\alpha'}^{-3/2}(z)J(z)\big(\gamma^2(z)+\gamma^{-2}(z)\big) \\ &= \frac{\vartheta_3^2(0)\vartheta_1^2(n\omega+\omega/2)}{\vartheta_2^2(\omega/2)\vartheta_4^2(n\omega)}\left[3(\beta-\alpha')+\frac{9}{5}(\beta'-\alpha')-\frac{9}{5}(\alpha'-\alpha)+\frac{6}{5}\frac{(\alpha'-\alpha)(\beta'-\alpha')}{\beta-\alpha'}\right] \\ &\qquad + \frac{3\pi^2}{4(\beta-\alpha')}\frac{\mathrm{d}^2}{\mathrm{d}\tilde{u}^2}J(z)\bigg|_{z=\alpha'},\end{aligned} \tag{7.6.49}$$

and

$$\begin{aligned}\operatorname*{Res}_{z=\beta'} & 3\psi_{\beta'}^{-3/2}(z)J(z)\big(\gamma^2(z)+\gamma^{-2}(z)\big) \\ &= \frac{\vartheta_3^2(0)\vartheta_2^2(n\omega+\omega/2)}{\vartheta_1^2(\omega/2)\vartheta_4^2(n\omega)}\left[-3(\beta'-\alpha)-\frac{9}{5}(\beta'-\alpha')+\frac{9}{5}(\beta-\beta')-\frac{6}{5}\frac{(\beta'-\alpha')(\beta-\beta')}{\beta'-\alpha}\right] \\ &\qquad - \frac{3\pi^2}{4(\beta'-\alpha)}\frac{\mathrm{d}^2}{\mathrm{d}\tilde{u}^2}J(z)\bigg|_{z=\beta'}.\end{aligned} \tag{7.6.50}$$

Also,

$$\begin{aligned}\operatorname*{Res}_{z=\alpha'} & \psi_{\alpha'}^{-3/2}(z)\big(\gamma^2(z)-\gamma^{-2}(z)\big)\vartheta_{11}(z)\vartheta_{12}(z) \\ &= \frac{\vartheta_3^2(0)\vartheta_1^2(n\omega+\omega/2)}{\vartheta_2^2(\omega/2)\vartheta_4^2(n\omega)}\left[-\frac{(\beta-\alpha')}{2}+\frac{3}{10}(\beta'-\alpha')-\frac{3}{10}(\alpha'-\alpha)+\frac{1}{5}\frac{(\alpha'-\alpha)(\beta'-\alpha')}{\beta-\alpha'}\right] \\ &\qquad + \frac{\pi^2}{4(\beta-\alpha')}\frac{\mathrm{d}^2}{\mathrm{d}\tilde{u}^2}(\vartheta_{11}\vartheta_{12})\bigg|_{z=\alpha'},\end{aligned} \tag{7.6.51}$$

and

$$\begin{aligned}\operatorname*{Res}_{z=\beta'} & \psi_{\beta'}^{-3/2}(z)\big(\gamma^2(z)-\gamma^{-2}(z)\big)\vartheta_{11}(z)\vartheta_{12}(z) \\ &= \frac{\vartheta_3^2(0)\vartheta_2^2(n\omega+\omega/2)}{\vartheta_1^2(\omega/2)\vartheta_4^2(n\omega)}\left[-\frac{(\beta'-\alpha)}{2}+\frac{3}{10}(\beta'-\alpha')-\frac{3}{10}(\beta-\beta')+\frac{1}{5}\frac{(\beta'-\alpha')(\beta-\beta')}{\beta'-\alpha}\right] \\ &\qquad + \frac{\pi^2}{4(\beta'-\alpha)}\frac{\mathrm{d}^2}{\mathrm{d}\tilde{u}^2}(\vartheta_{11}\vartheta_{12})\bigg|_{z=\beta'}.\end{aligned} \tag{7.6.52}$$

From (7.6.29), (7.6.40), (7.6.41), we have

$$\operatorname*{Res}_{z=\alpha'} 6\psi_{\alpha'}^{-3/2}(z)K(z) = 12\pi\frac{\vartheta_3^2(0)\vartheta_1^2(n\omega+\omega/2)}{\vartheta_2^2(\omega/2)\vartheta_4^2(n\omega)}\left[\frac{\vartheta_2'(\omega/2)}{\vartheta_2(\omega/2)} - \frac{\vartheta_1'(n\omega+\omega/2)}{\vartheta_1(n\omega+\omega/2)}\right]. \tag{7.6.53}$$

and

$$\underset{z=\beta'}{\operatorname{Res}}\, 6\psi_{\beta'}^{-3/2}(z)K(z) = 12\pi\frac{\vartheta_3^2(0)\vartheta_2^2(n\omega+\omega/2)}{\vartheta_1^2(\omega/2)\vartheta_4^2(n\omega)}\left[\frac{\vartheta_1'(\omega/2)}{\vartheta_1(\omega/2)} - \frac{\vartheta_2'(n\omega+\omega/2)}{\vartheta_2(n\omega+\omega/2)}\right]. \tag{7.6.54}$$

Combining (7.6.18), (7.6.19), (7.6.43), (7.6.45), and (7.6.47), we get that

$$\begin{aligned}\underset{z=\alpha}{\operatorname{Res}}[j_X^0(z)] &= \frac{\mathrm{i}}{96}\Bigg[\frac{\vartheta_3^2(0)\vartheta_4^2(n\omega+\omega/2)}{\vartheta_3^2(\omega/2)\vartheta_4^2(n\omega)}\Bigg(-\frac{7}{2}(\beta'-\alpha)-\frac{3}{2}(\beta-\alpha)-\frac{3}{2}(\alpha'-\alpha)\\ &\quad+\frac{(\alpha'-\alpha)(\beta-\alpha)}{\beta'-\alpha}-12\pi\left(\frac{\vartheta_3'(\omega/2)}{\vartheta_3(\omega/2)}-\frac{\vartheta_4'(n\omega+\omega/2)}{\vartheta_4(n\omega+\omega/2)}\right)\Bigg)\\ &\quad-\frac{\pi^2}{4(\beta'-\alpha)}\frac{\mathrm{d}^2}{\mathrm{d}\tilde{u}^2}[3(\vartheta_{11}^2+\vartheta_{12}^2)-\vartheta_{11}\vartheta_{12}]_{z=\alpha}\Bigg].\end{aligned} \tag{7.6.55}$$

Similarly, we have

$$\begin{aligned}\underset{z=\alpha'}{\operatorname{Res}}[j_X^0(z)] &= \frac{\mathrm{i}}{96}\Bigg[\frac{\vartheta_3^2(0)\vartheta_1^2(n\omega+\omega/2)}{\vartheta_2^2(\omega/2)\vartheta_4^2(n\omega)}\Bigg(\frac{7}{2}(\beta-\alpha')+\frac{3}{2}(\beta'-\alpha')-\frac{3}{2}(\alpha'-\alpha)\\ &\quad+\frac{(\alpha'-\alpha)(\beta'-\alpha')}{\beta-\alpha'}+12\pi\left(\frac{\vartheta_2'(\omega/2)}{\vartheta_2(\omega/2)}-\frac{\vartheta_1'(n\omega+\omega/2)}{\vartheta_1(n\omega+\omega/2)}\right)\Bigg)\\ &\quad+\frac{\pi^2}{4(\beta-\alpha')}\frac{\mathrm{d}^2}{\mathrm{d}\tilde{u}^2}[3J(z)-\vartheta_{11}\vartheta_{12}]_{z=\alpha'}\Bigg],\end{aligned} \tag{7.6.56}$$

$$\begin{aligned}\underset{z=\beta'}{\operatorname{Res}}[j_X^0(z)] &= \frac{\mathrm{i}}{96}\Bigg[\frac{\vartheta_3^2(0)\vartheta_2^2(n\omega+\omega/2)}{\vartheta_1^2(\omega/2)\vartheta_4^2(n\omega)}\Bigg(\frac{7}{2}(\beta'-\alpha)+\frac{3}{2}(\beta'-\alpha')-\frac{3}{2}(\beta-\beta')\\ &\quad+\frac{(\beta-\beta')(\beta'-\alpha')}{\beta'-\alpha}-12\pi\left(\frac{\vartheta_1'(\omega/2)}{\vartheta_1(\omega/2)}-\frac{\vartheta_2'(n\omega+\omega/2)}{\vartheta_2(n\omega+\omega/2)}\right)\Bigg)\\ &\quad+\frac{\pi^2}{4(\beta'-\alpha)}\frac{\mathrm{d}^2}{\mathrm{d}\tilde{u}^2}[3J(z)-\vartheta_{11}\vartheta_{12}]_{z=\beta'}\Bigg],\end{aligned} \tag{7.6.57}$$

and

$$(7.6.58)\quad \operatorname*{Res}_{z=\beta}[j_X^0(z)] = \frac{\mathrm{i}}{96}\Bigg[\frac{\vartheta_3^2(0)\vartheta_3^2(n\omega+\omega/2)}{\vartheta_4^2(\omega/2)\vartheta_4^2(n\omega)}\Bigg(-\frac{7}{2}(\beta-\alpha')-\frac{3}{2}(\beta-\alpha)-\frac{3}{2}(\beta-\beta') + \frac{(\beta-\beta')(\beta-\alpha)}{\beta-\alpha'}+12\pi\bigg(\frac{\vartheta_4'(\omega/2)}{\vartheta_4(\omega/2)}-\frac{\vartheta_3'(n\omega+\omega/2)}{\vartheta_3(n\omega+\omega/2)}\bigg)\Bigg) - \frac{\pi^2}{4(\beta-\alpha')}\frac{\mathrm{d}^2}{\mathrm{d}\tilde{u}^2}[3(\vartheta_{11}^2+\vartheta_{12}^2)-\vartheta_{11}\vartheta_{12}]_{z=\beta}\Bigg].$$

Using MAPLE for calculations, we get

$$(7.6.59)\quad \frac{\mathrm{d}^2}{\mathrm{d}\tilde{u}^2}[3(\vartheta_{11}^2+\vartheta_{12}^2)-\vartheta_{11}\vartheta_{12}]\bigg|_{z=\alpha} = 2\frac{\vartheta_3^2(0)\vartheta_4^2(n\omega+\omega/2)}{\vartheta_3^2(\omega/2)\vartheta_4^2(n\omega)}\bigg[5\frac{\vartheta_4''(n\omega+\omega/2)}{\vartheta_4(n\omega+\omega/2)}-5\frac{\vartheta_3''(\omega/2)}{\vartheta_3(\omega/2)}+7\bigg(\frac{\vartheta_4'(n\omega+\omega/2)}{\vartheta_4(n\omega+\omega/2)}\bigg)^2 + 17\bigg(\frac{\vartheta_3'(\omega/2)}{\vartheta_3(\omega/2)}\bigg)^2-24\frac{\vartheta_4'(n\omega+\omega/2)\vartheta_3'(\omega/2)}{\vartheta_4(n\omega+\omega/2)\vartheta_3(\omega/2)}\bigg],$$

$$(7.6.60)\quad \frac{\mathrm{d}^2}{\mathrm{d}\tilde{u}^2}[3(\vartheta_{11}^2+\vartheta_{12}^2)-\vartheta_{11}\vartheta_{12}]\bigg|_{z=\beta} = 2\frac{\vartheta_3^2(0)\vartheta_3^2(n\omega+\omega/2)}{\vartheta_4^2(\omega/2)\vartheta_4^2(n\omega)}\bigg[5\frac{\vartheta_3''(n\omega+\omega/2)}{\vartheta_3(n\omega+\omega/2)}-5\frac{\vartheta_4''(\omega/2)}{\vartheta_4(\omega/2)}+7\bigg(\frac{\vartheta_3'(n\omega+\omega/2)}{\vartheta_3(n\omega+\omega/2)}\bigg)^2 + 17\bigg(\frac{\vartheta_4'(\omega/2)}{\vartheta_4(\omega/2)}\bigg)^2-24\frac{\vartheta_3'(n\omega+\omega/2)\vartheta_4'(\omega/2)}{\vartheta_3(n\omega+\omega/2)\vartheta_4(\omega/2)}\bigg],$$

$$(7.6.61)\quad \frac{\mathrm{d}^2}{\mathrm{d}\tilde{u}^2}[3J(z)-\vartheta_{11}\vartheta_{12}]\bigg|_{z=\alpha'} = 2\frac{\vartheta_3^2(0)\vartheta_1^2(n\omega+\omega/2)}{\vartheta_2^2(\omega/2)\vartheta_4^2(n\omega)}\bigg[5\frac{\vartheta_1''(n\omega+\omega/2)}{\vartheta_1(n\omega+\omega/2)}-5\frac{\vartheta_2''(\omega/2)}{\vartheta_2(\omega/2)}+7\bigg(\frac{\vartheta_1'(n\omega+\omega/2)}{\vartheta_1(n\omega+\omega/2)}\bigg)^2 + 17\bigg(\frac{\vartheta_2'(\omega/2)}{\vartheta_2(\omega/2)}\bigg)^2-24\frac{\vartheta_1'(n\omega+\omega/2)\vartheta_2'(\omega/2)}{\vartheta_1(n\omega+\omega/2)\vartheta_2(\omega/2)}\bigg],$$

$$(7.6.62)\quad \frac{\mathrm{d}^2}{\mathrm{d}\tilde{u}^2}[3J(z)-\vartheta_{11}\vartheta_{12}]\bigg|_{z=\beta'} = 2\frac{\vartheta_3^2(0)\vartheta_2^2(n\omega+\omega/2)}{\vartheta_1^2(\omega/2)\vartheta_4^2(n\omega)}\bigg[5\frac{\vartheta_2''(n\omega+\omega/2)}{\vartheta_2(n\omega+\omega/2)}-5\frac{\vartheta_1''(\omega/2)}{\vartheta_1(\omega/2)}+7\bigg(\frac{\vartheta_2'(n\omega+\omega/2)}{\vartheta_2(n\omega+\omega/2)}\bigg)^2 + 17\bigg(\frac{\vartheta_1'(\omega/2)}{\vartheta_1(\omega/2)}\bigg)^2-24\frac{\vartheta_2'(n\omega+\omega/2)\vartheta_1'(\omega/2)}{\vartheta_2(n\omega+\omega/2)\vartheta_1(\omega/2)}\bigg].$$

These formulas, combined with (7.6.23), prove Lemma 7.6.1. □

7.6.2. Large n asymptotic formula for h_n. We evaluate the large n asymptotic behavior of h_{nn} and then we use formula (7.1.11) to obtain a formula for

h_n. By (7.5.6), $h_{nn} = [\mathbf{P}_1]_{12}$, and by (7.5.14),

$$[\mathbf{P}_1]_{12} = [\mathbf{R}_1]_{12}(-2\pi\mathrm{i}), \tag{7.6.63}$$

hence

$$h_{nn} = [\mathbf{R}_1]_{12}(-2\pi\mathrm{i}). \tag{7.6.64}$$

Furthermore, from (7.5.21) we obtain that

$$h_{nn} = \mathrm{e}^{nl}[\mathbf{T}_1]_{12}(-2\pi\mathrm{i}), \tag{7.6.65}$$

and from (7.5.22), that

$$h_{nn} = \mathrm{e}^{nl}[\mathbf{S}_1]_{12}(-2\pi\mathrm{i}). \tag{7.6.66}$$

It follows from (7.5.90) that

$$\mathbf{S}_1 = \mathbf{M}_1 + \mathbf{X}_1. \tag{7.6.67}$$

By (7.5.60),

$$[\mathbf{M}_1]_{12} = \frac{\mathrm{i}A\vartheta_4\big((n+1)\omega\big)}{\vartheta_4(n\omega)}, \qquad \omega = \frac{\pi(1+\zeta)}{2}, \quad A = \frac{\pi\vartheta_1'(0)}{2\vartheta_1(\omega)}, \tag{7.6.68}$$

and by (7.6.10),

$$[\mathbf{X}_1]_{12} = \frac{c(n)}{n} + \mathcal{O}(n^{-2}), \tag{7.6.69}$$

where

$$c(n) = X_\alpha + X_{\alpha'} + X_{\beta'} + X_\beta \tag{7.6.70}$$

is an explicit quasi-periodic function of n. Therefore,

$$h_{nn} = \mathrm{e}^{nl}\left[\frac{\mathrm{i}A\vartheta_4\big((n+1)\omega\big)}{\vartheta_4(n\omega)} + \frac{c(n)}{n} + \mathcal{O}(n^{-2})\right](-2\pi\mathrm{i}). \tag{7.6.71}$$

By (7.4.35),

$$e^{l/2} = \frac{\pi\vartheta_1'(0)}{2\mathrm{e}\vartheta_1(\omega)} = \frac{A}{e}, \tag{7.6.72}$$

hence

$$\begin{aligned} h_{nn} &= \left(\frac{A}{\mathrm{e}}\right)^{2n}\left[\mathrm{i}A\frac{\vartheta_4\big((n+1)\omega\big)}{\vartheta_4(n\omega)} + \frac{c(n)}{n} + \mathcal{O}(n^{-2})\right](-2\pi\mathrm{i}) \\ &= \frac{2\pi A^{2n+1}\vartheta_4\big((n+1)\omega\big)}{\mathrm{e}^{2n}\vartheta_4(n\omega)}\left(1 + \frac{c_1(n)}{n} + \mathcal{O}(n^{-2})\right), \end{aligned} \tag{7.6.73}$$

where

$$c_1(n) = \frac{c(n)\vartheta_4(n\omega)}{\mathrm{i}A\vartheta_4\big((n+1)\omega\big)}. \tag{7.6.74}$$

From (7.1.11) and the Stirling formula we obtain that

$$\frac{h_n}{(n!)^2} = \frac{n^{2n+1}h_{nn}}{(n!)^2(2\gamma)^{2n+1}} = \left(\frac{\mathrm{e}}{2\gamma}\right)^{2n+1}\frac{h_{nn}}{2\pi\mathrm{e}}\left(1 - \frac{1}{6n} + \mathcal{O}(n^{-2})\right), \tag{7.6.75}$$

hence by (7.6.73),

$$\frac{h_n}{(n!)^2} = \left(\frac{\mathrm{e}}{2\gamma}\right)^{2n+1} \frac{1}{2\pi\mathrm{e}} \frac{2\pi A^{2n+1}\vartheta_4\big((n+1)\omega\big)}{\gamma \mathrm{e}^{2n}\vartheta_4(n\omega)} \times \left(1 + \frac{c_1(n)}{n} - \frac{1}{6n} + \mathcal{O}(n^{-2})\right) \tag{7.6.76}$$
$$= G^{2n+1} \frac{\vartheta_4\big((n+1)\omega\big)}{\vartheta_4(n\omega)} \left(1 + \frac{c_2(n)}{n} + \mathcal{O}(n^{-2})\right),$$

where

$$G = \frac{A}{2\gamma} = \frac{\pi\vartheta_1'(0)}{4\gamma\vartheta_1(\omega)}, \quad c_2(n) = c_1(n) - \frac{1}{6}. \tag{7.6.77}$$

Observe that $c_1(n)$ has the form,

$$c_1(n) = f(n\omega, \omega), \tag{7.6.78}$$

where $f(x,\omega)$ is a real analytic function which is periodic with respect to both x and ω, of periods π and 2π, respectively, so that

$$f(x+\pi,\omega) = f(x,\omega), \quad f(x,\omega+2\pi) = f(x,\omega). \tag{7.6.79}$$

We can now summarize now the asymptotic formula for $h_n/(n!)^2$.

Proposition 7.6.2. *As $n \to \infty$,*

$$\frac{h_n}{(n!)^2} = G^{2n+1} \frac{\vartheta_4\big((n+1)\omega\big)}{\vartheta_4(n\omega)} \left[1 + \frac{f_0(n\omega,\omega)}{n} + \mathcal{O}(n^{-2})\right], \tag{7.6.80}$$

where

$$G = \frac{\pi\vartheta_1'(0)}{4\gamma\vartheta_1(\omega)}, \tag{7.6.81}$$

and

$$f_0(x,\omega) = f(x,\omega) - \tfrac{1}{6}, \tag{7.6.82}$$

is a real analytic function which satisfies the periodicity conditions

$$f_0(x+\pi,\omega) = f_0(x,\omega), \quad f_0(x,\omega+2\pi) = f_0(x,\omega). \tag{7.6.83}$$

By (7.6.74),

$$f(n\omega,\omega) = \frac{X(n\omega,\omega)\vartheta_4(n\omega)}{\mathrm{i}A\vartheta_4\big((n+1)\omega\big)}. \tag{7.6.84}$$

where

$$X \equiv X_\alpha + X_{\alpha'} + X_{\beta'} + X_\beta \tag{7.6.85}$$

and explicit expressions for $X_\alpha, X_{\alpha'}, X_{\beta'}, X_\beta$ are given in Lemma 7.6.1.

In the next two subsections, we carry out a concrete evaluation of $f(n\omega,\omega)$, obtaining that

$$f_0(x,\omega) \equiv 0, \tag{7.6.86}$$

which immediately improves Proposition 7.6.2 to Proposition 7.3.1. In order to prove (7.6.86), we will first show that $f(n\omega,\omega)$ does not depend on n.

7.6.3. The function $f(x,\omega)$ is constant in x. Denote

$$z = n\omega + \frac{\omega}{2}, \quad \tilde{f}(x,\omega) \equiv f\left(x - \frac{\omega}{2}, \omega\right), \tag{7.6.87}$$

so that

$$\tilde{f}(z,\omega) = f(n\omega,\omega). \tag{7.6.88}$$

To prove that $f(x,\omega)$ is constant in x we will prove the following lemmas.

Lemma 7.6.3. *The function $\tilde{f}(z,\omega)$ is doubly periodic in z.*

Lemma 7.6.4. *The function $\tilde{f}(z,\omega)$ is analytic throughout the z-plane.*

From these two lemmas, it follows immediately that $\tilde{f}(z,\omega)$ is constant in z, as it is a doubly periodic entire function of z, and it thus follows that $f(x,\omega)$ is constant in x.

To prove Lemma 7.6.3, we will check that $\tilde{f}(z+\pi\tau,\omega) = \tilde{f}(z,\omega)$. By (7.6.74), (7.6.70),

$$c_1(n) \equiv f(n\omega,\omega) = \tilde{f}(z,\omega) = Y_1 + Y_2 + Y_3 + Y_4, \tag{7.6.89}$$

where

$$\begin{aligned}
Y_1 &= \frac{X_\alpha \vartheta_4(n\omega)}{\mathrm{i}A\vartheta_4(n\omega+\omega)} \\
&= \frac{\vartheta_1(\omega)\vartheta_3^2(0)\vartheta_4^2(z)}{48\pi\vartheta_1'(0)\vartheta_3^2(\omega/2)\vartheta_4(z-\omega/2)\vartheta_4(z+\omega/2)}\left(C_\alpha + 12\pi\xi_\alpha + \frac{\pi^2\eta_\alpha}{2(\beta'-\alpha)}\right), \\
Y_2 &= \frac{X_{\alpha'} \vartheta_4(n\omega)}{\mathrm{i}A\vartheta_4(n\omega+\omega)} \\
&= \frac{\vartheta_1(\omega)\vartheta_3^2(0)\vartheta_1^2(z)}{48\pi\vartheta_1'(0)\vartheta_2^2(\omega/2)\vartheta_4(z-\omega/2)\vartheta_4(z+\omega/2)}\left(C_{\alpha'} + 12\pi\xi_{\alpha'} + \frac{\pi^2\eta_{\alpha'}}{2(\beta-\alpha')}\right), \\
Y_3 &= \frac{X_{\beta'} \vartheta_4(n\omega)}{\mathrm{i}A\vartheta_4(n\omega+\omega)} \\
&= \frac{\vartheta_1(\omega)\vartheta_3^2(0)\vartheta_2^2(z)}{48\pi\vartheta_1'(0)\vartheta_1^2(\omega/2)\vartheta_4(z-\omega/2)\vartheta_4(z+\omega/2)}\left(C_{\beta'} + 12\pi\xi_{\beta'} + \frac{\pi^2\eta_{\beta'}}{2(\beta'-\alpha)}\right), \\
Y_4 &= \frac{X_{\beta} \vartheta_4(n\omega)}{\mathrm{i}A\vartheta_4(n\omega+\omega)} \\
&= \frac{\vartheta_1(\omega)\vartheta_3^2(0)\vartheta_3^2(z)}{48\pi\vartheta_1'(0)\vartheta_4^2(\omega/2)\vartheta_4(z-\omega/2)\vartheta_4(z+\omega/2)}\left(C_{\beta} + 12\pi\xi_{\beta} + \frac{\pi^2\eta_{\beta}}{2(\beta-\alpha')}\right).
\end{aligned} \tag{7.6.90}$$

Observe that Y_1, Y_2, Y_3, Y_4 can be written in the form

$$\begin{aligned}
Y_1 &= Q_{11}h_{11}(z) + Q_{12}h_{12}(z) + Q_{13}h_{13}(z) + Q_{14}h_{14}(z), \\
Y_2 &= Q_{21}h_{21}(z) + Q_{22}h_{22}(z) + Q_{23}h_{23}(z) + Q_{24}h_{24}(z), \\
Y_3 &= Q_{31}h_{31}(z) + Q_{32}h_{32}(z) + Q_{33}h_{33}(z) + Q_{34}h_{34}(z), \\
Y_4 &= Q_{41}h_{41}(z) + Q_{42}h_{42}(z) + Q_{43}h_{43}(z) + Q_{44}h_{44}(z),
\end{aligned} \tag{7.6.91}$$

where

(7.6.92)
$$
\begin{aligned}
h_{11}(z) &= \frac{\vartheta_4^2(z)}{\vartheta_4(z-\omega/2)\vartheta_4(z+\omega/2)}, & h_{12}(z) &= \frac{\vartheta_4'(z)\vartheta_4(z)}{\vartheta_4(z-\omega/2)\vartheta_4(z+\omega/2)},\\
h_{13}(z) &= \frac{\vartheta_4'(z)^2}{\vartheta_4(z-\omega/2)\vartheta_4(z+\omega/2)}, & h_{14}(z) &= \frac{\vartheta_4''(z)\vartheta_4(z)}{\vartheta_4(z-\omega/2)\vartheta_4(z+\omega/2)},\\
h_{21}(z) &= \frac{\vartheta_1^2(z)}{\vartheta_4(z-\omega/2)\vartheta_4(z+\omega/2)}, & h_{22}(z) &= \frac{\vartheta_1'(z)\vartheta_1(z)}{\vartheta_4(z-\omega/2)\vartheta_4(z+\omega/2)},\\
h_{23}(z) &= \frac{\vartheta_1'(z)^2}{\vartheta_4(z-\omega/2)\vartheta_4(z+\omega/2)}, & h_{24}(z) &= \frac{\vartheta_1''(z)\vartheta_1(z)}{\vartheta_4(z-\omega/2)\vartheta_4(z+\omega/2)},\\
h_{31}(z) &= \frac{\vartheta_2^2(z)}{\vartheta_4(z-\omega/2)\vartheta_4(z+\omega/2)}, & h_{32}(z) &= \frac{\vartheta_2'(z)\vartheta_2(z)}{\vartheta_4(z-\omega/2)\vartheta_4(z+\omega/2)},\\
h_{33}(z) &= \frac{\vartheta_2'(z)^2}{\vartheta_4(z-\omega/2)\vartheta_4(z+\omega/2)}, & h_{34}(z) &= \frac{\vartheta_2''(z)\vartheta_2(z)}{\vartheta_4(z-\omega/2)\vartheta_4(z+\omega/2)},\\
h_{41}(z) &= \frac{\vartheta_3^2(z)}{\vartheta_4(z-\omega/2)\vartheta_4(z+\omega/2)}, & h_{42}(z) &= \frac{\vartheta_3'(z)\vartheta_3(z)}{\vartheta_4(z-\omega/2)\vartheta_4(z+\omega/2)},\\
h_{43}(z) &= \frac{\vartheta_3'(z)^2}{\vartheta_4(z-\omega/2)\vartheta_4(z+\omega/2)}, & h_{44}(z) &= \frac{\vartheta_3''(z)\vartheta_3(z)}{\vartheta_4(z-\omega/2)\vartheta_4(z+\omega/2)},
\end{aligned}
$$

and the numbers Q_{ij} do not depend on z. More specifically,
(7.6.93)
$$
\begin{aligned}
Q_{11} &= \frac{\vartheta_1(\omega)\vartheta_3^2(0)}{48\pi\vartheta_1'(0)\vartheta_3^2(\omega/2)}\\
&\quad\times\left[C_\alpha+12\pi\frac{\vartheta_3'(\omega/2)}{\vartheta_3(\omega/2)}+\frac{\pi^2}{2(\beta'-\alpha)}\left(-5\frac{\vartheta_3''(\omega/2)}{\vartheta_3(\omega/2)}+17\left(\frac{\vartheta_3'(\omega/2)}{\vartheta_3(\omega/2)}\right)^2\right)\right],\\
Q_{12} &= -\frac{\vartheta_1(\omega)\vartheta_3^2(0)}{4\vartheta_1'(0)\vartheta_3^2(\omega/2)}-\frac{\pi\vartheta_1(\omega)\vartheta_3^2(0)\vartheta_3'(\omega/2)}{4(\beta'-\alpha)\vartheta_1'(0)\vartheta_3^3(\omega/2)},\\
Q_{13} &= \frac{7\pi\vartheta_1(\omega)\vartheta_3^2(0)}{96(\beta'-\alpha)\vartheta_1'(0)\vartheta_3^2(\omega/2)},\\
Q_{14} &= \frac{5\pi\vartheta_1(\omega)\vartheta_3^2(0)}{96(\beta'-\alpha)\vartheta_1'(0)\vartheta_3^2(\omega/2)},
\end{aligned}
$$

and similar formulas hold for other Q_{ij}. In particular, notice that
(7.6.94)
$$
\begin{aligned}
Q_{13}+Q_{14} &= \frac{\pi\vartheta_1(\omega)\vartheta_3^2(0)}{8(\beta'-\alpha)\vartheta_1'(0)\vartheta_3^2(\omega/2)} = \frac{\vartheta_1(\omega)\vartheta_1(\omega/2)}{8\vartheta_1'(0)\vartheta_2(\omega/2)\vartheta_3(\omega/2)\vartheta_4(\omega/2)},\\
Q_{23}+Q_{24} &= -\frac{\pi\vartheta_1(\omega)\vartheta_3^2(0)}{8(\beta-\alpha')\vartheta_1'(0)\vartheta_2^2(\omega/2)} = -\frac{\vartheta_1(\omega)\vartheta_4(\omega/2)}{8\vartheta_1'(0)\vartheta_1(\omega/2)\vartheta_2(\omega/2)\vartheta_3(\omega/2)},\\
Q_{33}+Q_{34} &= -\frac{\pi\vartheta_1(\omega)\vartheta_3^2(0)}{8(\beta'-\alpha)\vartheta_1'(0)\vartheta_1^2(\omega/2)} = -\frac{\vartheta_1(\omega)\vartheta_3(\omega/2)}{8\vartheta_1'(0)\vartheta_1(\omega/2)\vartheta_2(\omega/2)\vartheta_4(\omega/2)},\\
Q_{43}+Q_{44} &= \frac{\pi\vartheta_1(\omega)\vartheta_3^2(0)}{8(\beta-\alpha')\vartheta_1'(0)\vartheta_4^2(\omega/2)} = \frac{\vartheta_1(\omega)\vartheta_2(\omega/2)}{8\vartheta_1'(0)\vartheta_1(\omega/2)\vartheta_3(\omega/2)\vartheta_4(\omega/2)}.
\end{aligned}
$$

Observe that all $Q_{ij} = Q_{ij}(\omega)$ are periodic functions of ω of period 2π. It follows from equations (7.2.2) that the functions

$$h_{j1}(z), \quad j = 1,2,3,4, \tag{7.6.95}$$

are doubly periodic, and

$$h_{j1}(z+\pi) = h_{j1}(z), \quad h_{j1}(z+\pi\tau) = h_{j1}(z), \tag{7.6.96}$$

while the functions $h_{j2}(z), h_{j3}(z), h_{j4}(z)$, for $j = 1,2,3,4$ satisfy the equations

$$\begin{aligned} h_{jk}(z+\pi) &= h_{jk}(z), \quad k = 2,3,4; \\ h_{j2}(z+\pi\tau) &= h_{j2}(z) - 2\mathrm{i}h_{j1}(z), \\ h_{j3}(z+\pi\tau) &= h_{j3}(z) - 4\mathrm{i}h_{j2}(z) - 4h_{j1}(z), \\ h_4(z+\pi\tau) &= h_{j4}(z) - 4\mathrm{i}h_{j2}(z) - 4h_{j1}(z). \end{aligned} \tag{7.6.97}$$

This implies that the functions $Y_j = Y_j(z)$ for $j = 1,2,3,4$, satisfy the equations

$$\begin{aligned} Y_j(z+\pi) &= Y_j(z), \\ Y_j(z+\pi\tau) &= Y_j(z) + (-2\mathrm{i}Q_{j2} - 4Q_{j3} - 4Q_{j4})h_{j1}(z) \\ &\quad + (-4\mathrm{i}Q_{j3} - 4\mathrm{i}Q_{j4})h_{j2}(z). \end{aligned} \tag{7.6.98}$$

The proof of Lemma 7.6.3 then follows immediately from (7.6.98) and the following identities:

$$\begin{aligned} (Q_{13}+Q_{14})h_{11}(z) + (Q_{23}+Q_{24})h_{21}(z) + (Q_{33}+Q_{34})h_{31}(z) \\ + (Q_{43}+Q_{44})h_{41}(z) \equiv 0 \end{aligned} \tag{7.6.99}$$

$$\begin{aligned} (Q_{13}+Q_{14})h_{12}(z) + (Q_{23}+Q_{24})h_{22}(z) + (Q_{33}+Q_{34})h_{32}(z) \\ + (Q_{43}+Q_{44})h_{42}(z) \equiv 0 \end{aligned} \tag{7.6.100}$$

$$Q_{12}h_{11}(z) + Q_{22}h_{21}(z) + Q_{32}h_{31}(z) + Q_{42}h_{41}(z) \equiv 0, \tag{7.6.101}$$

which are proven below. Introduce here the notation

$$\vartheta_j \equiv \vartheta_j\left(\frac{\omega}{2}\right), \quad \vartheta_j' \equiv \vartheta_j'\left(\frac{\omega}{2}\right), \quad \vartheta_j'' \equiv \vartheta_j''\left(\frac{\omega}{2}\right) \qquad \text{for } j = 1,2,3,4. \tag{7.6.102}$$

The sum in (7.6.99) can be written as

$$\frac{\vartheta_1(\omega)[\vartheta_1^2\vartheta_4^2(z) - \vartheta_4^2\vartheta_1^2(z) + \vartheta_2^2\vartheta_3^2(z) - \vartheta_3^2\vartheta_2^2(z)]}{8\vartheta_1'(0)\vartheta_1\vartheta_2\vartheta_3\vartheta_4\vartheta_4(z-\omega/2)\vartheta_4(z+\omega/2)}, \tag{7.6.103}$$

which is zero by the Jacobi identity (7.2.15).

The sum in (7.6.100) can be written as

$$\frac{\vartheta_1(\omega)\left[\vartheta_1^2\vartheta_4'(z)\vartheta_4(z) - \vartheta_4^2\vartheta_1'(z)\vartheta_1(z) + \vartheta_2^2\vartheta_3'(z)\vartheta_3(z) - \vartheta_3^2\vartheta_2'(z)\vartheta_2(z)\right]}{8\vartheta_1'(0)\vartheta_1\vartheta_2\vartheta_3\vartheta_4\vartheta_4(z-\omega/2)\vartheta_4(z+\omega/2)}. \tag{7.6.104}$$

Using the identities (7.2.8), we can write the expression in brackets in the numerator of (7.6.104) as

$$\begin{aligned} &\frac{\vartheta_1'(z)}{\vartheta_1(z)}[\vartheta_1^2\vartheta_4^2(z) - \vartheta_4^2\vartheta_1^2(z) - \vartheta_3^2\vartheta_2^2(z) + \vartheta_2^2\vartheta_3^2(z)] \\ &\qquad + \frac{\vartheta_2(z)\vartheta_3(z)\vartheta_4(z)}{\vartheta_1(z)}[\vartheta_3^2\vartheta_2^2(0) - \vartheta_1^2\vartheta_4^2(0) - \vartheta_2^2\vartheta_3^2(0)]. \end{aligned} \tag{7.6.105}$$

The first term in (7.6.105) vanishes by (7.2.15) and the second term vanishes by (7.2.16). Thus (7.6.100) is proven.

Finally, we can expand the sum in (7.6.101) and make the substitutions from identities (7.2.8), to obtain

$$\begin{aligned}(7.6.106)\quad & \frac{\vartheta_1(w)}{4\vartheta_1'(0)\vartheta_4(z-\omega/2)\vartheta_4(z+\omega/2)} \\ & \times\left[\frac{\vartheta_1'\left(\vartheta_4^2\vartheta_1^2(z)-\vartheta_1^2\vartheta_4^2(z)-\vartheta_2^2\vartheta_3^2(z)+\vartheta_3^2\vartheta_2^2(z)\right)}{\vartheta_1^2\vartheta_2\vartheta_3\vartheta_4}+\frac{\vartheta_3^2(z)}{\vartheta_4^2}\left(\vartheta_3^2(0)+\frac{\vartheta_2^2\vartheta_4^2(0)}{\vartheta_1^2}\right)\right. \\ & \qquad\left.+\frac{\vartheta_1^2(z)}{\vartheta_2^2}\left(\vartheta_3^2(0)-\frac{\vartheta_4^2\vartheta_2^2(0)}{\vartheta_1^2}\right)-\frac{\vartheta_3^2(0)}{\vartheta_1^2}\vartheta_2^2(z)\right].\end{aligned}$$

Once again, the first term vanishes by (7.2.15). If we substitute the identity (7.2.17) in the second and third terms, (7.6.106) becomes simply

$$(7.6.107)\quad \frac{\vartheta_1(w)}{4\vartheta_1'(0)\vartheta_1\vartheta_4(z-\omega/2)\vartheta_4(z+\omega/2)}[\vartheta_3^2(z)\vartheta_2^2(0)-\vartheta_2^2(z)\vartheta_3^2(0)-\vartheta_1^2(z)\vartheta_4^2(0)],$$

which is zero by (7.2.16). This proves (7.6.101) and thus Lemma 7.6.3.

We now turn to the proof of Lemma 7.6.4. Notice that in the fundamental rectangle

$$(7.6.108)\qquad \mathcal{R}=\left\{z\in\mathbb{C}:-\frac{\pi}{2}\le\operatorname{Re}z\le\frac{\pi}{2},0\le\operatorname{Im}z\le\pi\tau\right\},$$

the function $\tilde{f}(z,\omega)$ has the two possible simple poles,

$$(7.6.109)\qquad z_{1,2}=\frac{\pi\tau}{2}\pm\frac{\omega}{2}.$$

Because of the double-periodicity of $\tilde{f}$, we must have

$$(7.6.110)\qquad \operatorname*{Res}_{z=z_1}\tilde{f}(z,\omega)=-\operatorname*{Res}_{z=z_2}\tilde{f}(z,\omega).$$

Denote

$$(7.6.111)\qquad R_{ij}(\omega)=\operatorname*{Res}_{z=z_1}h_{ij}(z,\omega).$$

Then we have that

$$(7.6.112)\qquad \begin{aligned} R_{11} &= \frac{\vartheta_1^2}{\vartheta_1'(0)\vartheta_1(\omega)}, & R_{12} &= \frac{(\vartheta_1' - \mathrm{i}\vartheta_1)\vartheta_1}{\vartheta_1'(0)\vartheta_1(\omega)}, \\ R_{13} &= \frac{(\vartheta_1' - \mathrm{i}\vartheta_1)^2}{\vartheta_1'(0)\vartheta_1(\omega)}, & R_{14} &= \frac{(\vartheta_1'' - 2\mathrm{i}\vartheta_1' - \vartheta_1)\vartheta_1}{\vartheta_1'(0)\vartheta_1(\omega)}, \\ R_{21} &= \frac{\vartheta_4^2}{\vartheta_1'(0)\vartheta_1(\omega)}, & R_{22} &= \frac{(\vartheta_4' - \mathrm{i}\vartheta_4)\vartheta_4}{\vartheta_1'(0)\vartheta_1(\omega)}, \\ R_{23} &= \frac{(\vartheta_4' - \mathrm{i}\vartheta_4)^2}{\vartheta_1'(0)\vartheta_1(\omega)}, & R_{24} &= \frac{(\vartheta_4'' - 2\mathrm{i}\vartheta_4' - \vartheta_4)\vartheta_4}{\vartheta_1'(0)\vartheta_1(\omega)}, \\ R_{31} &= \frac{-\vartheta_3^2}{\vartheta_1'(0)\vartheta_1(\omega)}, & R_{32} &= \frac{-(\vartheta_3' - \mathrm{i}\vartheta_3)\vartheta_3}{\vartheta_1'(0)\vartheta_1(\omega)}, \\ R_{33} &= \frac{-(\vartheta_3' - \mathrm{i}\vartheta_3)^2}{\vartheta_1'(0)\vartheta_1(\omega)}, & R_{41} &= \frac{-\vartheta_2^2}{\vartheta_1'(0)\vartheta_1(\omega)}, \\ R_{42} &= \frac{-(\vartheta_2' - \mathrm{i}\vartheta_2)\vartheta_2}{\vartheta_1'(0)\vartheta_1(\omega)}, & R_{43} &= \frac{-(\vartheta_2' - \mathrm{i}\vartheta_2)^2}{\vartheta_1'(0)\vartheta_1(\omega)}, \\ R_{44} &= \frac{-(\vartheta_2'' - 2\mathrm{i}\vartheta_2' - \vartheta_2)\vartheta_2}{\vartheta_1'(0)\vartheta_1(\omega)}, & R_{34} &= \frac{-(\vartheta_3'' - 2\mathrm{i}\vartheta_3' - \vartheta_3)\vartheta_3}{\vartheta_1'(0)\vartheta_1(\omega)}, \end{aligned}$$

and

$$(7.6.113)\qquad \operatorname*{Res}_{z=z_1} \tilde{f}(z,\omega) = \sum_{j,k=1}^{4} Q_{jk} R_{jk}.$$

A priori, the sum in (7.6.113) is quite complicated, so we evaluate first the imaginary part . Multiplying out (7.6.113) and again making the substitutions from (7.2.8), we get

$$(7.6.114)\qquad \operatorname{Im} \operatorname*{Res}_{t} z = z_1 \tilde{f}(z,\omega) = \frac{(\vartheta_2^4 - \vartheta_4^4)\big(\vartheta_1^2\vartheta_3^2(0) + \vartheta_2^2\vartheta_4^2(0) - \vartheta_4^2\vartheta_2^2(0)\big)}{4\vartheta_1'(0)^2\vartheta_1^2\vartheta_2^2\vartheta_4^2},$$

which is zero by (7.2.17).

Substituting the identities (7.2.11), along with (7.2.8), into the sum (7.6.113) gives

$$(7.6.115)\qquad \operatorname*{Res}_{t} z = z_1 \tilde{f}(z,\omega) = \frac{24A + 17B + 12C + 10D + 7E + 5F}{96\vartheta_1'(0)^2\vartheta_1^2\vartheta_2^3\vartheta_3\vartheta_4^3},$$

where

$$(7.6.116)\qquad \begin{aligned} A &= \vartheta_3^2(\vartheta_2^4 + \vartheta_4^4)\big[\vartheta_2^2\vartheta_4^2\vartheta_2^2(0)\vartheta_4^2(0) - \vartheta_1^2\vartheta_3^2(0)\big(\vartheta_2^2\vartheta_4^2(0) - \vartheta_4^2\vartheta_2^2(0)\big)\big], \\ B &= -\vartheta_3^2\big(\vartheta_2^8\vartheta_4^4(0) + \vartheta_4^8\vartheta_2^4(0)\big), \\ C &= \vartheta_1^2\vartheta_2^2\vartheta_4^2(\vartheta_2^4 + \vartheta_4^4 - \vartheta_1^4 - \vartheta_3^4), \\ D &= -\vartheta_2^4\vartheta_3^2\vartheta_4^4\vartheta_3^4(0), \\ E &= -\vartheta_3^2\big[\vartheta_1^4\vartheta_3^4(0)\big(\vartheta_2^4 + \vartheta_4^4\big) + \vartheta_2^4\vartheta_4^4\big(\vartheta_2^4(0) + \vartheta_4^4(0)\big)\big], \\ F &= \vartheta_2^2\vartheta_4^2\vartheta_3^2(0)[\vartheta_2^2\vartheta_4^2(0)(\vartheta_3^4 - \vartheta_1^4 - \vartheta_4^4 + \vartheta_2^4) \\ &\qquad + \vartheta_4^2\vartheta_2^2(0)(\vartheta_4^4 - \vartheta_2^4 + \vartheta_3^4 - \vartheta_1^4)]. \end{aligned}$$

Note that none of these terms involve derivatives of theta functions. We immediately have $C = 0$ by the identity (7.2.14). We can also use this identity to write

$$F = 2\vartheta_2^2\vartheta_4^2\vartheta_3^2(0)[\vartheta_2^2\vartheta_4^2(0)(\vartheta_3^4 - \vartheta_4^4) + \vartheta_4^2\vartheta_2^2(0)(\vartheta_3^4 - \vartheta_2^4)], \tag{7.6.117}$$

and (7.2.17) to write

$$A = \vartheta_3^2(\vartheta_2^4 + \vartheta_4^4)[\vartheta_2^2\vartheta_4^2\vartheta_2^2(0)\vartheta_4^2(0) + \vartheta_1^4\vartheta_3^2(0)]. \tag{7.6.118}$$

We now combine the terms A, B, and E to obtain

$$\begin{aligned}24A + 17B + 7E &\\ = 24\vartheta_2^2\vartheta_3^2\vartheta_4^2&\big(\vartheta_4^2\vartheta_2^2(0) - \vartheta_2^2\vartheta_4^2(0)\big)\big(\vartheta_2^2\vartheta_2^2(0) - \vartheta_4^2\vartheta_4^2(0)\big) - 21\vartheta_2^4\vartheta_3^2\vartheta_4^4\vartheta_3^4(0)\\ &+ 7\vartheta_3^6(\vartheta_2^4 + \vartheta_4^4)\big(\vartheta_2^4(0) + \vartheta_4^4(0)\big) - 7\vartheta_3^2\big(\vartheta_4^8\vartheta_4^4(0) + \vartheta_2^8\vartheta_2^4(0)\big).\end{aligned} \tag{7.6.119}$$

By (7.2.17) and (7.2.19) we can write this sum as

$$\begin{aligned}24A + 17B + 7E &\\ = 24\vartheta_1^2\vartheta_2^2\vartheta_3^2\vartheta_4^2\vartheta_3^2(0)&\big(\vartheta_2^2\vartheta_2^2(0) - \vartheta_4^2\vartheta_4^2(0)\big) - 21\vartheta_2^4\vartheta_3^2\vartheta_4^4\vartheta_3^4(0)\\ &+ 7\vartheta_3^6\vartheta_3^4(0)(\vartheta_2^4 + \vartheta_4^4) - 7\vartheta_3^2\big(\vartheta_4^8\vartheta_4^4(0) + \vartheta_2^8\vartheta_2^4(0)\big).\end{aligned} \tag{7.6.120}$$

We now combine all terms and use (7.2.17) and (7.2.19) to write all terms solely in terms of ϑ_2, ϑ_4, and factors which are constant with respect to ω, yielding

$$\begin{aligned}24A + 17B + 10D + 7E + 5F &\\ = \left(41\frac{\vartheta_2^2\vartheta_4^2\vartheta_4^2(0)}{\vartheta_3^2(0)} + 24\frac{\vartheta_4^4\vartheta_4^4(0)}{\vartheta_2^2(0)\vartheta_3^2(0)} + 17\frac{\vartheta_2^4\vartheta_2^4(0)}{\vartheta_3^2(0)}\right)&\big(\vartheta_2^4(0) + \vartheta_4^4(0) - \vartheta_3^4(0)\big),\end{aligned} \tag{7.6.121}$$

which is zero by (7.2.19). Lemma 7.6.4 is thus proven, and it follows that $\tilde{f}(z,\omega)$ is constant in z. To evaluate the constant, we can take $z = 0$.

7.6.4. Evaluation of $\tilde{f}(0,\omega)$. We will evaluate $\tilde{f}(z,\omega)$ at $z = 0$. Notice that many of the functions $h_{jk}(z)$ vanish at $z = 0$. In fact we have

$$\begin{aligned}\tilde{f}(0,\omega) = Q_{11}h_{11}(0) + Q_{14}h_{14}(0) + Q_{23}h_{23}(0) + Q_{31}h_{31}(0) + Q_{34}h_{34}(0)&\\ + Q_{41}h_{41}(0) + Q_{44}h_{44}(0).&\end{aligned} \tag{7.6.122}$$

The identities (7.2.10) and (7.2.9) allow us to eliminate all derivatives of theta functions from the sum (7.6.122)) except $\vartheta_2', \vartheta_2''$. Making these substitutions and simplifying, (7.6.122) becomes

$$\begin{aligned}\tilde{f}(0,\omega) = &\frac{\vartheta_1(\omega)\big(\vartheta_1^2\vartheta_4^2(0) + \vartheta_2^2\vartheta_3^2(0) - \vartheta_3^2\vartheta_2^2(0)\big)}{96\vartheta_1'(0)}\\ &\times\left(\frac{5\vartheta_2''(0)}{\vartheta_1\vartheta_2\vartheta_3\vartheta_4^3} + \frac{17(\vartheta_2')^2}{\vartheta_1\vartheta_2^3\vartheta_3\vartheta_4^3} - \frac{5\vartheta_2''}{\vartheta_1\vartheta_2^2\vartheta_3\vartheta_4^3} + \frac{24\vartheta_2'\big(\vartheta_3^2\vartheta_2^2(0) + \vartheta_1^2\vartheta_4^2(0)\big)}{\vartheta_1^2\vartheta_2^3\vartheta_3^2\vartheta_4^2}\right)\\ &+ \frac{\vartheta_1(\omega)(24A + 17B + 7C + 5D + 3E + 2F)}{96\vartheta_1'(0)\vartheta_1^3\vartheta_2^3\vartheta_3^3\vartheta_4^3\vartheta_2^2(0)}\end{aligned} \tag{7.6.123}$$

where
(7.6.124)

$$A=\vartheta_2^2\vartheta_4^2\vartheta_2^2(0)\vartheta_3^2(0)\big(\vartheta_1^4\vartheta_4^4(0)+\vartheta_3^4\vartheta_2^4(0)\big),$$
$$B=\vartheta_4^2\vartheta_2^2(0)\big(\vartheta_1^2\vartheta_4^2(0)-\vartheta_3^2\vartheta_2^2(0)\big)\big((\vartheta_3^2\vartheta_2^2(0)+\vartheta_1^2\vartheta_4^2(0))^2-\vartheta_1^2\vartheta_3^2\vartheta_2^2(0)\vartheta_4^2(0)\big),$$
$$C=\vartheta_2^4\vartheta_4^2\vartheta_2^2(0)\vartheta_3^4(0)\big(\vartheta_1^2\vartheta_4^2(0)-\vartheta_3^2\vartheta_2^2(0)\big),$$
$$D=\vartheta_1^2\vartheta_3^2\Big[\vartheta_4^2\vartheta_2^4(0)\vartheta_4^2(0)\big(\vartheta_3^2\vartheta_2^2(0)-\vartheta_1^2\vartheta_4^2(0)\big)$$
$$+\vartheta_3^2(0)\Big(\vartheta_2^2(0)\big(\vartheta_3^4\vartheta_2^4(0)-\vartheta_1^4\vartheta_4^4(0)\big)+\vartheta_2^2\vartheta_1^2(0)\vartheta_4^2(0)\big(\vartheta_1^2(\vartheta_4^2(0)+\vartheta_2^2\vartheta_3^2(0))\big)\Big)\Big],$$
$$E=\vartheta_1^2\vartheta_2^2\vartheta_3^2\vartheta_2^2(0)\vartheta_3^2(0)\big[\vartheta_4^2(0)\big(\vartheta_1^2\vartheta_3^2(0)-\vartheta_4^2\vartheta_2^2(0)\big)-\vartheta_2^2\vartheta_2^4(0)\big],$$
$$F=\vartheta_1^2\vartheta_2^2\vartheta_3^2\vartheta_2^2(0)\big[4\vartheta_2^2\vartheta_3^2(0)\vartheta_4^4(0)-\vartheta_2^2(0)\big(\vartheta_1^2\vartheta_2^2(0)\vartheta_4^2(0)+\vartheta_3^2\vartheta_4^4(0)+\vartheta_3^2\vartheta_3^4(0)\big)\big].$$

The identity (7.2.16) implies that all terms in (7.6.123) involving derivatives of theta functions in the numerator vanish. Additionally, (7.2.16) – (7.2.19) allow us to simplify the numbers B, C, D and E. Namely, we have

$$(7.6.125)\qquad\begin{aligned}B&=-\vartheta_2^2\vartheta_4^2\vartheta_2^2(0)\vartheta_3^2(0)\big((\vartheta_3^2\vartheta_2^2(0)+\vartheta_1^2\vartheta_4^2(0))^2-\vartheta_1^2\vartheta_3^2\vartheta_2^2(0)\vartheta_4^2(0)\big),\\C&=-\vartheta_2^6\vartheta_4^2\vartheta_2^2(0)\vartheta_3^6(0),\\D&=\vartheta_1^2\vartheta_2^2\vartheta_3^2\vartheta_2^2(0)\vartheta_3^2(0)\big(\vartheta_1^2\vartheta_3^2(0)\vartheta_4^2(0)+\vartheta_3^2\vartheta_2^2(0)\vartheta_3^2(0)+\vartheta_4^2\vartheta_2^2(0)\vartheta_4^2(0)\big),\\E&=-\vartheta_1^2\vartheta_2^4\vartheta_3^2\vartheta_2^2(0)\vartheta_3^6(0).\end{aligned}$$

Combining these terms gives us

$$(7.6.126)\qquad\begin{aligned}&24A+17B+7C+5D+3E+2F\\&=\vartheta_2^2\vartheta_2^2(0)\big[-12\vartheta_1^2\vartheta_3^2\vartheta_4^2\vartheta_2^2(0)\vartheta_3^2(0)\vartheta_4^2(0)\\&\quad+7\vartheta_4^2\vartheta_3^2(0)\big(\vartheta_3^4\vartheta_2^4(0)+\vartheta_1^4\vartheta_4^4(0)-\vartheta_2^4\vartheta_3^4(0)\big)+8\vartheta_1^2\vartheta_2^2\vartheta_3^2\vartheta_3^2(0)\vartheta_4^4(0)\\&\quad+5\vartheta_1^4\vartheta_3^2\vartheta_3^4(0)\vartheta_4^2(0)+3\vartheta_1^2\vartheta_3^2\vartheta_3^4(0)\big(\vartheta_3^2\vartheta_2^2(0)-\vartheta_2^2\vartheta_3^2(0)\big)\\&\quad-2\vartheta_1^2\vartheta_3^2\vartheta_2^2(0)\vartheta_4^2(0)\big(\vartheta_1^2\vartheta_2^2(0)+\vartheta_3^2\vartheta_4^2(0)\big)\big].\end{aligned}$$

Again using (7.2.16) – (7.2.19), this expression simplifies to

$$(7.6.127)\qquad 24A+17B+7C+5D+3E+2F=8\vartheta_1^2\vartheta_2^2\vartheta_3^2\vartheta_4^2\vartheta_2^2(0)\vartheta_3^2(0)\vartheta_4^2(0).$$

Inserting this into (7.6.123), we get

$$(7.6.128)\qquad \tilde f(0,\omega)\equiv\tilde f(z,\omega)\equiv f(n\omega,\omega)=\frac{\vartheta_1(\omega)\vartheta_2^2(0)\vartheta_3^2(0)\vartheta_4^2(0)}{12\vartheta_1'(0)\vartheta_1\vartheta_2\vartheta_3\vartheta_4}=\frac{1}{6}$$

by (7.2.7) and (7.2.12). It then follows that

$$(7.6.129)\qquad f_0(n\omega,\omega)=f(n\omega,\omega)-\tfrac{1}{6}=0.$$

This proves (7.6.86) and therefore Proposition 7.3.1. □

7.7. The constant term

To complete the proof of Theorem 7.3.2, let us show that the constant C in the asymptotic expansion of Z_n does not depend on t. From (7.1.6) and (7.1.6), we have

$$(7.7.1)\qquad \log Z_n=n^2\log(ab)+\log\tau_n-2\sum_{k=0}^{n-1}\log k!,\qquad \tau_n=2^{n^2}\prod_{k=0}^{n-1}h_k.$$

The function τ_n satisfies the Toda equation,

$$(\log \tau_n)'' = \frac{\tau_{n+1}\tau_{n-1}}{\tau_n^2} \qquad \left(' = \frac{\partial}{\partial t}\right) \tag{7.7.2}$$

(cf. (4.4.15)), and we therefore have

$$(\log Z_n)'' = n^2(\log ab)'' + \frac{4h_n}{h_{n-1}}. \tag{7.7.3}$$

It follows from the RH analysis that formula (7.3.3) can be written up to the term of the order of $\mathcal{O}(n^{-3})$ as

$$\frac{h_n}{(n!)^2} = G^{2n+1}\frac{\vartheta_4\big((n+1)\omega\big)}{\vartheta_4(n\omega)}\left(1+\frac{c(n\omega)}{n^2}+\mathcal{O}(n^{-3})\right), \tag{7.7.4}$$

where $c(x)$ is a periodic function of period 2π, hence

$$\begin{aligned}\frac{4h_n}{h_{n-1}} = n^2\left(\frac{\pi\vartheta_1'(0)}{2\gamma}\right)^2 &\frac{\vartheta_4\big((n+1)\omega\big)\vartheta_4\big((n-1)\omega\big)}{\vartheta_1^2(\omega)\vartheta_4^2(n\omega)}\\ &\times\left(1+\frac{c(n\omega)-c((n-1)\omega)}{n^2}+\mathcal{O}(n^{-3})\right).\end{aligned} \tag{7.7.5}$$

To evaluate the expression on the right, we use the addition formula (see [76]),

$$\vartheta_4(y+z)\vartheta_4(y-z)\vartheta_4^2(0) = \vartheta_4^2(y)\vartheta_4^2(z) - \vartheta_1^2(y)\vartheta_1^2(z), \tag{7.7.6}$$

which gives that

$$\vartheta_4\big((n+1)\omega\big)\vartheta_4\big((n-1)\omega\big)\vartheta_4^2(0) = \vartheta_4^2(n\omega)\vartheta_4^2(\omega) - \vartheta_1^2(n\omega)\vartheta_1^2(\omega), \tag{7.7.7}$$

hence

$$\begin{aligned}\frac{4h_n}{h_{n-1}} = n^2\left(\frac{\pi\vartheta_1'(0)}{2\gamma\vartheta_4(0)}\right)^2 &\left[\frac{\vartheta_4^2(\omega)}{\vartheta_1^2(\omega)} - \frac{\vartheta_1^2(n\omega)}{\vartheta_4^2(n\omega)}\right]\\ &\times\left(1+\frac{c(n\omega)-c((n-1)\omega)}{n^2}+\mathcal{O}(n^{-3})\right).\end{aligned} \tag{7.7.8}$$

We have the following identity:

$$\frac{\vartheta_4^2(z)}{\vartheta_1^2(z)} - \frac{\vartheta_1^2(y)}{\vartheta_4^2(y)} = -\left(\frac{\vartheta_4(0)}{\vartheta_1'(0)}\right)^2\left[\frac{\mathrm{d}^2\log\vartheta_1(z)}{\mathrm{d}z^2} - \frac{\mathrm{d}^2\log\vartheta_4(y)}{\mathrm{d}y^2}\right]. \tag{7.7.9}$$

Indeed, introduce the function

$$f(z) = \frac{\vartheta_4^2(z)}{\vartheta_1^2(z)} + \left(\frac{\vartheta_4(0)}{\vartheta_1'(0)}\right)^2\frac{\mathrm{d}^2\log\vartheta_1(z)}{\mathrm{d}z^2}. \tag{7.7.10}$$

As $z \to 0$,

$$\frac{\vartheta_4^2(z)}{\vartheta_1^2(z)} = \frac{\vartheta_4^2(0)}{z^2\vartheta_1'(0)^2} + \mathcal{O}(1), \qquad \frac{\mathrm{d}^2\log\vartheta_1(z)}{\mathrm{d}z^2} = -\frac{1}{z^2} + \mathcal{O}(1), \tag{7.7.11}$$

hence $f(z) = \mathcal{O}(1)$. Since f is doubly periodic, this implies that $f(z)$ is constant in z. Since

$$f\left(z - \frac{\pi\tau}{2}\right) = \frac{\vartheta_1^2(z)}{\vartheta_4^2(z)} + \left(\frac{\vartheta_4(0)}{\vartheta_1'(0)}\right)^2\frac{\mathrm{d}^2\log\vartheta_4(z)}{\mathrm{d}z^2}, \tag{7.7.12}$$

we obtain identity (7.7.9). Applying identity (7.7.9) to (7.7.8), we obtain that

$$\frac{4h_n}{h_{n-1}} = -n^2\left(\frac{\pi}{2\gamma}\right)^2\left[\frac{\mathrm{d}^2 \log\vartheta_1(\omega)}{\mathrm{d}\omega^2} - \frac{\mathrm{d}^2\log\vartheta_4(n\omega)}{\mathrm{d}(n\omega)^2}\right] \times\left(1+\frac{c(n\omega)-c((n-1)\omega)}{n^2}+\mathcal{O}(n^{-3})\right). \tag{7.7.13}$$

Since $\omega = \pi(1+t/\gamma)/2$, we obtain that

$$\begin{aligned}\frac{4h_n}{h_{n-1}} &= \left[-n^2\big(\log\vartheta_1(\omega)\big)'' + \big(\log\vartheta_4(n\omega)\big)''\right] \\ &\quad\times\left(1+\frac{c(n\omega)-c((n-1)\omega)}{n^2}+\mathcal{O}(n^{-3})\right) \quad \left({}'=\frac{\partial}{\partial t}\right) \\ &= -n^2\big(\log\vartheta_1(\omega)\big)'' + \big(\log\vartheta_4(n\omega)\big)'' \\ &\quad - \big(\log\vartheta_1(\omega)\big)''\Big(c(n\omega)-c\big((n-1)\omega\big)\Big)+\mathcal{O}(n^{-1}).\end{aligned} \tag{7.7.14}$$

Thus by (7.7.3),

$$\begin{aligned}(\log Z_n)'' &= n^2(\log ab)'' - n^2\big(\log\vartheta_1(\omega)\big)'' + \big(\log\vartheta_4(n\omega)\big)'' \\ &\quad - \big(\log\vartheta_1(\omega)\big)''\Big(c(n\omega)-c\big((n-1)\omega\big)\Big)+\mathcal{O}(n^{-1}).\end{aligned} \tag{7.7.15}$$

We would like to integrate (7.7.15) twice to obtain a formula for Z_n. Notice first that we can write the function $c(x)$ as

$$c(x) = a + c_0(x), \qquad \int_0^{2\pi} c_0(x)\,\mathrm{d}x = 0, \tag{7.7.16}$$

and by the Riemann – Lebesgue lemma, for any $t_1, t_2 \in (-\gamma,\gamma)$,

$$\int_{t_1}^{t_2}\big(\log\vartheta_1(\omega)\big)'' c_0(n\omega)\,\mathrm{d}\omega = \mathcal{O}(n^{-1}). \tag{7.7.17}$$

It follows that after integrating, the quasi-periodic term in (7.7.15) is $\mathcal{O}(n^{-1})$, and thus integrating twice gives

$$\log Z_n = n^2\log\frac{ab}{\vartheta_1(\omega)} + \log\vartheta_4(n\omega) + d_1(\gamma,n)t + d_0(\gamma,n) + \mathcal{O}(n^{-1}), \tag{7.7.18}$$

where $d_1(\gamma,n)$ and $d_0(\gamma,n)$ do not depend on t. On the other hand, from (7.3.5) we have

$$\log Z_n = n^2\log\frac{ab}{\vartheta_1(\omega)} + n^2\log\frac{\pi\vartheta_1'(0)}{2\gamma} + \log\vartheta_4(n\omega) + \log C + \mathcal{O}(n^{-1}). \tag{7.7.19}$$

Combining (7.7.18) and (7.7.19), we find

$$\log C = d_1(\gamma,n)t + d_2(\gamma,n) + \mathcal{O}(n^{-1}), \tag{7.7.20}$$

where

$$d_2(\gamma,n) = d_0(\gamma,n) - n^2\log\frac{\pi\vartheta_1'(0)}{2\gamma}, \tag{7.7.21}$$

also does not depend on t. Taking $n\to\infty$, we find that

$$C = \mathrm{e}^{C_0+C_1 t}, \tag{7.7.22}$$

where C_0 and C_1 do not depend on t or n, but may depend on γ. By the symmetry of the partition function with respect to $t \mapsto (-t)$, we have that $C_1 = 0$, hence $C = \mathrm{e}^{C_0}$ is independent of t.

CHAPTER 8

Ferroelectric Phase

8.1. Introduction and formulation of the main results

Here we discuss the ferroelectric phase, and we will use parametrization (4.4.2) for the weights:

$$a = \sinh(t-\gamma), \quad b = \sinh(t+\gamma), \quad c = \sinh(2|\gamma|), \qquad 0 < |\gamma| < t. \tag{8.1.1}$$

Without loss of generality we may assume that

$$\gamma > 0, \tag{8.1.2}$$

which corresponds to the region

$$b > a + c. \tag{8.1.3}$$

The parameter Δ in the ferroelectric phase reduces to

$$\Delta = \cosh(2\gamma) > 1. \tag{8.1.4}$$

In this phase, the function $\phi(t)$ in the Hankel determinant τ_n is in fact the Laplace transform of a discrete measure lying on the positive integers. We have that

$$\begin{aligned} \phi(t) = \frac{\sinh(2\gamma)}{\sinh(t+\gamma)\sinh(t-\gamma)} &= 2\left[\frac{e^{-2t+2\gamma} - e^{-2t-2\gamma}}{(1-e^{-2t+2\gamma})(1-e^{-2t-2\gamma})}\right] \\ &= 2\left[\frac{1}{1-e^{-2t+2\gamma}} - \frac{1}{1-e^{-2t-2\gamma}}\right] \\ &= 4\sum_{l=1}^{\infty} e^{-2tl}\sinh(2\gamma l). \end{aligned} \tag{8.1.5}$$

Introduce now discrete monic polynomials $P_j(x) = x^j + \cdots$ orthogonal on the set $\mathbb{N} = \{l = 1, 2, \dots\}$ with respect to the weight,

$$w(l) = 2e^{-2tl}\sinh(2\gamma l) = e^{-2tl+2\gamma l} - e^{-2tl-2\gamma l}, \tag{8.1.6}$$

so that

$$\sum_{l=1}^{\infty} P_j(l)P_k(l)w(l) = h_k\delta_{jk}. \tag{8.1.7}$$

Then it follows from (8.1.5) and Proposition 1.2.2 that

$$\tau_n = 2^{n^2}\prod_{k=0}^{n-1} h_k. \tag{8.1.8}$$

We will prove the following asymptotics of h_k.

Theorem 8.1.1. *For any $\varepsilon > 0$, as $k \to \infty$,*

$$h_k = \frac{(k!)^2 q^{k+1}}{(1-q)^{2k+1}} \left(1 + \mathcal{O}(\mathrm{e}^{-k^{1-\varepsilon}})\right), \tag{8.1.9}$$

where

$$q = \mathrm{e}^{2\gamma - 2t}. \tag{8.1.10}$$

The error term in (8.1.9) *is uniform on any compact subset of the set*

$$\{(t, \gamma) : 0 < \gamma < t\}. \tag{8.1.11}$$

Our main result in this chapter is the following theorem.

Theorem 8.1.2. *In the ferroelectric phase with $t > \gamma > 0$, for any $\varepsilon > 0$, as $n \to \infty$,*

$$Z_n = CG^n F^{n^2}[1 + \mathcal{O}(\mathrm{e}^{-n^{1-\varepsilon}})], \tag{8.1.12}$$

where $C = 1 - \mathrm{e}^{-4\gamma}$, $G = \mathrm{e}^{\gamma - t}$, and $F = \sinh(t + \gamma)$. The error term in (8.1.9) *is uniform on any compact subset of the set* (8.1.11).

Up to the constant factor, this result follows directly from Theorem 8.1.1. In principle, we could prove Theorem 8.1.1 using a modification of the Riemann–Hilbert approach outlined in Chapter 3. In this case, however, we can avoid this by noticing that the weight of orthogonality for the polynomials defined in (8.1.7) is the difference of two exponential terms, and that the first term dominates for large l. This motivates comparison of these polynomials with polynomials orthogonal with respect to a simple exponential weight on $\mathbb{N}$. Such polynomials are a special case of the classical *Meixner polynomials* (see, e.g., [49]), and their properties are well known. We review the properties of the Meixner polynomials in Section 8.2 and compare them with the polynomials (8.1.7) in Section 8.4. To find the constant factor C we will use the Toda equation, combined with the asymptotics of C as $t \to \infty$. This work is shown in Section 8.5.

8.2. Meixner polynomials

In addition to the weight $w(l)$ defined in (8.1.6), consider the exponential weight on $\mathbb{N}$,

$$w^{\mathrm{Q}}(l) = q^l, \quad l \in \mathbb{N}; \qquad q = \mathrm{e}^{2\gamma - 2t} < 1, \tag{8.2.1}$$

which can be viewed as an approximation to $w(l)$ for large l. The orthogonal polynomials with the weight $w^{\mathrm{Q}}(l)$ are expressed in terms of the Meixner polynomials with $\beta = 1$, see [49], which are defined by the formula,

$$M_k(z; q) = \sum_{j=0}^{k} \frac{(1 - q^{-1})^j \prod_{i=0}^{j-1}(k - i) \prod_{i=0}^{j-1}(z - i)}{(j!)^2}. \tag{8.2.2}$$

They satisfy the orthogonality condition,

$$\sum_{l=0}^{\infty} M_j(l; q) M_k(l; q) q^l = \frac{q^{-k}\delta_{jk}}{1 - q}. \tag{8.2.3}$$

For the corresponding monic polynomials,

$$P_k^{\mathrm{M}}(z) = \frac{k!}{(1 - q^{-1})^k} M_k(z; q) \tag{8.2.4}$$

(M in P_k^{M} stands for Meixner), the orthogonality condition reads

$$\sum_{l=0}^{\infty} P_j^{\mathrm{M}}(l)P_k^{\mathrm{M}}(l)q^j = h_k^{\mathrm{M}}\delta_{jk}, \quad h_k^{\mathrm{M}} = \frac{(k!)^2 q^k}{(1-q)^{2k+1}}. \tag{8.2.5}$$

They satisfy the three term recurrence relation,

$$zP_k^{\mathrm{M}}(z) = P_{k+1}^{\mathrm{M}}(z) + \frac{kq+k+q}{1-q}\,P_k^{\mathrm{M}}(z) + \frac{k^2 q}{(1-q)^2}\,P_{k-1}^{\mathrm{M}}(z). \tag{8.2.6}$$

According to (8.2.1), we take $q = \mathrm{e}^{2\gamma-2t}$.

For our purposes it is convenient to introduce a shifted Meixner polynomial,

$$Q_k(z) = P_k^{\mathrm{M}}(z-1) = \frac{(-1)^k k! q^k}{(1-q)^k} M_k(z-1;q), \tag{8.2.7}$$

which is a monic polynomial as well. Equation (8.2.5) implies the orthogonality condition,

$$\sum_{l=1}^{\infty} Q_j(l)Q_k(l)q^l = h_k^{\mathrm{Q}}\delta_{jk}, \quad h_k^{\mathrm{Q}} = \frac{(k!)^2 q^{k+1}}{(1-q)^{2k+1}}. \tag{8.2.8}$$

By analogy with (8.1.8), define

$$\tau_n^{\mathrm{Q}} = 2^{n^2}\prod_{k=0}^{n-1} h_k^{\mathrm{Q}}. \tag{8.2.9}$$

From (8.2.5) and (8.2.8) we obtain that

$$\tau_n^{\mathrm{Q}} = 2^{n^2}\prod_{k=0}^{n-1}\frac{(k!)^2 q^{k+1}}{(1-q)^{2k+1}} = \frac{2^{n^2} q^{(n+1)n/2}}{(1-q)^{n^2}}\prod_{k=0}^{n-1}(k!)^2. \tag{8.2.10}$$

By analogy with (4.4.5), define also

$$Z_n^{\mathrm{Q}} = \frac{[\sinh(\gamma+t)\sinh(\gamma-t)]^{n^2}}{\prod_{k=0}^{n-1}(k!)^2}\,\tau_n^{\mathrm{Q}}. \tag{8.2.11}$$

Then from (8.2.10) we obtain that

$$Z_n^{\mathrm{Q}} = F^{n^2}G^n, \tag{8.2.12}$$

where

$$\begin{aligned} F &= \frac{2\sinh(t-\gamma)\sinh(t+\gamma)q^{1/2}}{1-q} \\ &= \frac{2\sinh(t-\gamma)\sinh(t+\gamma)e^{\gamma-t}}{1-\mathrm{e}^{2\gamma-2t}} = \sinh(t+\gamma), \end{aligned} \tag{8.2.13}$$

and

$$G = q^{1/2} = \mathrm{e}^{\gamma-t}. \tag{8.2.14}$$

8.3. Two interpolation problems

In order to compare the orthogonal polynomials (8.1.7) and (8.2.8), it can be useful to compare their associated interpolation problems. For the polynomials (8.1.7), the problem is as follows. For a given $k = 0, 1, \dots$, find a 2×2 matrix-valued function $\mathbf{P}_k(z) = (\mathbf{P}_k(z)_{ij})_{1\le i,j\le 2}$ with the following properties:

Analyticity: $\mathbf{P}_k(z)$ is an analytic function of z for $z \in \mathbb{C}\setminus\mathbb{N}$.

Residues at poles: At each node $l \in \mathbb{N}$, the elements $\mathbf{P}_k(z)_{11}$ and $\mathbf{P}_k(z)_{21}$ of the matrix $\mathbf{P}_k(z)$ are analytic functions of z, and the elements $\mathbf{P}_k(z)_{12}$ and $\mathbf{P}_k(z)_{22}$ have a simple pole with the residues,

$$\operatorname*{Res}_{z=l} \mathbf{P}_k(z)_{j2} = w(l)\mathbf{P}_k(l)_{j1}, \quad j = 1, 2. \tag{8.3.1}$$

Asymptotics at infinity: As $z \to \infty$, $\mathbf{P}_k(z)$ admits the asymptotic expansion,

$$\mathbf{P}_k(z) \sim \left(I + \frac{\mathbf{P}_1}{z} + \frac{\mathbf{P}_2}{z^2} + \cdots\right)\begin{pmatrix} z^k & 0 \\ 0 & z^{-k}\end{pmatrix}, \quad z \in \mathbb{C}\setminus\left[\bigcup_{l=1}^{\infty} D(l, r_l)\right], \tag{8.3.2}$$

where $D(z, r)$ is a disk of radius $r > 0$ centered at $z \in \mathbb{C}$ and

$$\lim_{l\to\infty} r_l = 0. \tag{8.3.3}$$

The unique solution to this interpolation problem is

$$\mathbf{P}_k(z) = \begin{pmatrix} P_k(z) & C(wP_k)(z) \\ (h_{k-1})^{-1}P_{k-1}(z) & (h_{k-1})^{-1}C(wP_{k-1})(z)\end{pmatrix}, \tag{8.3.4}$$

where C is the discrete Cauchy transformation on $\mathbb{N}$, defined by the formula,

$$C(f)(z) = \sum_{l=1}^{\infty} \frac{f(l)}{z-l}, \tag{8.3.5}$$

and $P_k(z) = z^k + \cdots$ are the monic polynomials orthogonal with respect to the weight $w(l)$, given in (8.1.7). It follows from (8.3.4) that

$$h_k = [\mathbf{P}_1]_{21}, \tag{8.3.6}$$

where $[\mathbf{P}_1]_{21}$ is the (21)-element of the matrix $\mathbf{P}_1$ on the right in (8.3.2). In order to compare these polynomials with the shifted Meixner polynomials (8.2.8), introduce also the 2×2 matrix

$$\mathbf{Q}_k(z) = \begin{pmatrix} Q_k(z) & C(w^{\mathrm{Q}}Q_k)(z) \\ (h^{\mathrm{Q}}_{k-1})^{-1}Q_{k-1}(z) & (h^{\mathrm{Q}}_{k-1})^{-1}C(w^{\mathrm{Q}}Q_{k-1})(z)\end{pmatrix}. \tag{8.3.7}$$

This matrix function satisfies an interpolation problem identical to the one for $\mathbf{P}_k(z)$, but with $w(l)$ replaced by $w^{\mathrm{Q}}(l)$.

Consider the quotient matrix,

$$\mathbf{X}_k(z) := \mathbf{P}_k(z)[\mathbf{Q}_k(z)]^{-1}. \tag{8.3.8}$$

Observe that $\det \mathbf{Q}_k(z)$ has no poles and it approaches 1 as $z \to \infty$ outside of the disks $D(l, r_l)$, $l = 1, 2, \dots$, hence,

$$\det \mathbf{Q}_k(z) \equiv 1, \tag{8.3.9}$$

The elements of the matrix $\mathbf{X}(z)$ are thus

$$(8.3.10)\qquad \begin{aligned} X_k(z)_{11} &= P_k(z)(h_{k-1}^{\mathrm{Q}})^{-1}C(w^{\mathrm{Q}}Q_{k-1})(z) \\ &\qquad - C(wP_k)(z)(h_{k-1}^{\mathrm{Q}})^{-1}Q_{k-1}(z), \\ X_k(z)_{12} &= -P_k(z)C(w^{\mathrm{Q}}Q_k)(z) + C(wP_k)(z)Q_k(z), \\ X_k(z)_{21} &= (h_{k-1})^{-1}P_{k-1}(z)(h_{k-1}^{\mathrm{Q}})^{-1}C(w^{\mathrm{Q}}Q_{k-1})(z) \\ &\qquad - (h_{k-1})^{-1}C(wP_{k-1})(z)(h_{k-1}^{\mathrm{Q}})^{-1}Q_{k-1}(z), \\ X_k(z)_{22} &= -(h_{k-1})^{-1}P_{k-1}(z)C(w^{\mathrm{Q}}Q_k)(z) \\ &\qquad + (h_{k-1})^{-1}C(wP_{k-1})(z)Q_k(z). \end{aligned}$$

Also,

$$(8.3.11)\qquad \mathbf{X}_k(z) \to I \quad \text{as } z \to \infty \text{ outside of the disks } D(l, r_l),\ l = 1, 2, \ldots$$

This implies that the matrix $\mathbf{X}_k$ can be written as

$$(8.3.12)\qquad \mathbf{X}_k(z) = I + C[(w^{\mathrm{Q}} - w)\mathbf{R}],$$

where

$$(8.3.13)\qquad \mathbf{R}(z) = \begin{pmatrix} (h_{k-1}^{\mathrm{Q}})^{-1}P_k(z)Q_{k-1}(z) & -P_k(z)Q_k(z) \\ (h_{k-1}h_{k-1}^{\mathrm{Q}})^{-1}P_{k-1}(z)Q_{k-1}(z) & -(h_{k-1})^{-1}P_{k-1}(z)Q_k(z) \end{pmatrix}.$$

The formula (8.3.12) can be further used to evaluate the large n asymptotics of the orthogonal polynomials $P_n(z)$, but we will not pursue this approach. Instead, we note simply that from formula (8.3.6) and (8.3.12) we obtain that

$$(8.3.14)\qquad h_k - h_k^{\mathrm{Q}} = -\sum_{l=1}^{\infty} P_k(l)Q_k(l)\,[w^{\mathrm{Q}}(l) - w(l)].$$

We can use this identity to estimate $|h_k - h_k^{\mathrm{Q}}|$ directly.

8.4. Evaluation of the ratio h_k/h_k^{Q}

In this section we will prove Theorem 8.1.1. By applying the Cauchy-Schwarz inequality to identity (8.3.14), we obtain that

$$(8.4.1)\quad |h_k - h_k^{\mathrm{Q}}| \le \left[\sum_{l=1}^{\infty} P_k(l)^2|w(l) - w^{\mathrm{Q}}(l)|\right]^{1/2}\left[\sum_{l=1}^{\infty} Q_k(l)^2|w(l) - w^{\mathrm{Q}}(l)|\right]^{1/2},$$

so that

$$(8.4.2)\quad \left|\frac{h_k}{h_k^{\mathrm{Q}}} - 1\right| \le \left[\frac{1}{h_k^{\mathrm{Q}}}\sum_{l=1}^{\infty} P_k(l)^2|w(l) - w^{\mathrm{Q}}(l)|\right]^{1/2} \times \left[\frac{1}{h_k^{\mathrm{Q}}}\sum_{l=1}^{\infty} Q_k(l)^2|w(l) - w^{\mathrm{Q}}(l)|\right]^{1/2},$$

From (8.1.6),

$$(8.4.3)\qquad |w(l) - w^{\mathrm{Q}}(l)| = \mathrm{e}^{-(2t+2\gamma)l} \le C_0 w(l), \quad l \ge 1; \qquad C_0 = \frac{1}{\mathrm{e}^{4\gamma} - 1},$$

hence

$$\frac{1}{h_k^{\mathrm{Q}}}\sum_{l=1}^{\infty} P_k(l)^2\,|w(l)-w^{\mathrm{Q}}(l)| \le C_0\frac{1}{h_k^{\mathrm{Q}}}\sum_{l=1}^{\infty} P_k(l)^2 w(l) = \frac{C_0 h_k}{h_k^{\mathrm{Q}}} \le C_0(1+\varepsilon_k), \tag{8.4.4}$$

where

$$\varepsilon_k = \left|\frac{h_k}{h_k^{\mathrm{Q}}} - 1\right|. \tag{8.4.5}$$

Thus, by (8.4.2),

$$\varepsilon_k^2 \le C_0(1+\varepsilon_k)\delta_k, \tag{8.4.6}$$

where

$$\delta_k = \frac{1}{h_k^{\mathrm{Q}}}\sum_{l=1}^{\infty} Q_k(l)^2|w(l)-w^{\mathrm{Q}}(l)|. \tag{8.4.7}$$

By (8.4.3),

$$\delta_k = \frac{1}{h_k^{\mathrm{Q}}}\sum_{l=1}^{\infty} Q_k(l)^2\, q_0^l, \qquad q_0 = \mathrm{e}^{-2(t+\gamma)}. \tag{8.4.8}$$

Let us evaluate δ_k.

We partition the sum in (8.4.8) into two parts:

$$\delta_k' = \frac{1}{h_k^{\mathrm{Q}}}\sum_{l=1}^{L} Q_k(l)^2\, q_0^l, \tag{8.4.9}$$

and

$$\delta_k'' = \frac{1}{h_k^{\mathrm{Q}}}\sum_{l=L+1}^{\infty} Q_k(l)^2\, q_0^l, \tag{8.4.10}$$

where

$$L = \lceil k^\lambda \rceil, \quad 0<\lambda<1. \tag{8.4.11}$$

We first estimate δ_k'. We have from (8.2.7), (8.2.8) that

$$\frac{Q_k(l)}{(h_k^{\mathrm{Q}})^{1/2}} = \frac{(-1)^k(1-q)^{1/2}q^{k/2}}{q^{1/2}} M_k(l-1;q). \tag{8.4.12}$$

By (8.2.2),

$$\begin{aligned} M_k(l-1;q) = 1 &+ (1-q^{-1})k(l-1) + (1-q^{-1})^2\frac{k(k-1)(l-1)(l-2)}{(2!)^2} \\ &+ (1-q^{-1})^3\frac{k(k-1)(k-2)(l-1)(l-2)(l-3)}{(3!)^2} + \cdots . \end{aligned} \tag{8.4.13}$$

If $l<k$, then the latter sum consists of l nonzero terms. For $l \le L$ it is estimated as

$$M_k(l-1;q) = \mathcal{O}(k^L L^{L+1}) = \mathcal{O}\big(\mathrm{e}^{L\log k + (L+1)\log L}\big), \tag{8.4.14}$$

hence

$$\frac{Q_k(l)}{(h_k^{\mathrm{Q}})^{1/2}} = \mathcal{O}\big(\mathrm{e}^{(k\log q)/2 + L\log k + (L+1)\log L}\big). \tag{8.4.15}$$

Due to our choice of L in (8.4.11), this implies the estimate,

$$\frac{Q_k(l)}{(h_k^{\mathrm{Q}})^{1/2}} = \mathcal{O}\big(\mathrm{e}^{(k\log q)/2+2k^\lambda \log k}\big). \tag{8.4.16}$$

Since $0 < q < 1$ and $0 < \lambda < 1$, the expression on the right is exponentially small as $k \to \infty$.

From (8.4.9) we obtain now that

$$\delta_k' = \mathcal{O}(\mathrm{e}^{k\log q + 4k^\lambda \log k}). \tag{8.4.17}$$

Since $\lambda < 1$ and $q < 1$, we obtain that

$$\delta_k' = \mathcal{O}(\mathrm{e}^{-c_0 k}), \quad c_0 = -\frac{\log q}{2} > 0. \tag{8.4.18}$$

Now let us estimate δ_k''. By (8.2.8),

$$\frac{1}{h_k^{\mathrm{Q}}} \sum_{l=1}^{\infty} Q_k(l)^2 q^l = 1, \tag{8.4.19}$$

hence

(8.4.20)

$$\delta_k'' = \frac{1}{h_k^{\mathrm{Q}}} \sum_{l=L+1}^{\infty} Q_k(l)^2\, q_0^l < \left(\frac{q_0}{q}\right)^L \frac{1}{h_k^{\mathrm{Q}}} \sum_{l=L+1}^{\infty} Q_k(l)^2\, q^l < \left(\frac{q_0}{q}\right)^L = \mathrm{e}^{-4\gamma L}.$$

Thus,

$$\delta_k'' < e^{-4\gamma(k^\lambda - 1)}. \tag{8.4.21}$$

Since $0 < \lambda < 1$ is an arbitrary number, we obtain from (8.4.18) and (8.4.21) that for any $\eta > 0$,

$$\delta_k = \mathcal{O}\big(\mathrm{e}^{-k^{1-\eta}}\big). \tag{8.4.22}$$

Let us return back to inequality (8.4.6). Consider two cases: (1) $\varepsilon_k > 1$ and (2) $\varepsilon_k \le 1$. In the first case (8.4.6) implies that

$$\varepsilon_k \le 2C_0\delta_k, \tag{8.4.23}$$

which is impossible, because of (8.4.22). Hence, for large k, $\varepsilon_k \le 1$, in which case (8.4.6) gives that

$$\varepsilon_k^2 \le 2C_0\delta_k. \tag{8.4.24}$$

Estimate (8.4.22) implies now that for any $\eta > 0$,

$$\varepsilon_k = \mathcal{O}\big(\mathrm{e}^{-k^{1-\eta}}\big), \tag{8.4.25}$$

so that as $k \to \infty$,

$$h_k = h_k^{\mathrm{Q}}(1 + \tilde{\varepsilon}_k), \quad |\tilde{\varepsilon}_k| = \varepsilon_k = \mathcal{O}\big(\mathrm{e}^{-k^{1-\eta}}\big). \tag{8.4.26}$$

From (8.4.26) we obtain that for any $\eta > 0$,

$$Z_n = Z_n^{\mathrm{Q}} \prod_{k=0}^{n} (1 + \tilde{\varepsilon}_k) = CZ_n^{\mathrm{Q}}\big[1 + \mathcal{O}\big(\mathrm{e}^{-n^{1-\eta}}\big)\big], \tag{8.4.27}$$

where

$$C = \prod_{k=0}^{\infty} (1 + \tilde{\varepsilon}_k) > 0. \tag{8.4.28}$$

Thus, we have proved the following result.

Proposition 8.4.1. *For any* $\varepsilon > 0$, *as* $n \to \infty$,

$$Z_n = CF^{n^2}G^n\big[1 + \mathcal{O}(\mathrm{e}^{-n^{1-\varepsilon}})\big], \tag{8.4.29}$$

where $C > 0$, $F = \sinh(t+\gamma)$, *and* $G = \mathrm{e}^{\gamma - t}$.

8.5. Evaluation of the constant factor

Let us now compute the exact value of the constant C in formula (8.4.29). This is be done in two steps: first, with the help of the Toda equation, we find the form of the dependence of C on t; and second, we find the large t asymptotics of C. By combining these two steps, we obtain the exact value of C.

Recall that the function τ_n satisfies the Toda equation

$$(\log \tau_n)'' = \frac{\tau_{n+1}\tau_{n-1}}{\tau_n^2}, \qquad \left(' = \frac{\partial}{\partial t}\right), \tag{8.5.1}$$

cf. (4.4.15). From (8.1.8) we then obtain that

$$(\log \tau_n)'' = \frac{4h_n}{h_{n-1}}. \tag{8.5.2}$$

From (8.1.9) we obtain that

$$\frac{4h_n}{h_{n-1}} = \frac{4n^2 q}{(1-q)^2} + \mathcal{O}\big(\mathrm{e}^{-n^{1-\varepsilon}}\big). \tag{8.5.3}$$

We have that

$$\frac{4q}{(1-q)^2} = \frac{4\mathrm{e}^{2\gamma-2t}}{(1-\mathrm{e}^{2\gamma-2t})^2} = \left[\frac{(-2)}{1-\mathrm{e}^{2\gamma-2t}}\right]' = [-\log(1-\mathrm{e}^{2\gamma-2t})]'', \tag{8.5.4}$$

hence from (8.5.2), (8.5.3) we obtain that

$$(\log \tau_n)'' = n^2[-\log(1-\mathrm{e}^{2\gamma-2t})]'' + \mathcal{O}\big(\mathrm{e}^{-n^{1-\varepsilon}}\big). \tag{8.5.5}$$

By (4.4.5) this implies that

$$(\log Z_n)'' = n^2\left[\log \frac{\sinh(t-\gamma)\sinh(t+\gamma)}{1-\mathrm{e}^{2\gamma-2t}}\right]'' + \mathcal{O}\big(\mathrm{e}^{-n^{1-\varepsilon}}\big). \tag{8.5.6}$$

Since

$$\log\left[\frac{\sinh(t-\gamma)\sinh(t+\gamma)}{1-\mathrm{e}^{2\gamma-2t}}\right] = \log[\sinh(t+\gamma)] + (t-\gamma) - \log 2, \tag{8.5.7}$$

we can simplify (8.5.6) to

$$(\log Z_n)'' = n^2[\log\sinh(t+\gamma)]'' + \mathcal{O}\big(\mathrm{e}^{-n^{1-\varepsilon}}\big). \tag{8.5.8}$$

Observe that the error term in the last formula is uniform when t belongs to a compact set on (γ, ∞), hence by integrating it we obtain that

$$\log Z_n = n^2\log\sinh(t+\gamma) + c_1 t + c_0 + \mathcal{O}\big(\mathrm{e}^{-n^{1-\varepsilon}}\big), \tag{8.5.9}$$

where c_0, c_1 do not depend on t. In general, c_0 and c_1 depend on γ and n. By substituting formula (8.4.29) into the preceding equation, we obtain that

$$\log C + n(\gamma - t) = c_1 t + c_0 + \mathcal{O}\big(\mathrm{e}^{-n^{1-\varepsilon}}\big). \tag{8.5.10}$$

Denote

$$d_0 = c_0 - n\gamma, \quad d_1 = c_1 + n. \tag{8.5.11}$$

Then equation (8.5.10) reads

$$\log C = d_1 t + d_0 + \mathcal{O}\big(e^{-n^{1-\varepsilon}}\big). \tag{8.5.12}$$

Observe that $C = C(\gamma, t)$ does not depend on n, while $d_{0,1} = d_{0,1}(\gamma, n)$ does not depend on t. Take any $0 < \gamma < t_1 < t_2$. Then

$$\log C(\gamma, t_2) - \log C(\gamma, t_1) = d_1(t_2 - t_1) + \mathcal{O}\big(e^{-n^{1-\varepsilon}}\big). \tag{8.5.13}$$

From this formula we obtain that the limit,

$$\lim_{n\to\infty} d_1(\gamma, n) = d_1(\gamma), \tag{8.5.14}$$

exists. This in turn implies that the limit,

$$\lim_{n\to\infty} d_0(\gamma, n) = d_0(\gamma), \tag{8.5.15}$$

exists. By taking the limit $n \to \infty$ in (8.5.12), we obtain that

$$\log C = d_1(\gamma) t + d_0(\gamma). \tag{8.5.16}$$

Now consider the following regime:

$$\gamma \text{ is fixed}, \quad t \to \infty, \tag{8.5.17}$$

and let us evaluate the asymptotics of C in this regime. By (8.1.6) we have that

$$h_0 = \sum_{l=1}^{\infty} w(l) = \sum_{l=1}^{\infty} \big(e^{-2tl+2\gamma l} - e^{-2tl-2\gamma l}\big) = \frac{e^{-2t+2\gamma}}{1 - e^{-2t+2\gamma}} - \frac{e^{-2t-2\gamma}}{1 - e^{-2t-2\gamma}}. \tag{8.5.18}$$

Similarly, by (8.2.8),

$$h_0^{\mathrm{Q}} = \frac{e^{-2t+2\gamma}}{1 - e^{-2t+2\gamma}}, \tag{8.5.19}$$

hence

$$\frac{h_0}{h_0^{\mathrm{Q}}} = 1 - e^{-4\gamma} + \mathcal{O}(e^{-2t}), \quad t \to \infty. \tag{8.5.20}$$

Let us evaluate $\varepsilon_k = |h_k / h_k^{\mathrm{Q}} - 1|$ for $k \ge 1$.

By (8.4.6),

$$\varepsilon_k^2 \le C_0 (1 + \varepsilon_k) \delta_k, \quad C_0 = \frac{1}{e^{4\gamma} - 1}. \tag{8.5.21}$$

In the partition of δ_k as $\delta_k' + \delta_k''$ in (8.4.9), (8.4.10), let us choose

$$L = [k^{2/3} + t^{2/3}]. \tag{8.5.22}$$

From (8.4.12), (8.4.13) we obtain that for $l \le L$,

$$\frac{|Q_k(l)|}{(h_k^{\mathrm{Q}})^{1/2}} \le q^{(k-1)/2} k^L L^{L+1}, \quad q = e^{2\gamma - 2t}, \tag{8.5.23}$$

hence

$$\delta_k' \le \frac{q_0 q^{k-1} k^L L^{L+1}}{1 - q_0} \le \frac{q^k k^L L^{L+1}}{1 - q_0}, \quad q_0 = e^{-2\gamma - 2t}. \tag{8.5.24}$$

In addition, by (8.4.20),

$$\delta_k'' \le \mathrm{e}^{-4\gamma L}. \tag{8.5.25}$$

Our choice of L in (8.5.22) ensures that there exists $t_0 > 0$ such that for any $t \ge t_0$ and any $k \ge 1$,

$$\delta_k = \delta_k' + \delta_k'' \le \mathrm{e}^{-k^{1/2}-t^{1/2}}. \tag{8.5.26}$$

From (8.5.21) we obtain now that for $k \ge 1$ and large t,

$$\varepsilon_k \le C_1 \mathrm{e}^{-k^{1/2}/2-t^{1/2}/2}, \quad C_1 = (2C_0)^{1/2}. \tag{8.5.27}$$

By (8.4.28),

$$\log C = \sum_{k=0}^{\infty} \log(1+\tilde{\varepsilon}_k), \quad |\tilde{\varepsilon}_k| = \varepsilon_k. \tag{8.5.28}$$

From equations (8.5.20) and (8.5.27) we obtain now that

$$\log C = \log(1-\mathrm{e}^{-4\gamma}) + \mathcal{O}(\mathrm{e}^{-t^{1/2}/2}), \quad t \to \infty. \tag{8.5.29}$$

On the other hand, by (8.5.12)

$$\log C = d_1(\gamma)t + d_0(\gamma) \tag{8.5.30}$$

This implies that

$$d_1(\gamma) = 0, \quad d_0(\gamma) = \log(1-\mathrm{e}^{-4\gamma}), \tag{8.5.31}$$

so that

$$C = 1 - \mathrm{e}^{-4\gamma}. \tag{8.5.32}$$

By substituting expression (8.5.32) into formula (8.4.29), we prove Theorem 8.1.2. □

As a final note in this chapter, we compare the asymptotics of the free energy in the ferroelectric phase with the energy of the ground state.

8.6. Ground state configuration

The ground state in the ferroelectric phase region is the configuration

$$\sigma^{\mathrm{gs}}(x) = \begin{cases} \sigma_5 & \text{if } x \text{ is on the diagonal} \\ \sigma_3 & \text{if } x \text{ is above the diagonal} \\ \sigma_4 & \text{if } x \text{ is below the diagonal,} \end{cases} \tag{8.6.1}$$

see Figure 8.1. The weight of the ground state configuration is

$$w(\sigma^{\mathrm{gs}}) = b^{n^2}\left(\frac{c}{b}\right)^n = F^{n^2} G_0^n, \tag{8.6.2}$$

where

$$F = \sinh(t+\gamma), \quad G_0 = \frac{\sinh(2\gamma)}{\sinh(t+\gamma)}. \tag{8.6.3}$$

The ratio $Z_n / w(\sigma^{\mathrm{gs}})$ is evaluated as

$$\frac{Z_n}{w(\sigma^{\mathrm{gs}})} = G_1^n, \tag{8.6.4}$$

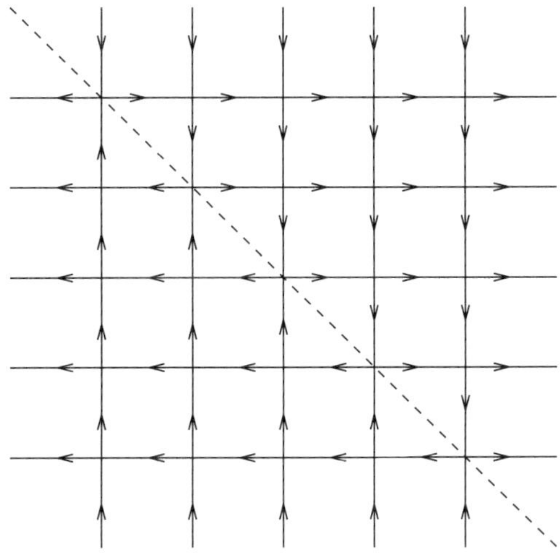

FIGURE 8.1. A ground state configuration in the ferroelectric phase region.

where

$$G_1 = \frac{G}{G_0} = \frac{e^{\gamma-t}\sinh(t+\gamma)}{\sinh 2\gamma} = \frac{e^{2\gamma}-e^{-2t}}{e^{2\gamma}-e^{-2\gamma}} > 1. \tag{8.6.5}$$

Observe that

$$\lim_{n\to\infty}\frac{\log Z_n}{n^2} = \lim_{n\to\infty}\frac{\log w(\sigma^{\mathrm{gs}})}{n^2} = \log F, \tag{8.6.6}$$

so that the free energy is determined by the free energy of the ground state configuration. This can be explained by the fact that low energy excited states are local perturbations of the ground state around the diagonal. Namely, it is impossible to create a new configuration by perturbing the ground state locally away of the diagonal: the conservation law $N_3(\sigma) = N_4(\sigma)$ forbids such a configuration. Therefore, typical configurations of the six-vertex model in the ferroelectric phase are frozen outside of a relatively small neighborhood of the diagonal.

CHAPTER 9

Between the Phases

In the preceding three chapters, we have discussed the large n asymptotic formula for the partition function for the six-vertex model in the three main regions of the phase diagram (see Figure 4.8). That is, we have found different asymptotic formulas for Z_n in the disordered, antiferroelectric, and ferroelectric phase regions, which correspond respectively to $-1 < \Delta < 1$, $\Delta < -1$, and $\Delta > 1$, where the anisotropy parameter Δ is defined in (4.4.1). These formulas do not extend to the boundaries between the phase regions. To complete the description of the asymptotics of the partition function, we must consider separately the cases $\Delta = 1$ and $\Delta = -1$. In fact, for $|\Delta| = 1$, the Izergin – Korepin formula does not apply directly, and we must consider a limiting case of our orthogonal polynomials formula. This is described below.

9.1. The critical line between the ferroelectric and disordered phases

There are two critical lines between the ferroelectric and disordered phases (see Figure 4.8). In this section we consider the critical line

$$\frac{b}{c} - \frac{a}{c} = 1. \tag{9.1.1}$$

The line $a/c - b/c = 1$ is similar. Specifically, we fix a point,

$$\frac{a}{c} = \frac{\alpha - 1}{2}, \quad \frac{b}{c} = \frac{\alpha + 1}{2}; \qquad \alpha > 1, \tag{9.1.2}$$

on this line, and we are interested in the large n asymptotics of the partition function

$$Z_n = Z_n\left(\frac{\alpha - 1}{2}, \frac{\alpha - 1}{2}, \frac{\alpha + 1}{2}, \frac{\alpha + 1}{2}, 1, 1\right). \tag{9.1.3}$$

As mentioned above, the Izergin – Korepin formula is not directly applicable, and we must consider a limit as the critical line is approached from either the disordered or ferroelectric phase region. We will consider the limit from the ferroelectric side. That is, starting with ferroelectric parametrization of the Izergin – Korepin formula (4.4.5) – (4.4.7), consider the limit as

$$t, \gamma \to +0, \quad \frac{t}{\gamma} \to \alpha. \tag{9.1.4}$$

Observe that in this limit,

$$\frac{a}{c} = \frac{\sinh(t - \gamma)}{\sinh(2\gamma)} \to \frac{\alpha - 1}{2}, \quad \frac{b}{c} = \frac{\sinh(t + \gamma)}{\sinh(2\gamma)} \to \frac{\alpha + 1}{2}. \tag{9.1.5}$$

By (4.2.18),

$$Z_n\left(\frac{a}{c},\frac{a}{c},\frac{b}{c},\frac{b}{c},1,1\right)=\frac{Z_n(a,a,b,b,c,c)}{c^{n^2}}, \tag{9.1.6}$$

hence by (4.4.5), and (8.1.8),

$$Z_n\left(\frac{a}{c},\frac{a}{c},\frac{b}{c},\frac{b}{c},1,1\right)=\left[\frac{2\sinh(t-\gamma)\sinh(t+\gamma)}{\sinh(2\gamma)}\right]^{n^2}\prod_{k=0}^{n-1}\frac{h_k}{(k!)^2}, \tag{9.1.7}$$

where h_k are the normalizing coefficients associated with the orthogonal polynomials $\{P_k(l)\}_{k=0}^{\infty}$ defined in (8.1.7), so that

$$\sum_{l=1}^{\infty}P_j(l)P_k(l)w(l)=h_k\delta_{jk};\quad w(l)=\mathrm{e}^{-2tl+2\gamma l}-\mathrm{e}^{-2tl-2\gamma l}. \tag{9.1.8}$$

To deal with limit (9.1.4) we need to rescale the orthogonal polynomials $P_k(l)$. Introduce the rescaled variable,

$$x=2tl-2\gamma l, \tag{9.1.9}$$

and the rescaled limiting weight,

$$w_\alpha(x)=\lim_{\substack{t,\gamma\to+0\\ t/\gamma\to\alpha}}(\mathrm{e}^{-2tl+2\gamma l}-\mathrm{e}^{-2tl-2\gamma l})=\mathrm{e}^{-x}-\mathrm{e}^{-rx},\quad r=\frac{\alpha+1}{\alpha-1}>1. \tag{9.1.10}$$

Consider monic orthogonal polynomials $P_j(x;\alpha)$ satisfying the orthogonality condition,

$$\int_0^\infty P_j(x;\alpha)P_k(x;\alpha)w_\alpha(x)\,\mathrm{d}x=h_{k,\alpha}\delta_{jk}. \tag{9.1.11}$$

To find a relation between $P_k(l)$ and $P_k(x;\alpha)$, introduce the monic polynomials

$$\widetilde{P}_k(x)=\delta^kP_k(x/\delta), \tag{9.1.12}$$

where

$$\delta=2t-2\gamma, \tag{9.1.13}$$

and rewrite orthogonality condition (9.1.8) in the form

$$\sum_{l=1}^{\infty}\widetilde{P}_j(l\delta)\widetilde{P}_k(l\delta)w_\alpha(l\delta)\delta=\delta^{2k+1}h_k\delta_{jk}, \tag{9.1.14}$$

which is a Riemann sum for the integral in orthogonality condition (9.1.11). Therefore,

$$\lim_{\substack{t,\gamma\to+0\\ t/\gamma\to\alpha}}\widetilde{P}_k(x)=P_k(x;\alpha), \tag{9.1.15}$$

and

$$\lim_{\substack{t,\gamma\to+0\\ t/\gamma\to\alpha}}\delta^{2k+1}h_k=h_{k,\alpha}. \tag{9.1.16}$$

Let us rewrite formula (9.1.7) as

$$Z_n\left(\frac{a}{c},\frac{a}{c},\frac{b}{c},\frac{b}{c},1,1\right)=\left[\frac{2\sinh(t-\gamma)\sinh(t+\gamma)}{\sinh(2\gamma)\delta}\right]^{n^2}\prod_{k=0}^{n-1}\frac{\delta^{2k+1}h_k}{(k!)^2}, \tag{9.1.17}$$

and take limit (9.1.4). In the limit we obtain that

$$Z_n = Z_n\left(\frac{\alpha-1}{2}, \frac{\alpha-1}{2}, \frac{\alpha+1}{2}, \frac{\alpha+1}{2}, 1, 1\right) = \left(\frac{\alpha+1}{2}\right)^{n^2} \prod_{k=0}^{n-1} \frac{h_{k,\alpha}}{(k!)^2}. \tag{9.1.18}$$

We thus see that we need an asymptotic formula for the normalizing constants $h_{k,\alpha}$ for the system of orthogonal polynomials (9.1.11). These polynomials are not in the class of orthogonal polynomials described in Chapter 2, as the measure of orthogonality is supported on the half-line rather than on the full real line. Such orthogonal polynomials may be referred to as *Laguerre type* polynomials. The asymptotic solution to the corresponding Riemann – Hilbert problem can still be constructed using the Deift – Zhou steepest descent method, but there is a different local solution at the origin due to the presence of the "hard edge" at zero. This local solution is built from Bessel functions. The general asymptotic analysis of systems of Laguerre type orthogonal polynomials was carried out by Vanlessen in [74]. The analysis of that paper was adapted and applied specifically to the polynomials (9.1.11) in the paper [14]. The main technical result of that paper is the following asymptotic formula for $h_{k,\alpha}$.

Theorem 9.1.1. *As $k \to \infty$,*

$$\log\left[\frac{h_{k,\alpha}}{(k!)^2}\right] = -\frac{\zeta(\frac{3}{2})}{2\sqrt{\pi(r-1)}k^{1/2}} + \frac{1}{4k} + \mathcal{O}(k^{-3/2}), \quad r = \frac{\alpha+1}{\alpha-1}, \tag{9.1.19}$$

where

$$\zeta(s) = 1 + \frac{1}{2^s} + \frac{1}{3^s} + \cdots, \quad \operatorname{Re} s > 1, \tag{9.1.20}$$

is the Riemann zeta function.

From the above Theorem 9.1.1 and the formula (9.1.18), we immediately obtain the the following asymptotics of Z_n on the critical line between ferroelectric and disordered phases.

Theorem 9.1.2. *As $n \to \infty$,*

$$Z_n\left(\frac{\alpha-1}{2}, \frac{\alpha-1}{2}, \frac{\alpha+1}{2}, \frac{\alpha+1}{2}, 1, 1\right) = Cn^{\kappa} G^{\sqrt{n}} F^{n^2}[1 + \mathcal{O}(n^{-1/2})], \tag{9.1.21}$$

where $C > 0$,

$$\kappa = \frac{1}{4}, \qquad G = \exp\left[-\zeta\left(\frac{3}{2}\right)\sqrt{\frac{\alpha-1}{2\pi}}\right], \tag{9.1.22}$$

and

$$F = \frac{\alpha+1}{2}. \tag{9.1.23}$$

PROOF. From formula (9.1.18) and asymptotics (9.1.19) we obtain that

$$\log\left[\frac{Z_n\big((\alpha-1)/2,(\alpha-1/2),(\alpha+1)/2,(\alpha+1)/2,1,1\big)}{\big((\alpha+1)/2\big)^{n^2}}\right] \tag{9.1.24}$$
$$=\sum_{k=0}^{n-1}\log\left[\frac{h_{k,\alpha}}{(k!)^2}\right]=\sum_{k=0}^{n-1}\left[-\frac{\zeta(\frac{3}{2})}{2\sqrt{\pi(r-1)}\,k^{1/2}}+\frac{1}{4k}+\mathcal{O}(k^{-3/2})\right]$$
$$=-\zeta\left(\frac{3}{2}\right)\sqrt{\frac{(\alpha-1)}{2\pi}}\,n^{1/2}+\frac{\log n}{4}+C_0+\mathcal{O}(n^{-1/2}),$$

which implies Theorem 9.1.2. □

9.2. The critical line between the antiferroelectric and disordered phases

We now consider the case $\Delta=-1$, which is the line $a+b=c$ between the antiferroelectric and disordered phases. We can use the Izergin–Korepin formula to approach this line from either the antiferroelectric and disordered phase. We will approach it from the antiferroelectric phase, and begin with the orthogonal polynomial formula (7.1.7). On the critical line

$$\frac{a}{c}+\frac{b}{c}=1, \tag{9.2.1}$$

we fix a point,

$$\frac{a}{c}=\frac{1-\alpha}{2},\quad \frac{b}{c}=\frac{1+\alpha}{2},\qquad -1<\alpha<1, \tag{9.2.2}$$

and consider the partition function

$$Z_n=Z_n\left(\frac{1-\alpha}{2},\frac{1-\alpha}{2},\frac{1+\alpha}{2},\frac{1+\alpha}{2},1,1\right). \tag{9.2.3}$$

This corresponds to taking a limit of the Izergin–Korepin formula in the antiferroelectric phase as $t,\gamma\to 0$, and $t/\gamma=\alpha$. Introduce the rescaled variable,

$$x=-2tl+2\gamma l, \tag{9.2.4}$$

and the rescaled limiting weight,

$$w_\alpha(x)=\lim_{\substack{t,\gamma\to+0\\ t/\gamma\to\alpha}}\mathrm{e}^{2tl-2\gamma|l|}=\begin{cases}\mathrm{e}^{-x}, & x\geq 0\\ \mathrm{e}^{rx}, & x<0,\end{cases} \tag{9.2.5}$$

where

$$r=\frac{1+\alpha}{1-\alpha}>0. \tag{9.2.6}$$

Consider monic orthogonal polynomials $P_j(x;\alpha)$ satisfying the orthogonality condition,

$$\int_{\mathbb{R}}P_j(x;\alpha)P_k(x;\alpha)w_\alpha(x)\,\mathrm{d}x=h_{k,\alpha}\delta_{jk}, \tag{9.2.7}$$

which can be obtained from the polynomials (7.1.5) by taking the appropriate scaling limit as $t, \gamma \to 0$, and $t/\gamma = \alpha$. Similar to (9.1.18), we obtain

$$Z_n = Z_n\left(\frac{\alpha-1}{2}, \frac{\alpha-1}{2}, \frac{\alpha+1}{2}, \frac{\alpha+1}{2}, 1, 1\right) = \left(\frac{1+\alpha}{2}\right)^{n^2} \prod_{k=0}^{n-1} \frac{h_{k,\alpha}}{(k!)^2}. \tag{9.2.8}$$

The orthogonal polynomials (9.2.7) can easily be rescaled with n in order to apply the steepest descent analysis of Chapter 2 and obtain an asymptotic formula for the constants $h_{k,\alpha}$, and thus for Z_n. The analysis is complicated by the fact the the weight $w_\alpha(x)$ is singular at the origin, and therefore the nonlinear steepest descent method requires a special local analysis close to that point. The weight $w_\alpha(x)$ is somewhat similar to the class of *Freud weights* studied by Kriecherbauer and McLaughlin in their paper [52]. In that paper, the authors were able to prove the existence of a local solution to the corresponding Riemann–Hilbert problem, but in general were unable to formulate the solution explicitly. Such explicit information would in general be necessary to obtain the subleading terms in the asymptotic expansion of $h_{k,\alpha}$, which are in turn necessary to evaluate the product in (9.2.8). A somewhat different approach was employed by Bleher and Bothner in the paper [8]. Using the Riemann–Hilbert approach, they were able to obtain the following asymptotic formula for $h_{k,\alpha}$.

Proposition 9.2.1. *As $k \to \infty$,*

$$\frac{h_{k,\alpha}}{k!} = \left(\frac{\pi(1-\alpha)}{2\cos(\pi\alpha/2)}\right)^{2k+1}\left(1 + \frac{1}{12k} + \varepsilon_k\right). \tag{9.2.9}$$

where

$$\varepsilon_k = \frac{(-1)^k \cos\big(\pi\alpha(k+\frac{1}{2})\big)}{2k(\log k)^2} \tag{9.2.10}$$

$$\times\left(1 - \frac{2\log\log k}{\log k} + \frac{1-2\gamma_{\mathrm{e}} - 4\log 2 - 2\log\big(\cos(\pi\alpha/2)\big)}{\log k} + \mathcal{O}\left(\left(\frac{\log\log k}{k}\right)^2\right)\right),$$

and γ_{e} is the Euler constant. The error term is uniform in α on compact sets of $\alpha \in (-1,1)$.

Note that we have used a slightly different scaling here than the one in [8]. From this proposition and the formula (9.2.8) we obtain the following asymptotic formula for Z_n on the critical line separating the antiferroelectric and disordered phases.

Theorem 9.2.2. *On the critical line separating the antiferroelectric and disordered phases, as $n \to \infty$,*

$$Z_n\left(\frac{\alpha-1}{2}, \frac{\alpha-1}{2}, \frac{\alpha+1}{2}, \frac{\alpha+1}{2}, 1, 1\right) = CF^{n^2} n^{1/12}\big(1 + \mathcal{O}(n^{-1})\big), \tag{9.2.11}$$

where

$$F = \frac{\pi(1-\alpha^2)}{4\cos(\pi\alpha/2)}, \quad C = \left(\cos\frac{\pi\alpha}{2}\right)^{1/12} C_0, \tag{9.2.12}$$

and $C_0 > 0$ does not depend on α. The error term is uniform in α on compact sets of $\alpha \in (-1,1)$.

PROOF. The free energy F and the power-like term $n^{1/12}$ are obtained immediately from Proposition 9.2.1 and formula (9.2.8). For the dependence of the constant term on α, one can use the Toda equation. This is similar to the calculation of the constant term in the disordered phase, and we leave it to the reader. □

Let us remark that the leading term in the formula (9.2.11) first appeared in the physics literature in the paper [17] of Bogoliubov, Kitaev, and Zvonarev. Their result implied a much weaker error estimate, and said nothing about the constant term C. The stronger result above is due to Bleher and Bothner [8].

9.3. The order of the phase transitions

The limiting free energy per site in the six-vertex model can be defined as

$$\lim_{n\to\infty}\frac{\log Z_n}{n^2}. \tag{9.3.1}$$

In the main results of the preceding three chapters, we have proven that within each phase region, the free energy is an analytic function of the parameters. In this section we discuss the behavior of the free energy as the parameters pass between the phase regions. As we will see, the free energy remains a continuous function of the parameters, but is not analytic on the critical lines between phases.

9.3.1. The transition between the F and D regions. Let us first compare the free energy in the disordered phase and in the ferroelectric phase when we approach a point of phase transition. Consider first the ferroelectric phase $b > a+c$, using the parametrization (4.4.2). In this phase region, according to (8.1.12), the free energy is simply the parameter b, and we consider the regime,

$$t,\gamma \to +0,\qquad \frac{t}{\gamma}\to \alpha > 1. \tag{9.3.2}$$

In this regime,

$$\lim_{\gamma\to 0}\frac{b}{c}=\lim_{\gamma\to 0}\frac{\sinh(t+\gamma)}{\sinh(2\gamma)}=\frac{\alpha+1}{2},\qquad \lim_{\gamma\to 0}\frac{a}{c}=\lim_{\gamma\to 0}\frac{\sinh(t-\gamma)}{\sinh(2\gamma)}=\frac{\alpha-1}{2}. \tag{9.3.3}$$

Observe that $t,\gamma\to 0$ as we approach the line of phase transition,

$$\frac{b}{c}=\frac{a}{c}+1. \tag{9.3.4}$$

hence $a,b,c\to 0$ in parametrization (4.4.2). We therefore rescale formula (8.1.12) according to (4.2.18),

$$Z_n\left(\frac{a}{c},\frac{a}{c},\frac{b}{c},\frac{b}{c},1,1\right)=c^{-n^2}Z_n(a,a,b,b,c,c)=CG^nF_{\mathrm{F}}^{n^2}\big[1+\mathcal{O}\big(e^{-n^{1-\varepsilon}}\big)\big], \tag{9.3.5}$$

where

$$F_{\mathrm{F}}=\frac{F}{c}=\frac{b}{c}. \tag{9.3.6}$$

Similarly in the disordered phase, from Theorem 6.1.3 we obtain

$$Z_n\left(\frac{a}{c},\frac{a}{c},\frac{b}{c},\frac{b}{c},1,1\right)=Cn^{\kappa}F_{\mathrm{D}}^{n^2}[1+\mathcal{O}(n^{-1})], \tag{9.3.7}$$

where

$$F_{\mathrm{D}}=\frac{F}{c}=\frac{\pi ab}{2\gamma c\cos\big(\pi t/(2\gamma)\big)}, \tag{9.3.8}$$

and the parameters t and γ are according to the parametrization (4.4.4). As we approach critical line (9.3.4), the parametrization (4.4.4) corresponds to the limit

$$t, \gamma \to \frac{\pi}{2} - 0, \quad \frac{\pi/2 - t}{\pi/2 - \gamma} \to \alpha > 1. \tag{9.3.9}$$

It is convenient to shift the parametrization by $t \mapsto \pi/2 - t$ and $\gamma \mapsto \pi/2 - \gamma$. This gives the parametrization,

$$a = \sin(t - \gamma), \quad b = \sin(t + \gamma), \quad c = \sin(2|\gamma|), \quad |\gamma| < t, \tag{9.3.10}$$

and the approach to critical line (9.3.4) is also described by the regime (9.3.2). In the new t, γ the formula (9.3.8) for the free energy in the disordered phase reads

$$F_{\text{D}} = \frac{\pi ab}{(\pi - 2\gamma)c \cos\left[\pi(\pi/2 - t)/\left(2(\pi/2 - \gamma)\right)\right]}. \tag{9.3.11}$$

We consider the free energy on the line

$$\frac{a + b}{c} = \alpha, \tag{9.3.12}$$

which crosses the critical line (9.3.4) at a right angle in the coordinates a/c and b/c. We will also use the parameter

$$\beta = \frac{b - a}{c}, \tag{9.3.13}$$

on this line. Note that $\beta = 1$ corresponds to the point of phase transition. In variables α and β, the free energy is given by

$$F = \begin{cases} F_{\text{F}} = \dfrac{\alpha + \beta}{2} & \text{for } \beta \geq 1 \\ F_{\text{D}} = \dfrac{(\alpha + \beta)g(t, \gamma)}{2} & \text{for } \beta \leq 1, \end{cases} \tag{9.3.14}$$

where

$$g(t, \gamma) = \frac{\pi \sin(t - \gamma)}{(\pi - 2\gamma) \sin[\pi(t - \gamma)/(\pi - 2\gamma)]}, \tag{9.3.15}$$

in the parametrization (9.3.10). A straightforward calculation shows that on the line $(a + b)/c = \alpha$ in the disordered phase, as $\beta \to 1 - 0$,

$$g(t, \gamma) = 1 + \frac{2(\alpha - 1)^{3/2}(1 - \beta)^{3/2}}{3\pi(\alpha + 1)^{1/2}} + \mathcal{O}((1 - \beta)^2). \tag{9.3.16}$$

This implies that the free energy F exhibits a phase transition of the order $\frac{3}{2}$ with respect to the parameter β at the point $\beta = 1$. Figure 9.1 depicts the graph of $F = F(\beta)$ (the left graph) and its derivative, $F'(\beta)$ (the right graph), as a function of $\beta = (b - a)/c$ on the line $(b + a)/c = 2$. Observe the square root singularities of F_0' at $\beta = 1$, which corresponds to the phase transition of order $\frac{3}{2}$. Since

$$\Delta = \frac{a^2 + b^2 - c^2}{2ab} = \frac{\alpha^2 + \beta^2 - 2}{\alpha^2 - \beta^2} = 1 + \frac{4(\beta - 1)}{\alpha^2 - 1} + \mathcal{O}((\beta - 1)^2), \tag{9.3.17}$$

it is a phase transition of the order $\frac{3}{2}$ with respect to the parameter Δ at the point $\Delta = 1$ as well.

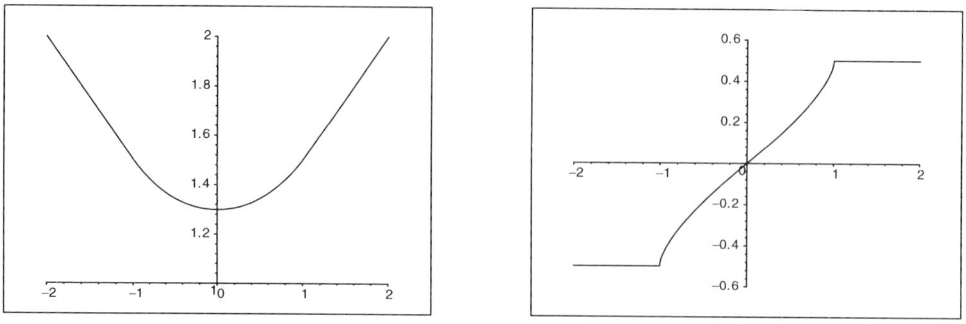

FIGURE 9.1. Free energy $F = F(\beta)$ (the left graph) and its derivative (the right graph), as functions of $\beta = (b-a)/c$ on the line $(b+a)/c = 2$. This shows the phase transition from the disordered phase region ($|\beta| < 1$) to the ferroelectric region ($|\beta| > 1$).

9.3.2. The transition between the AF and D regions. Let us now consider the free energy as we pass between the antiferroelectric and disordered phases. In the antiferroelectric phase, Theorem 7.3.2 and equation (4.2.18) give that

$$Z_n\left(\frac{a}{c},\frac{a}{c},\frac{b}{c},\frac{b}{c},1,1\right) = c^{-n^2} Z_n(a,a,b,b,c,c) = C\vartheta_4(n\omega)F_{\mathrm{AF}}^{n^2}[1+\mathcal{O}(n^{-1})], \tag{9.3.18}$$

where

$$F_{\mathrm{AF}} = \frac{\pi a\, b\, \vartheta_1'(0)}{2\gamma\, c\, \vartheta_1(\omega)}. \tag{9.3.19}$$

In this formula, as in Chapter 7, the parameters t and γ are according to the parametrization (4.4.3), $\omega = (\pi/2)(1 + t/\gamma)$, and the elliptic nome for the Jacobi theta functions is $q = \mathrm{e}^{-\pi^2/(2\gamma)}$.

The critical line between the AF and D regions is given by

$$\frac{a}{c} + \frac{b}{c} = 1, \tag{9.3.20}$$

and we can parametrize that line by taking

$$\frac{a}{c} = \frac{1-\alpha}{2}, \quad \frac{b}{c} = \frac{1+\alpha}{2}, \qquad -1 < \alpha < 1. \tag{9.3.21}$$

In the parametrization (4.4.3), the critical line is approached when

$$t \to 0, \quad \gamma \to 0^+, \qquad \frac{t}{\gamma} \to \alpha. \tag{9.3.22}$$

In particular when γ approaches zero, so does q, the elliptic nome for the Jacobi theta functions. It is immediate from the definition of ϑ_1 in (7.2.1) that as $q \to 0^+$,

$$\begin{aligned} \vartheta_1(z) &= \vartheta_1(z \mid q) = 2q^{1/4}\sin z + \mathcal{O}(q^{9/4}), \\ \vartheta_1'(0) &= \vartheta_1'(0 \mid q) = 2q^{1/4} + \mathcal{O}(q^{9/4}), \end{aligned} \tag{9.3.23}$$

and therefore the ratio of theta functions in (9.3.19) is

$$\frac{\vartheta_1'(0)}{\vartheta_1(\omega)} = \frac{1}{\sin\omega} + \mathcal{O}(q^2) = \frac{1}{\cos\bigl(\pi t/(2\gamma)\bigr)} + \mathcal{O}(q^2). \tag{9.3.24}$$

Let us write the free energy (9.3.19) as the sum

$$F_{\rm AF} = F_{\rm AF}^{\rm Reg} + F_{\rm AF}^{\rm Sing}, \tag{9.3.25}$$

where

$$F_{\rm AF}^{\rm Reg} = \frac{\pi ab}{2\gamma c\cos\big(\pi t/(2\gamma)\big)}, \quad F_{\rm AF}^{\rm Sing} = \frac{\pi ab}{2\gamma c}\left(\frac{\vartheta_1'(0)}{\vartheta_1(\omega)} - \frac{1}{\cos\big(\pi t/(2\gamma)\big)}\right). \tag{9.3.26}$$

The term $F_{\rm AF}^{\rm Reg}$ looks identical to the formula (9.3.8) for the free energy in the disordered region, but we must remember that the parameterizations are slightly different in the two phase regions.

To compare $F_{\rm AF}^{\rm Reg}$ and $F_{\rm D}$, consider the new coordinates

$$\frac{a}{c} = \frac{1-\alpha}{2} + \beta, \quad \frac{b}{c} = \frac{1+\alpha}{2} + \beta. \tag{9.3.27}$$

Notice that $\beta < 0$ corresponds to the antiferroelectric region $\beta > 0$ corresponds to the disordered region. The critical line is crossed at $\beta = 0$. To consider the free energy in these coordinates, let us take

$$F_{\rm AF}^{\rm Reg}(\alpha,\beta) \equiv F_{\rm AF}^{\rm Reg}\big(t(\alpha,\beta),\gamma(\alpha,\beta)\big), \tag{9.3.28}$$

where the transformation $(t,\gamma) \mapsto (\alpha,\beta)$ is given by (4.4.3) and (9.3.27). We also take

$$F_{\rm D}(\alpha,\beta) \equiv F_{\rm D}^{\rm Reg}\big(t(\alpha,\beta),\gamma(\alpha,\beta)\big), \tag{9.3.29}$$

where the transformation $(t,\gamma) \mapsto (\alpha,\beta)$ is given by (4.4.4) and (9.3.27). To see the phase transition, we consider these formulas for $\alpha \in (-1,1)$ fixed, and for β in a neighborhood of zero.

Notice that in the disordered phase region,

$$\alpha = \frac{b-a}{c} = \frac{\sin(\gamma+t) - \sin(\gamma-t)}{\sin(2\gamma)} = \frac{\sin t}{\sin\gamma}, \tag{9.3.30}$$

and in the antiferroelectric phase region,

$$\alpha = \frac{b-a}{c} = \frac{\sinh(\gamma+t) - \sinh(\gamma-t)}{\sinh(2\gamma)} = \frac{\sinh t}{\sinh\gamma}. \tag{9.3.31}$$

Furthermore, in the disordered phase region corresponding to $\beta > 0$,

$$\gamma_{\rm D}(\beta) = \arcsin\left(2\sqrt{\frac{\beta(1+\beta)}{(1+2\beta)^2 - \alpha^2}}\right), \tag{9.3.32}$$

and in the antiferroelectric phase region corresponding to $\beta < 0$,

$$\gamma_{\rm AF}(\beta) = \operatorname{arc\,sinh}\left(2\sqrt{\frac{-\beta(1+\beta)}{(1+2\beta)^2 - \alpha^2}}\right). \tag{9.3.33}$$

Similarly the parameter t is given in the disordered phase region by

$$t_{\rm D}(\beta) = \arcsin\left(2\alpha\sqrt{\frac{\beta(1+\beta)}{(1+2\beta)^2 - \alpha^2}}\right), \tag{9.3.34}$$

and in the antiferroelectric phase region by

$$t_{\rm AF}(\beta) = \operatorname{arc\,sinh}\left(2\alpha\sqrt{\frac{-\beta(1+\beta)}{(1+2\beta)^2 - \alpha^2}}\right). \tag{9.3.35}$$

For a fixed α, let us define the free energy in a neighborhood of $\beta = 0$ as

$$F(\beta) = \frac{\pi ab}{c^2} \times \begin{cases} \dfrac{\sin(2\gamma_{\mathrm{D}}(\beta))}{2\gamma_{\mathrm{D}}(\beta)\cos\big((\pi/2)(t_{\mathrm{D}}(\beta)/\gamma_{\mathrm{D}}(\beta))\big)}, & \beta \geq 0 \\ \dfrac{\sinh(2\gamma_{\mathrm{AF}}(\beta))}{2\gamma_{\mathrm{AF}}(\beta)\cos\big((\pi/2)(t_{\mathrm{AF}}(\beta)/\gamma_{\mathrm{AF}}(\beta))\big)}, & \beta \leq 0, \end{cases} \tag{9.3.36}$$

which coincides with F_{D} for $\beta > 0$, and with $F_{\mathrm{AF}}^{\mathrm{Reg}}$ for $\beta < 0$. This function is well defined at $\beta = 0$ for a fixed α. Indeed we have

$$F(0) = \frac{\pi(1-\alpha^2)}{4\cos(\pi\alpha/2)}, \tag{9.3.37}$$

in agreement with the result of Theorem 9.2.2 for the free energy on the critical line. We claim that the function $F(\beta)$ is an analytic function of β at $\beta = 0$. Indeed as functions of a complex parameter β in a neighborhood of the origin, both $\gamma_{\mathrm{D}}(\beta)$ and $t_{\mathrm{D}}(\beta)$ have a branch cut at the origin, but their ratio does not. Taking the branch of the square root which maps the negative real axis to the positive imaginary axis, we have for β close to 0,

$$\begin{aligned} \gamma_{\mathrm{D}}(-\beta) &= \arcsin\left(2\sqrt{\frac{-\beta(1-\beta)}{(1-2\beta)^2-\alpha^2}}\right) \\ &= \arcsin\left(2\mathrm{i}\sqrt{\frac{\beta(1-\beta)}{(1-2\beta)^2-\alpha^2}}\right) \\ &= \mathrm{i}\,\mathrm{arc\,sinh}\left(2\sqrt{\frac{\beta(1-\beta)}{(1-2\beta)^2-\alpha^2}}\right), \end{aligned} \tag{9.3.38}$$

and similarly

$$t_{\mathrm{D}}(-\beta) = \mathrm{i}\,\mathrm{arc\,sinh}\left(2\alpha\sqrt{\frac{\beta(1-\beta)}{(1-2\beta)^2-\alpha^2}}\right). \tag{9.3.39}$$

Combining the equations (9.3.38) and (9.3.39) above gives the analytic continuation of the ratio $t_{\mathrm{D}}/\gamma_{\mathrm{D}}$ to the negative real axis. Combining them with (9.3.33) and (9.3.35), we find that for $0 < \beta < \varepsilon$ for some $\varepsilon > 0$,

$$\frac{t_{\mathrm{D}}(-\beta)}{\gamma_{\mathrm{D}}(-\beta)} = \frac{t_{\mathrm{AF}}(-\beta)}{\gamma_{\mathrm{AF}}(-\beta)}. \tag{9.3.40}$$

It is a similar exercise to show that for $0 < \beta < \varepsilon$,

$$\frac{\sin\big(2\gamma_{\mathrm{D}}(-\beta)\big)}{2\gamma_{\mathrm{D}}(-\beta)} = \frac{\sin\big(2\gamma_{\mathrm{AF}}(-\beta)\big)}{2\gamma_{\mathrm{AF}}(-\beta)}, \tag{9.3.41}$$

where the quantity on the left again represents the analytic continuation to the negative real axis. This proves that the function (9.3.36) is indeed an analytic function of β, and so the only phase transition comes from the term $F_{\mathrm{AF}}^{\mathrm{Sing}}$ in (9.3.25). According to (9.3.24), this term approaches zero exponentially fast as $\gamma \to 0^+$, and thus as $\beta \to 0$. This indicates an infinite order phase transition.

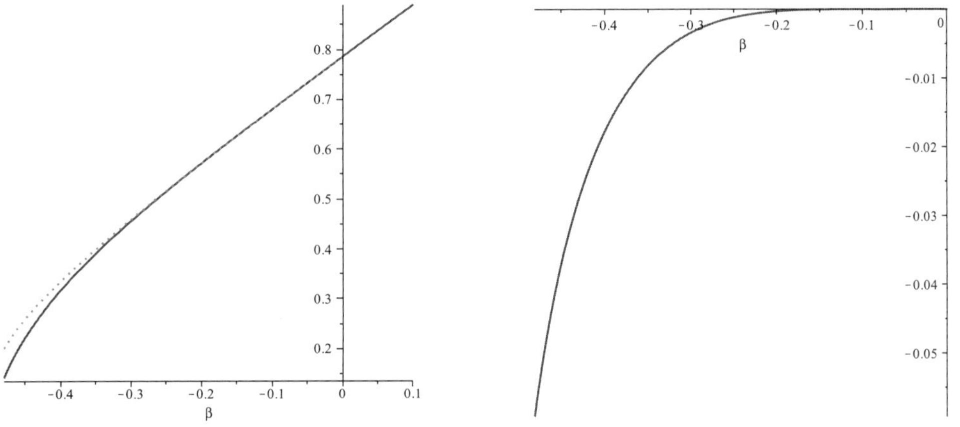

FIGURE 9.2. The solid line in the left graph gives the free energy $F = F(\beta)$ as a function of $\beta = (a+b-c)/(2c)$ on the diagonal $a = b$. The smooth phase transition from the antiferroelectric phase region to the disordered region occurs at $\beta = 0^-$. The dotted line in the left graph gives the analytic continuation of the free energy from the disordered region. The right graph is the singular term $F_{\mathrm{AF}}^{\mathrm{Sing}}$ as a function of $\beta < 0$, which gives the deviation of the free energy from an analytic function.

Bibliography

1. G. Akemann, J. Baik, and P. Di Francesco (eds.), *The Oxford handbook of random matrix theory*, Oxford Univ. Press, Oxford, 2011.
2. G. W. Anderson, *Spectral statistics of unitary ensembles*, The Oxford Handbook of Random Matrix Theory (G. Akemann, J. Baik, and P. Di Francesco, eds.), Oxford Univ. Press, Oxford, 2011, pp. 66–85.
3. G. W. Anderson, A. Guionnet, and O. Zeitouni, *An introduction to random matrices*, Cambridge Stud. Adv. Math., vol. 118, Cambridge Univ. Press, Cambridge, 2010.
4. Z. Bai and J. W. Silverstein, *Spectral analysis of large dimensional random matrices*, 2nd ed., Springer Ser. Statist., Springer, New York, 2010.
5. J. Baik, T. Kriecherbauer, K. T.-R. McLaughlin, and P. D. Miller, *Discrete orthogonal polynomials: Asymptotics and applications*, Ann. of Math. Stud., vol. 164, Princeton Univ. Press, Princeton, NJ, 2007.
6. M. T. Batchelor, R. J. Baxter, M. J. O'Rourke, and C. M. Yung, *Exact solution and interfacial tension of the six-vertex model with anti-periodic boundary conditions*, J. Phys. A **28** (1995), no. 10, 2759–2770.
7. R. J. Baxter, *Exactly solved models in statistical mechanics*, Academic Press, London, 1982.
8. P. Bleher and T. Bothner, *Exact solution of the six-vertex model with domain wall boundary conditions: Critical line between disordered and antiferroelectric phases*, Random Matrices Theory Appl. **1** (2012), no. 4, 1250012.
9. P. Bleher and V. V. Fokin, *Exact solution of the six-vertex model with domain wall boundary conditions. Disordered phase*, Comm. Math. Phys. **268** (2006), no. 1, 223–284.
10. P. Bleher and A. Its, *Semiclassical asymptotics of orthogonal polynomials, Riemann–Hilbert problem, and universality in the matrix model*, Ann. of Math. (2) **150** (1999), no. 1, 185–266.
11. ———, *Double scaling limit in the random matrix model: the Riemann-Hilbert approach*, Comm. Pure Appl. Math. **56** (2003), no. 4, 433–516.
12. ——— (ed.), *Random matrix models and their applications*, Math. Sci. Res. Inst. Publ., vol. 40, Cambridge Univ. Press, Cambridge, 2001.
13. P. Bleher and K. Liechty, *Exact solution of the six-vertex model with domain wall boundary conditions: Ferroelectric phase*, Comm. Math. Phys. **286** (2009), no. 2, 777–801.
14. ———, *Exact solution of the six-vertex model with domain wall boundary conditions: Critical line between ferroelectric and disordered phases*, J. Stat. Phys. **134** (2009), no. 3, 463–485.
15. ———, *Exact solution of the six-vertex model with domain wall boundary conditions: Antiferroelectric phase*, Comm. Pure Appl. Math. **63** (2010), no. 6, 779–829.
16. ———, *Uniform asymptotics for discrete orthogonal polynomials with respect to varying exponential weights on a regular infinite lattice*, Int. Math. Res. Not. IMRN **2** (2011), 342–386.
17. N. M. Bogoliubov, A. V. Kitaev, and M. B. Zvonarev, *Boundary polarization in the six-vertex model*, Phys. Rev. E (3) **65** (2002), no. 2, 026126, 4.
18. A. Borodin, *Determinantal point processes*, The Oxford Handbook of Random Matrix Theory (G. Akemann, J. Baik, and P. Di Francesco, eds.), Oxford Univ. Press, Oxford, 2011, pp. 231–249.
19. A. Borodin and D. Boyarchenko, *Distribution of the first particle in discrete orthogonal polynomial ensembles*, Comm. Math. Phys. **234** (2003), no. 2, 287–338.
20. A. Borodin and A. Soshnikov, *Janossy densities*. I: *Determinantal ensembles*, J. Statist. Phys. **113** (2003), no. 12, 595–610.
21. D. M. Bressoud, *Proofs and confirmations: The story of the alternating sign matrix conjecture*, MAA Spectrum, Math. Assoc. America, Washington, DC, 1999.

22. R. C. Brower, N. Deo, S. Jain, and C. I Tan, *Symmetry breaking in the double-well Hermitian matrix models*, Nuclear Phys. B **405** (1993), no. 1, 166–187.
23. F. Colomo and A. G. Pronko, *On some representations of the six vertex model partition function*, Phys. Lett. A **315** (2003), no. 3-4, 231–236.
24. ———, *On the partition function of the six-vertex model with domain wall boundary conditions*, J. Phys. A **37** (2004), no. 6, 1987–2002.
25. ———, *Square ice, alternating sign matrices, and classical orthogonal polynomials*, J. Stat. Mech. Theory Exp. **1** (2005), 005, 33 pp. (electronic).
26. ———, *The role of orthogonal polynomials in the six-vertex model and its combinatorial applications*, J. Phys. A **39** (2006), no. 28, 9015–9033.
27. P. Deift, *Orthogonal polynomials and random matrices: a Riemann – Hilbert approach*, Courant Lect. Notes Math., vol. 3, New York Univ., Courant Inst. Math. Sci., New York, 1999.
28. P. Deift, T. Kriecherbauer, and K. T.-R. McLaughlin, *New results on the equilibrium measure for logarithmic potentials in the presence of an external field*, J. Approx. Theory **95** (1998), no. 3, 388–475.
29. P. Deift, T. Kriecherbauer, K. T.-R. McLaughlin, S. Venakides, and X. Zhou, *Uniform asymptotics for polynomials orthogonal with respect to varying exponential weights and applications to universality questions in random matrix theory*, Comm. Pure Appl. Math. **52** (1999), no. 11, 1335–1425.
30. ———, *Strong asymptotics of orthogonal polynomials with respect to exponential weights*, Comm. Pure Appl. Math. **52** (1999), no. 12, 1491–1552.
31. P. Deift and K. T-R McLaughlin, *A continuum limit of the Toda lattice*, Mem. Amer. Math. Soc. **131** (1998), no. 624, x+216.
32. P. Deift and X. Zhou, *A steepest descent method for oscillatory Riemann – Hilbert problems. Asymptotics for the MKdV equation*, Ann. of Math. (2) **137** (1993), no. 2, 295–368.
33. P. Di Francesco, *2D quantum gravity, matrix models and graph combinatorics*, Applications of Random Matrices in Physics (Les Houches, 2004) (É. Brézin, V. Kazakov, D. Serban, P. Wiegmann, and A. Zabrodin, eds.), NATO Sci. Ser. II Math. Phys. Chem., vol. 221, Springer, Dordrecht, 2006, pp. 33–88.
34. P. Di Francesco, P. Ginsparg, and J. Zinn-Justin, *2D gravity and random matrices*, Phys. Rep. **254** (1995), no. 1-2, 133.
35. P. D. Dragnev and E. B. Saff, *Constrained energy problems with applications to orthogonal polynomials of a discrete variable*, J. Anal. Math. **72** (1997), 223–259.
36. F. J. Dyson, *Statistical theory of the energy levels of complex systems*. I, J. Mathematical Phys. **3** (1962), 140–156; II, 157–165; III, 166–175.
37. B. Eynard, *A concise expression for the ODE's of orthogonal polynomials*, `arXiv:math-ph/0109018`.
38. ———, *Counting surfaces: Combinatorics, matrix models and algebraic geometry*, Prog. Math. Phys., vol. 114, Birkhäuser, Basel, 2012.
39. H. M. Farkas and I. Kra, *Riemann surfaces*, 2nd ed., Grad. Texts in Math., vol. 71, Springer, New York, 1992.
40. P. L. Ferrari and H. Spohn, *Domino tilings and the six-vertex model at its free-fermion point*, J. Phys. A **39** (2006), no. 33, 10297–10306.
41. A. S. Fokas, A. R. Its, and A. V. Kitaev, *The isomonodromy approach to matrix models in 2D quantum gravity*, Comm. Math. Phys. **147** (1992), no. 2, 395–430.
42. P. J. Forrester, *Log-gases and random matrices*, London Math. Soc. Monogr. Ser., vol. 34, Princeton Univ. Press, Princeton, NJ, 2010.
43. J. Harnad (ed.), *Random matrices, random processes and integrable systems*, CRM Ser. Math. Phys, Springer, New York, 2011.
44. J. B. Hough, M. Krishnapur, Y. Peres, and B. Virág, *Determinantal processes and independence*, Probab. Surv. **3** (2006), 206–229.
45. ———, *Zeros of Gaussian analytic functions and determinantal point processes*, Univ. Lecture Ser., vol. 51, Amer. Math. Soc., Providence, RI, 2009.
46. A. G. Izergin, *Partition function of a six-vertex model in a finite volume*, Dokl. Akad. Nauk SSSR **297** (1987), no. 2, 331–333 (Russian); English transl., Soviet Phys. Dokl. **32** (1987), no. 11, 878–879.

47. A. G. Izergin, D. A. Coker, and V. E. Korepin, *Determinant formula for the six-vertex model*, J. Phys. A **25** (1992), no. 16, 4315–4334.
48. S. Kamvissis, K. D. T.-R. McLaughlin, and P. D. Miller, *Semiclassical soliton ensembles for the focusing nonlinear Schrödinger equation*, Ann. of Math. Stud., vol. 154, Princeton Univ. Press, Princeton, NJ, 2003.
49. R. Koekoek, P. A. Lesky, and R. F. Swarttouw, *Hypergeometric orthogonal polynomials and their q-analogues*, with a foreword by T. H. Koornwinder, Springer Monogr. Math., Springer, Berlin, 2010.
50. V. E. Korepin, *Calculation of norms of Bethe wave functions*, Comm. Math. Phys. **86** (1982), no. 3, 391–418.
51. V. Korepin and P. Zinn-Justin, *Thermodynamic limit of the six-vertex model with domain wall boundary conditions*, J. Phys. A **33** (2000), no. 40, 7053–7066.
52. T. Kriecherbauer and K. T.-R. McLaughlin, *Strong asymptotics of polynomials orthogonal with respect to Freud weights*, Internat. Math. Res. Notices **6** (1999), 299–333.
53. A. B. J. Kuijlaars, *On the finite-gap ansatz in the continuum limit of the Toda lattice*, Duke Math. J. **104** (2000), no. 3, 433–462.
54. A. B. J. Kuijlaars and K. T-R McLaughlin, *Generic behavior of the density of states in random matrix theory and equilibrium problems in the presence of real analytic external fields*, Comm. Pure Appl. Math. **53** (2000), no. 6, 736–785.
55. G. Kuperberg, *Another proof of the alternating-sign matrix conjecture*, Internat. Math. Res. Notices **3** (1996), 139–150.
56. E. H. Lieb, *Exact solution of the problem of the entropy of two-dimensional ice*, Phys. Rev. Lett. **18** (1967), no. 17, 692–694.
57. ______, *Exact solution of the F model of an antiferroelectric*, Phys. Rev. Lett. **18** (1967), no. 24, 1046–1048.
58. ______, *Exact solution of the two-dimensional Slater KDP model of a ferroelectric*, Phys. Rev. Lett. **19** (1967), no. 3, 108–110.
59. ______, *Residual entropy of square ice*, Phys. Rev. **162** (1967), no. 1, 162–172.
60. E. H. Lieb and F.-Y. Wu, *Two dimensional ferroelectric models*, Phase Transitions and Critical Phenomena (C. Domb and M. S. Green, eds.), Vol. 1, Academic Press, New York, 1972, pp. 331–490.
61. M. L. Mehta, *Random matrices*, 2nd ed., Academic Press Inc., Boston, MA, 1991.
62. W. H. Mills, D. P. Robbins, and H. Rumsey Jr., *Alternating sign matrices and descending plane partitions*, J. Combin. Theory Ser. A **34** (1983), no. 3, 340–359.
63. P. D. Miller, *Asymptotics of semiclassical soliton ensembles: rigorous justification of the WKB approximation*, Int. Math. Res. Not. **8** (2002), 383–454.
64. N. I. Muskhelishvili, *Singular integral equations. Boundary problems of function theory and their application to mathematical physics*, translated by J. R. M. Radok, P. Noordhoff N. V., Groningen, 1953.
65. F. W. J. Olver, *Asymptotics and special functions*, AKP Class., A K Peters Ltd., Wellesley, MA, 1997. Reprint of the 1974 original.
66. L. Pastur and M. Shcherbina, *Eigenvalue distribution of large random matrices*, Math. Surveys Monogr., vol. 171, Amer. Math. Soc., Providence, RI, 2011.
67. N. Reshetikhin, *Lectures on the integrability of the six-vertex model*, Exact Methods in Low-Dimensional Statistical Physics and Quantum Computing (Les Houches, 2008) (J. Jacobsen, Ouvry Stéphane, V. Pasquier, D. Serban, and L. F. Cugliandolo, eds.), Oxford Univ. Press, Oxford, 2010, pp. 197–266.
68. E. B. Saff and V. Totik, *Logarithmic potentials with external fields*, with Appendix B by T. Bloom, Grundlehren Math. Wiss., vol. 316, Springer, Berlin, 1997.
69. J. C. Slater, *Theory of the transition of* KH_2PO_4, J. Chem. Phys. **9** (1941), no. 1, 16–33.
70. K. Sogo, *Toda molecule equation and quotient-difference method*, J. Phys. Soc. Japan **62** (1993), no. 4, 1081–1084.
71. B. Sutherland, *Exact solution of a two-dimensional model for hydrogen-bonded crystals*, Phys. Rev. Lett. **19** (1967), 103-104.
72. G. Szegő, *Orthogonal polynomials*, 4th ed., Amer. Math. Soc. Colloq. Publ., vol. 23, Amer. Math. Soc., Providence, RI, 1975.
73. T. Tao, *Topics in random matrix theory*, Grad. Stud. Math., vol. 132, Amer. Math. Soc., Providence, RI, 2012.

74. M. Vanlessen, *Strong asymptotics of Laguerre-type orthogonal polynomials and applications in random matrix theory*, Constr. Approx. **25** (2007), no. 2, 125–175.
75. F.-Y. Wu and K. Y. Lin, *Staggered ice-rule vertex model — The Pfaffian solution*, Phys. Rev. B **12** (1975), no. 1, 419–428.
76. E. T. Whittaker and G. N. Watson, *A course of modern analysis*, Cambridge Univ. Press, Cambridge, 2000. Reprint of the fourth (1927) edition.
77. D. Zeilberger, *Proof of the refined alternating sign matrix conjecture*, New York J. Math. **2** (1996), 59–68, electronic.
78. P. Zinn-Justin, *Six-vertex model with domain wall boundary conditions and one-matrix model*, Phys. Rev. E (3) **62** (2000), no. 3, 3411–3418.
79. ———, *Six-vertex, loop and tiling models: Integrability and combinatorics*, LAP Lambert Acad. Publ., 2010.

Published Titles in This Series

32 **Pavel Bleher and Karl Liechty,** Random Matrices and the Six-Vertex Model, 2013

31 **Jean-Pierre Labesse and Jean-Loup Waldspurger,** La Formule des Traces Tordue d'après le Friday Morning Seminar, 2013

30 **Joseph H. Silverman,** Moduli Spaces and Arithmetic Dynamics, 2012

29 **Marcelo Aguiar and Swapneel Mahajan,** Monoidal Functors, Species and Hopf Algebras, 2010

28 **Saugata Ghosh,** Skew-Orthogonal Polynomials and Random Matrix Theory, 2009

27 **Jean Berstel, Aaron Lauve, Christophe Reutenauer, and Franco V. Saliola,** Combinatorics on Words, 2008

26 **Victor Guillemin and Reyer Sjamaar,** Convexity Properties of Hamiltonian Group Actions, 2005

25 **Andrew J. Majda, Rafail V. Abramov, and Marcus J. Grote,** Information Theory and Stochastics for Multiscale Nonlinear Systems, 2005

24 **Dana Schlomiuk, AndreĭA. Bolibrukh, Sergei Yakovenko, Vadim Kaloshin, and Alexandru Buium,** On Finiteness in Differential Equations and Diophantine Geometry, 2005

23 **J. J. M. M. Rutten, Marta Kwiatkowska, Gethin Norman, and David Parker,** Mathematical Techniques for Analyzing Concurrent and Probabilistic Systems, 2004

22 **Montserrat Alsina and Pilar Bayer,** Quaternion Orders, Quadratic Forms, and Shimura Curves, 2004

21 **Andrei Tyurin,** Quantization, Classical and Quantum Field Theory and Theta Functions, 2003

20 **Joel Feldman, Horst Knörrer, and Eugene Trubowitz,** Riemann Surfaces of Infinite Genus, 2003

19 **L. Lafforgue,** Chirurgie des grassmanniennes, 2003

18 **G. Lusztig,** Hecke Algebras with Unequal Parameters, 2003

17 **Michael Barr,** Acyclic Models, 2002

16 **Joel Feldman, Horst Knörrer, and Eugene Trubowitz,** Fermionic Functional Integrals and the Renormalization Group, 2002

15 **José I. Burgos Gil,** The Regulators of Beilinson and Borel, 2002

14 **Eyal Z. Goren,** Lectures on Hilbert Modular Varieties and Modular Forms, 2002

13 **Michael Baake and Robert V. Moody, Editors,** Directions in Mathematical Quasicrystals, 2000

12 **Masayoshi Miyanishi,** Open Algebraic Surfaces, 2000

11 **Spencer J. Bloch,** Higher Regulators, Algebraic K-Theory, and Zeta Functions of Elliptic Curves, 2000

10 **James D. Lewis and B. Brent Gordon,** A Survey of the Hodge Conjecture, Second Edition, 1999

9 **Yves Meyer,** Wavelets, Vibrations and Scalings, 1998

8 **Ioannis Karatzas,** Lectures on the Mathematics of Finance, 1997

7 **John Milton,** Dynamics of Small Neural Populations, 1996

6 **Eugene B. Dynkin,** An Introduction to Branching Measure-Valued Processes, 1994

5 **Andrew Bruckner,** Differentiation of Real Functions, 1994

4 **David Ruelle,** Dynamical Zeta Functions for Piecewise Monotone Maps of the Interval, 1994

3 **V. Kumar Murty,** Introduction to Abelian Varieties, 1993

2 **M. Ya. Antimirov, A. A. Kolyshkin, and Rémi Vaillancourt,** Applied Integral Transforms, 1993

1 **Dan Voiculescu, Kenneth J. Dykema, and Alexandru Nica,** Free Random Variables, 1992